D1267237

Yale Western Americana Series, 22

"Ye Mining Engineer." From EMJ, 109 *(May 15, 1920), between pp. 116–17.*

MINING ENGINEERS & THE AMERICAN WEST

The Lace-Boot Brigade, 1849–1933

BY CLARK C. SPENCE

New Haven and London, Yale University Press, 1970

Published with assistance from the foundation established in memory of James Wesley Cooper, of the Class of 1865, Yale College.

Library of Congress catalog card number: 74-104621
Standard book number: 300-01224-1
Designed by Sally Sullivan, set in Linotype Caslon type, and printed in the United States of America by Vail-Ballou Press, Inc., Binghamton, N.Y.
Distributed in Great Britain, Europe, Asia, and Africa by Yale University Press Ltd., London; in Canada by McGill-Queen's University Press, Montreal; and in Mexico by Centro Interamericano de Libros Académicos, Mexico City.

To Christian Edward Spence

CONTENTS

ILLUSTRATIONS

MAPS

TABLES

PREFACE

In one sense the historian's approach to the history of the mineral West has been like that of early Nevada miners toward the materials with which they worked. For years prospectors toiled away in the valley of the Carson under the shadow of Mt. Davidson, gaining a modest amount of gold, but sweating and cursing as they cast aside the "damned blue stuff" that clogged their primitive equipment. Finally in 1859, when assays were run, this "damned blue stuff" proved to be worth nearly $5,000 per ton in silver—and the rush was on.

In like fashion, the historian has traditionally been preoccupied with working the ground where gold obviously glittered and until recent years has ignored the unfamiliar, more complex metal deposits. Until the new generation of historians—men like Rodman Paul, W. Turrentine Jackson, and Russell Elliott—emphasis has too often been on the color and drama of the mining frontiers, rather than on the more prosaic but equally important mineral industry. In part, this has been a function of the Turnerian focus on the pre-1890 era; in part, it has been an outgrowth of the pragmatic belief that glamorous "rush" periods will pay larger dividends on smaller investments where publishers are concerned. Moreover, the nature of the field has tended to discourage all but the most committed: serious research in the mineral industries requires some technical knowledge, either self-acquired or otherwise, of mining, milling, economics, or even accounting and law.

This study of the mining engineer is an effort to give an important body of men their deserved place in the written history of the West. It has not been a matter of striking rich bonanza paydirt, but rather of locating and working veins of low-grade ore in quantity. But if the process of locating, extracting, separating, and refining, has produced a product that helps in understanding western development the effort has been worthwhile.

Many people aided in bringing this project to fruition. Among others, I am indebted to the staffs of the Bancroft Library, the Huntington Library, the Denver Public Library, the New York Public Library, the Beinecke and Sterling libraries at Yale University, Columbia University, the New-York Historical Society, the State Historical Society of Colorado, the Idaho State Historical Society, and the Montana Historical Society. I owe a special debt to John Brennan of the Western Historical Collection, University of Colorado, and especially to Gene Gressley of the Western History Research Center, University of Wyoming, for his real zeal in collecting mining materials and making them readily usable. I should also like to thank a number of individuals who took time out of busy schedules to answer queries, make valuable suggestions, and often to make private materials available. Here I would include L. K. Requa of Salt Lake City; Mrs. Spencer C. Browne of Kensington, California; George O. Argall, George Hurst, Ira Joralemon, James Parks Bradley, Donald M. McLaughlin, and the late Henry C. Carlisle—all of San Francisco; Mr. and Mrs. Otis Marston, Francis Seeley Foote, Philip R. Bradley, Jr., and F. Milton Sizer—all of Berkeley.

I wish also to thank Daniel Barthell, John Cordulack, Joseph E. King, David Schob, and Barry Hoffman, who as graduate students at the University of Illinois made substantial contributions. For typing much of the manuscript, I am grateful to Ruth Lewis, Glenna Johnson, and Jeane Cooper, all of the office staff of the Department of History at the University of Illinois; and for her role as critic and proofreader I am indebted to my wife, Mary Lee.

I also wish to acknowledge the generous financial support, both direct and indirect, of the University of Illinois Research Board, the Ford Foundation, the Social Science Research Council, the American Philosophical Society, and the American Academy of Arts and Sciences.

C. C. S.

University of Illinois
August 1, 1969

ABBREVIATIONS

AJM	*American Journal of Mining, Milling, Ore Boring, Geology, Mineralogy, Metallurgy, etc.* New York, 1866–July 1869.
AIME *Trans.*	*Transactions of the American Institute of Mining Engineers.* New York, 1871–1932.
DAB	Allen Johnson & Dumas Malone, eds. *Dictionary of American Biography.* 22 vols. New York, 1943–1958.
EMJ	*The Engineering and Mining Journal.* New York, 1869–1932.
MI&R	*The Mining Industry and Review.* Denver, 1895–98.
M&M	*Mining and Metallurgy.* New York, 1920–40.
M&SP	*Mining and Scientific Press.* San Francisco, 1860–1922.
NCAB	*National Cyclopedia of American Biography.* 49 vols. New York, 1898–1966.
SMQ	*The School of Mines Quarterly.* New York, 1879–1914.

CHAPTER 1

Introduction

A toast to the adventurous!
Explorers, miners, engineers!
Who blazed the trail ahead for us,
The pioneers! the pioneers!
—Thomas A. Rickard [1]

"Mining engineering, in its most general sense, is the application of science to the discovery and working of mines, and to the subsequent treatment of ores." Thus did an early member of the profession, William Ashburner, define his field in 1874.[2] A more common and succinct definition was the one current in the 1920s: "An engineer is a man who can do for one dollar what any fool can do for two." [3] Both definitions are accurate, perhaps, but neither takes into account the complexity of the field. Geologist George Becker in 1877 thought mining engineering "the most complex of all the mechanic arts, involving almost all the rest." [4] Another in 1910 believed mining engineering "the most polyglot of all the professions. . . . Geologist, surveyor, lawyer, mechanic, chemist, metallurgist, mineralogist, electrician, this was the mining engineer of the old school." [5] The mining engineer of the latter part of the nineteenth century was a general practitioner—a jack-of-all-

1. "To the Memory of the Pioneers of American Mining," *EMJ, 79* (January 5, 1905), 1.
2. William Ashburner, "On the Profession of Mining Engineering," *Bulletin of the University of California,* no. 5 (November 1874), p. 3.
3. W. H. Shockley to editor, Palo Alto, February 14, 1922, *M&SP, 124* (March 4, 1922), 285.
4. George E. Becker, "Annual Report of the College of Mines," University of California, *Reports to the President of the University, from the Colleges of Agriculture and the Mechanic Arts,* p. 66.
5. Benjamin B. Lawrence, "Engineering Profession," *SMQ, 21* (April 1910), 207.

trades. His task included locating, developing, exposing, measuring, and removing ore from the ground; he was charged with reducing metal from the ore and often with marketing it. He devised intricate machinery, became an expert in mining litigation, and frequently engaged in corporate promotion and investment. Moreover, he carried his profession to the four corners of the world, becoming in the process an instrument for the diffusion of technological advance, ever borrowing, innovating, and adapting ideas and equipment to new environments. So long as technical knowledge had not advanced too far, the engineer was able to remain a general practitioner, but ultimately—even before the twentieth century—he began to develop specializations that might take him in any of half a dozen different directions.

It is with the mining engineer, both as a generalist and as a specialist, that this study is concerned, with a particular focus on his work and his impact in the trans-Mississippi West prior to the 1930s. Until recently, the mining engineer and the geologist (both often performed the same tasks) have remained shadowy and relatively obscure historical figures, eclipsed by an emphasis (perhaps even an overemphasis) on the earlier Forty-Niners, Fifty-Niners, or captains of industry on the western scene. It is only fitting, as historians think more and more of the mineral west in terms of a developing mineral industry with its own problems of capital formation, labor-management relations, and the application of technology, that this important cog in the industrial machine receive special attention.

Not a pathfinder, not a pioneer in the traditional sense, the mining engineer to date has been best remembered for his occasional contributions in other, especially the political, realms. Few monuments or biographies and only a handful of place names on the map [6] recall his role in the West, yet his was the

6. Raymond Peak in California was named by the State Geological Survey in 1865 for Rossiter W. Raymond, one of the most influential of the early mining engineers; Blake's Ravine and Blake's Sea, also in California, commemorate William P. Blake, and Hoover Lake was named for Theodore J. Hoover, then mine manager at Bodie and a brother of Herbert Hoover. Daggett County in

Mining centers of California and Nevada

hand instrumental in guiding the rapid exploitation of underground resources when the day of the short-lived mineral frontier was gone. It is hoped that this study can view the engineer for what he was, picture him against the background of his work, describe his actual professional role and accomplishments, delineate the problems he faced and the life he led, and assess the imprint that he made on the western environment during the years from the California gold rush down to the years of the Great Depression.

The mining engineer does not stand alone. He was a product of his times and his surroundings, and he was at once a causal factor and an outgrowth of the wide-ranging changes taking place around him. His role was small in California's early unsophisticated placer years, when mining was dominated by a pragmatic trial-and-error approach, by wasted effort, and often by capital loss. But his importance became clear after the late 1850s, when large-scale, deep-level operations began on the Comstock and when this approach and hydraulic methods brought both maturity and stability to the mines of California. With the decline of the Comstock by the late 1870s, he became a central figure in the production of Colorado's silver-lead ores, the silver-lead-zinc combinations of the Coeur d'Alene, and the copper of Butte. Wherever mining expanded, from Ajo to the Black Hills, from Cripple Creek to the Yukon, his advice was eagerly solicited and his ideas put into practice. He was instrumental in introducing two radical innovations, effective gold-dredging machinery and low-grade copper techniques, that revolutionized the industry after the turn of the century. In short, in the transition from unsystematic to scientific mining under way since the mid-nineteenth century, the engineer was the focal figure.

Who was he and what was his background? Perhaps a point of departure in answering this question is to indicate

---

northeastern Utah bears the name of Ellsworth Daggett, a prominent mine engineer who was instrumental in bringing irrigation to that portion of the state. Edwin Gudde, *California Place Names,* pp. 30, 136, 249; *M&M, 4* (August 1923), 430.

that the term "mining engineer" traditionally was used loosely and that for the purposes of this study a mining engineer was anyone—technically trained or otherwise—who did the work normally done by the profession. Modern writers particularly are prone to use the designation indiscriminately, applying it to almost anyone connected with the mineral field. Thus, Jesse R. Grant, son of President U. S. Grant, Ambrose Bierce, who for a time worked as an agent for a mining company, and even highwayman "Black Bart" have been so dignified,[7] although their careers do not warrant the use of the title.

It seems fairly clear that the mining engineer profession was in general white, masculine, and middle class. Despite W. Sherman Savage's comment that some Negroes in the West "became experts in the technical phases of mining," [8] there is little evidence that the Negro played any real part in this field. An editorial reference in 1870 mentions a mine near Hornitos, California, "where a white superintendent, a black foreman, and a force of yellow miners seem to do very well together." "Indeed," commented the editor, who was also an engineer, "one might expect distinctions to disappear underground, since there is no difference of color in the dark." [9] Henry O. Flipper, the first Negro graduate of West Point, has been called "America's first Negro mining engineer," [10] but Flipper was the exception. There may have been others of his race in the profession but they are not evident, and it has to be concluded

7. See *MI&R, 17* (March 19, 1896), 428; Don Ashbaugh, *Nevada's Turbulent Yesterday,* p. 307; Andrew Rolle, *California,* p. 290.

8. W. Sherman Savage, "The Negro on the Mining Frontier," *The Journal of Negro History, 30* (January 1945), 45.

9. *EMJ, 10* (November 22, 1870), 329.

10. Flipper graduated from the academy and served with the army in the West until 1882, when he was court-martialed and dismissed for alleged carelessness with funds, though to his death in 1940 he proclaimed his innocence. After 1882, according to his biographer, he spent thirty-seven years as a civil and mining engineer, including one year as "resident engineer" for the Balvanera Mining Company in Mexico. He was employed by Albert Fall on the legal staff of another concern and ultimately went with Fall to Washington, D.C., where in 1921 he became assistant to the Secretary of the Interior. Theodore D. Harris, ed., *Negro Frontiersman: the Western Memoirs of Henry O. Flipper,* pp. vii–ix, 42–43; Wesley A. Brown, "Eleven Men of West Point," *The Negro History Bulletin, 19* (April 1956), 149.

that never, throughout the era, was the Negro in the main current of mining engineering.

This branch of engineering was, of course, a man's world, into which but a few of the gentler sex intruded. Women occasionally invested in western mines, and the San Francisco Exchange had its "Mud Hens," feminine speculators in mining shares,[11] and here and there a woman was associated with the technical side of mining, but, as a San Francisco editor remarked in 1888, "a woman, as a skilled writer on metallurgy and mining, or as a superintendent of mines, would be a new thing under the sun."[12] In the 1890s, a few of the fair sex, including the wife of John P. St. John, Prohibitionist ex-governor of Kansas, supervised western mine operations, but the news that a woman had been appointed to manage a property in Australia was met with real skepticism. "To meet one of the gentler sex walking with heavy tread and grim determined gait through a level, with a pick in one hand and a tallow candle in the other, would be a great blow to our feelings," said a Denver observer. "Similarly a lady mining engineer is a revolutionary and anarchical departure from established precedent."[13]

The U.S. Census of 1900 showed only three female mining engineers,[14] and there is little evidence that many enrolled in the mining schools. In more than three quarters of a century, the Colorado School of Mines graduated only four women, and the first of them (in 1898) was in civil rather than mining engineering.[15] A survey of mining graduates of the University

11. Joseph L. King, *History of the San Francisco Stock and Exchange Board,* p. 224.

12. *M&SP, 42* (March 19, 1888), 178.

13. *M&SP, 67* (September 30, 1893), 210, *71* (August 25, 1895), 119, *76* (January 15, 1898), 62; *MI&R, 20* (February 3, 1898), 346.

14. In addition, 11 were listed under the heading "Surveyors" and 248 under "Chemists, Assayers and Metallurgists." U.S. Bureau of the Census, *Occupations at the Twelfth Census,* p. 7.

15. Two of the four took positions with the National Bureau of Standards. At least one of the others, Ninetta Davis, class of 1920, held several eningeering positions, including one with Shell Oil in Denver. *MI&R, 20* (June 9, 1898), 538; Jesse R. Morgan, *A World School: The Colorado School of Mines,* pp. 110–11.

of California, 1865 to 1936, includes no women at all.[16] Nor is there any indication that Columbia Mines enrolled any, at least until 1911, when Mrs. F. B. Flower, a "lady miner" at Porcupine in Canada, was reported to have passed the examinations and matriculated.[17] However, several women were members of the American Institute of Mining Engineers (AIME),[18] and it was announced in 1907 that Mrs. J. C. Gregory, "a lady mining engineer who will open an office in Columbus, Ohio," was inspecting the Coeur d'Alene mining region.[19] William Shockley, whose engineering career took him all over the West and as far afield as China and Abyssinia, married a "charming country-woman," May Bradford, described as "an engineer (formerly a U.S. Deputy Mineral Surveyor) and artist."[20] Another, Doska Monical, who for a time was with the U.S. Geological Survey, in 1920 was in charge of the inquiry department for Augustus Locke, mining geologist in San Francisco.[21] That women played no appreciable direct role as practicing mining engineers in the West is borne out not only by the technical literature, but also by such veteran engineers as Ira Joralemon of San Francisco, who recalls that Anaconda Copper for a time had a very competent woman engineer in charge of its Salt Lake office, but that, of the few women in the profession, almost none attempted to follow the normal opportunities of the career.[22]

Mining engineers in the West were a cosmopolitan lot, drawn from nearly every quarter of the globe, although undoubtedly the majority were American. Foreign engineers came either for a specific job, then returned home, or, as was the case with many, they migrated permanently to the West.

16. E. T. H. Bunje, F. J. Schmitz, and D. I. Wainright, *Careers of University of California Mining Engineers, 1865–1936.*

17. *EMJ, 91* (January 14, 1911), 108.

18. Robert Hallowell Richards, *Robert Hallowell Richards: His Mark,* pp. 198–99.

19. *Northwest Mining News* (Spokane), *1* (October 1907), 8.

20. Edward T. McCarthy, *Further Incidents in the Life of a Mining Engineer,* p. 367; *EMJ, 110* (August 14, 1920), 313.

21. *EMJ, 109* (January 31, 1920), 369.

22. Interview with Ira B. Joralemon, San Francisco, April 6, 1964.

English firms, such as John Taylor and Sons, Bewick Moreing, or Bainbridge, Seymour & Rathbone, did much engineering work in the western United States, and sometimes even maintained branch offices in San Francisco or Salt Lake City.[23] For them or for other clients came a steady parade of British engineers to inspect property or to manage it for English investors. William Petherick, a well-known English engineer whose "integrity was proverbial," died in the Cosmopolitan Hotel in San Francisco while on such a mission.[24] J. Arthur Phillips, a copper specialist and "one of the best metallurgists of the day," made several inspection trips to California and Nevada in the 1860s;[25] Ernest Woakes, Royal School of Mines graduate, included California and Oregon mine examinations in an active and varied global career;[26] and Edgar P. Rathbone, father of movie actor Basil Rathbone and a prominent engineer in the Transvaal, was one of the first mining engineers in the Klondyke.[27]

Other British mining engineers migrated and played a more permanent part in the unfolding drama of western mine exploitation. Philip Argall, for example, commenced his career in the Irish mines, built up a reputation in Australia–New Zealand and in France, and then came to Colorado for an English firm, there making a brilliant record.[28] Richard Pearce was born in Cornwall. At the age of fourteen, he went to work in the tin-dressing plants, but managed to study at the Royal

23. Between 1883 and 1900, John Taylor and Sons inspected at least a dozen western mines. Bewick Moreing set up a branch office in San Francisco, took Edward Hooper into partnership, and in 1891 shifted their American headquarters to Salt Lake City because of its central location. *EMJ, 46* (July 21, 1888), 49, *52* (November 28, 1891), 620; *M&SP, 60* (April 5, 1890), 236; *Who's Who in Mining and Metallurgy,* p. 42; John Taylor and Sons, MS Index Book of Inspections (1883–1900).

24. *M&SP, 13* (July 21, 1866), 33.

25. *M&SP, 11* (December 9, 1865), 361, *13* (October 20, 1866), 241; London *Times,* January 7, 1887; J. Arthur Phillips to Adolph Sutro, London, August 16, 1866, Adolph Sutro MSS, Box 9.

26. Margaret Reeks, *Register of the Associates and Old Students of the Royal School of Mines and History of the Royal School of Mines,* pp. 207–08.

27. *EMJ, 41* (June 12, 1886), 423, *46* (July 28, 1888), 68, *65* (June 11, 1898), 708, *114* (August 19, 1922), 316; *M&SP, 79* (September 9, 1899), 292. See also Basil Rathbone, *In and Out of Character,* pp. 19–22, 34–35.

28. *EMJ, 15* (Decmeber 5, 1891), 739–40; *DAB, 1,* 344–45.

School of Mines and at the Königliche Sächsische Bergakademie at Freiberg. Sent by clients to Colorado in 1871, he subsequently joined hands with Nathaniel P. Hill to establish one of the first successful smelters in the Rockies. Until he returned to England in 1902, Pearce was one of the best-known engineers in America: he was British vice-consul in Denver, twice president of the Colorado Scientific Society, and president of the American Institute of Mining Engineers.[29] J. H. Ernest Waters, also a Royal School of Mines man, with a dash of Freiberg, had followed his profession in Colorado, Japan, and China before returning to the Rockies in the early 1880s to manage the Sheridan and Mendota, mines owned by a group of Englishmen in Shanghai.[30] Arthur Buck Kitchener of the Denver engineering firm of Walters Brothers & Kitchener in 1900 was a brother of the famed British field marshall.[31]

Along with these and numerous other English engineers in the West were countless Germans, for in the 1860s Germany led the way in technical mining training. Philip Deidesheimer, whose "square set" timbering technique revolutionized western mining, was among them, as were Baron von Richthofen, later to achieve world fame as a geologist;[32] Carl Stetefeldt, son of a Lutheran minister and subsequently one of the pioneer metallurgists in California; and Albert Arents, who led the way in lead-silver smelting in the West.[33]

European engineers followed French capital across the Mississippi, though when the Société Anonyme des Mines de Lexington appointed Giovanni Lavagnino as its engineer-manager in 1886, an American editor believed that "if the company were to appoint the entire French corps of mining engineers, it

29. *EMJ, 37* (June 14, 1884), 438; *DAB, 14,* 353–54; William N. Byers, *Encyclopedia of Biography of Colorado,* p. 234.

30. *EMJ, 52* (August 22, 1891), 213, *54* (July 9, 1892), 36.

31. *EMJ, 69* (April 14, 1900), 446; Edwin Sharpe Grew, *Field-Marshall Lord Kitchener, 1,* 50.

32. Josiah D. Whitney to William Brewer, San Francisco, April 14, 1866, Brewer-Whitney MSS, Box 2.

33. *DAB, 1,* 343–44; *17,* 595; *Helena Herald,* March 19, 1869, November 1, 1871.

could not make its mines pay, nor bring back its badly invested money." [34] One Frenchman, George Marie Chartier, had fifteen years of professional experience in Colorado prior to his appointment as consular agent for France and Belgium in California in 1901, and André Chevanne, an École des Mines graduate, amassed a fortune during the forty years he practiced his profession on the Pacific Coast.[35] Anthony Luchich, a Dalmatian and graduate of the Polytechnic at Gratz, left the Austrian navy to become an American citizen in 1885, followed a mining engineering career in Colorado under his new name, Lucas, and eventually brought in the gusher at Spindletop that made him famous.[36] P. H. Van Diest, born in the Netherlands, came to Colorado in 1872 from the Dutch East Indies for his health, and carved out a distinguished engineering career for himself.[37] A Swiss, Alex Trippel, made a reputation for himself in the mines of Arizona in the late nineteenth century; Per Gustaf Linder, graduate of the Royal Mining Institute at Stockholm, superintended mines for James Ben Ali Haggin and others for many years after he came to the United States in 1882; and John O. Norbom, a Norwegian engineer, spent twenty years on the Pacific Coast before he was killed on the Berkeley–San Francisco ferry in 1911, by an explosion of some chemical in his pocket.[38] A Russian engineer was at Tombstone as early as 1880, and another, Boris Gorow, reportedly of royal blood, became a naturalized American and was killed by Mexican rebels in 1913.[39]

Gelasio Caetani, an Italian with an English mother, received his technical training at Columbia Mines, his baptism under fire at the Bunker Hill & Sullivan in Idaho, and when

34. *EMJ, 42* (August 14, 1886), 117.

35. *EMJ, 64* (November 13, 1897), 582; Wellington C. Wolfe, ed., *Men of California, 1900–1902,* p. 132.

36. Thomas A. Rickard, ed., *Interviews with Mining Engineers,* p. 310; *M&M, 2* (November 1921), 42.

37. *EMJ, 22* (August 5, 1876), v *75* (January 10, 1903), 93.

38. *M&M, 1* (July 1920), 28; *EMJ, 91* (January 28, 1911), 234; Ira B. Joralemon, *Romantic Copper,* p. 208.

39. Entry for April 7, 1880, George W. Parsons, *The Private Journal of George Whitwell Parsons,* p. 111; EMJ, *96* (July 12, 1913), 83.

World War I broke was a member of an engineering partnership in San Francisco. Subsequently, he joined the Italian army, became a national hero for his wartime exploits, returned as Italian ambassador to the United States, and was eventually an aide of Premier Mussolini and mayor of Rome.[40] Alexander Del Mar, a Spaniard, was trained in Madrid, but early migrated to California; Juan Feliz Brandes, born in Buenos Aires and educated in Berlin and at Columbia, practiced after 1899 in Mexico and along the Pacific Coast.[41] About the same time, a young Peruvian engineer named Parra put an end to his grief with a bullet in his head and died on the grave of his wife at Butte.[42] D'Arcy Weatherbe, son of a former chief justice of Nova Scotia, was a Canadian, a global mining engineer and an expert on western dredging, and Robert Postlethwaite, perhaps the first and the most important of the California dredge engineers, was a New Zealander by birth and training.[43] Tadaatsu Matsudaira, believed to have been the first Japanese in Colorado, was of noble birth, and served two years as assistant to the state inspector of mines, before his untimely death in the late 1880s.[44]

On the other hand, despite these and many other exceptions, the majority of western engineers were American by birth, but were members of a highly cosmopolitan fraternity. This became even more apparent because many of the Americans were themselves trained abroad in the 1860s and 1870s and because their services would be in great demand in foreign mining fields during the 1890s and thereafter.

The typical native-born mining engineer might be almost anyone. He might be drawn from almost any walk of life, but the odds favored his coming from a background of at least

40. *EMJ, 93* (February 3, 1912), 280; *M&SP, 112* (January 1, 1916), 28; *Autobiography of William Francis Kett*, pp. 147–48.

41. *Who's Who in Engineering, 1922–23*, p. 181; J. R. Robertson, *The Life of Hon. Alex. Del Mar, M.E.*, p. 2.

42. Kett, *Autobiography*, p. 76.

43. *EMJ, 65* (May 14, 1898), 588, *116* (August 4, 1923), 203.

44. *Empire* (Denver), *18* (July 2, 1967).

some means. He might be the son of financier William C. Ralston of California,[45] or of the Chief Justice of Rhode Island.[46] Like Daniel M. Barringer, he might be the son of a minister to Spain;[47] like Wilbur E. Sanders, he might be the son of a U.S. senator from Montana.[48] Karl Krug came from an old California wine-making family;[49] Robert Bunsen, a successful Colorado and New Mexico engineer of the 1880s, was a "near relative of the great chemist of that name."[50] Sherman Day, an engineer whose reputation in the early 1860s "probably stands second to none on the Pacific Coast," was the son of a president of Yale College, and Robert G. Brown's father was president of Hamilton College.[51] Others were sons of lawyers, merchants, physicians, publishers, or cotton brokers.[52]

Noticeable, too, was the tendency for a son to follow his father's footsteps in the profession. Thomas A. Rickard was of a family that at one time had eight members in the AIME. His grandfather, one of the earliest accredited mining engineers in the West, brought one of the first stamp mills to California, and his father was one of five brothers, all following the same calling.[53] Joseph H. Collins was known as "the grandfather of technology in Cornwall and the father of four good mining engineers": Arthur, the eldest son, had a global reputation by the time he was assassinated in Colorado in 1902, Henry was the author of a standard textbook on silver-lead metallurgy, and both Edgar and George had managed

45. *W. C. Ralston: A Business Man for Governor,* 1–3.

46. *NCAB, 14,* 440.

47. *NCAB, 22,* 317.

48. Franklin Harper, ed., *Who's Who on the Pacific Coast,* p. 496.

49. *EMJ, 93* (January 20, 1912), 184; William Durbrow Interview, University of California Regional Cultural History Project, typescript, pp. 31–32.

50. *EMJ, 44* (October 1, 1887), 237.

51. *M&SP, 9* (November 26 & December 3, 1864), 344–45, 359; William Brewer to George Brush (New Almaden, August 18, 1861), Brush family MSS; Eben Olcott to Henry Munroe, May 5, 1891, copy, Letterbook 17, Ebenezer E. Olcott MSS (N.Y.).

52. See biographical sketches of Eben Olcott, Louis Janin, N. P. Emmons, Albert Holden, and Albert F. Schneider for examples taken at random. *NCAB, 5,* 265, *18,* 11–12, *20,* 163, *22,* 230, *24,* 70.

53. Thomas A. Rickard, *Retrospect,* pp. 8–9.

mines throughout the West and as far afield as Russia.[54] Similarly, the Janins constituted another mining engineering dynasty in the West. Originally from Louisiana, the three brothers—Louis, Henry, and Alexis—came to be known wherever mining men congregated during the last half of the nineteenth century, and several of Louis' sons carried on the tradition: Eugene, the youngest, died of typhoid in 1894, just seven weeks after commencing his first professional work; Louis, Jr., began a brilliant career but retired early "on account of ill health"; but Charles carried the family name well into the twentieth century.[55] Frederick W. Bradley and Philip R. Bradley, sons of a pioneer California civil engineer, both attended the University of California at Berkeley, became top engineers, and passed the baton on to their sons. Fred Bradley wrote a business associate, Bernard Baruch, in 1927:

> Yes, my boys are growing up and the oldest one now has charge of a quicksilver mine in Lake County, this State. The next boy graduates from the College of Mining at Berkeley next May and then is going to a mining property in the Yellow Pine District, Central Idaho; and I have other mining properties in preparation for the next two boys to practice on.[56]

Charles F. Hoffmann, a Freiberg-trained pioneer, who was with the Lander expedition and the Whitney Geological Survey in California in his early years, had four sons who followed in his footsteps and were generally scattered "from Dawson to Johannesburg." Moreover, the family of J. Ross Browne, with whom the Hoffmanns intermarried, produced two generations of mining engineers.[57] Nor were these isolated instances; sons of numerous well-known engineers—

54. *M&SP, 112* (April 29, 1916), 617, *116* (June 8, 1918), 806; *Who's Who in Mining,* p. 19.

55. *M&SP, 95* (August 10, 1907), 158; *EMJ, 58* (September 15, 1894), 252, *97* (March 21, 1914), 633, *109* (May 29, 1920), 1215.

56. Frederick W. Bradley to Bernard Baruch, San Francisco, September 21, 1927, copy, Frederick Worthen Bradley MSS.

57. *EMJ, 96* (July 12, 1913), 87; *M&SP, 95* (August 10, 1907), 158; interview with Mrs. Spencer C. Browne, Kensington, California, March 8, 1964.

James D. Hague, Arthur D. Foote, William A. Dennis, Philip Argall, John Farish, and Richard Pearce, to name a few—carved out niches for themselves in the same profession.[58]

A number of the earlier engineers came up in the school of practical experience from poor backgrounds, and some of the later group, such as bluff Daniel Cowan Jackling, worked their way up through mining school from poverty,[59] but probably most came at least from a middle-class environment, and some from well-to-do families. Technical education was expensive, and as Thomas Rickard noted in 1916, a good start was vital —"such as a father able to pay for the necessary education, a kind uncle to give the graduate a job, and friends glad to give a push when most needed." [60]

Familiarity with the work and the life undoubtedly helps explain why sons of mining engineers took up the profession. In some instance, the pressure of parents who were not in the field was decisive. Louis Janin, Sr., a lawyer by trade, sent three of his sons to Freiberg for technical training in the 1850s and 1860s because he believed, as a result of his connection with California legal cases, that engineering was a field of rapidly increasing importance and opulence.[61] Alexander Del Mar was given a mining education because his father, a Spaniard, wished him to supervise silver mines owned by the family.[62]

Young Robert Livermore first wanted to be a sailor or a cowboy and tried his hand at both, but mine-sampling experience in Wyoming persuaded him to take up engineering.[63] Others were undoubtedly influenced by contacts with teachers and with established engineers. Roy McLaughlin made the de-

58. *EMJ, 82* (July 28, 1906), 174, *86* (August 29, 1908), 437, *89* (January 15 & 22, 1910), 186, 236, *95* (May 31, 1913), 1117; Arthur D. Foote, "William Hague," copy, Arthur DeWint Foote MSS.

59. Rickard, ed., *Interviews*, pp. 193–94.

60. *M&SP, 112* (May 20, 1916), 735.

61. Rossiter W. Raymond, "Biographical Notice of Louis Janin," AIME *Trans., 49* (1914), 833.

62. Robertson, *Life of Del Mar*, p. 2.

63. Gene M. Gressley, ed., *Bostonians and Bullion: The Journal of Robert Livermore, 1892–1915*, pp. 4–28, 45–69.

cision after reading a U.S. Geological Survey publication on Leadville and discussing it with his favorite high school instructor, a former classmate of Herbert Hoover at Stanford. Ernest A. Wiltsee was prepared to enroll at Harvard, where his mother hoped he would study for the ministry, but a chance meeting with Frederick Corning, an engineer en route to Leadville in the late 1870s, so impressed Wiltsee that he went instead to the Columbia School of Mines.[64] William Durbrow registered at the University of California in the 1890s, intending to study law, but was so infected with the enthusiasm of a mining man at dinner one night that he changed his curriculum.[65] Others commenced their training in the field of civil or electrical engineering, then moved to the mining course in college because they found it more interesting.[66]

No doubt some were influenced by the stereotype of the mining engineer—the dashing, daring, two-fisted, he-man adventure—depicted by Richard Harding Davis and other turn-of-the-century novelists, or by the success stories of such eminent and highly publicized engineers as John Hays Hammond and Herbert Hoover. And supply followed demand: the profession was an expanding one, and when editors commented on the need for more skilled engineers, as they did repeatedly in the 1880s and 1890s,[67] this surely helped attract recruits.

Moreover, mining engineering in America was accorded a higher status than in other countries, especially England. At the end of a busy career, Herbert Hoover looked back and commented:

> It was the American universities that took engineering away from rule-of-thumb surveyors, mechanics, and Cornish foremen and lifted it into the realm of application of

64. Ernest A. Wiltsee, "Reminiscences," 4b, typescript, Ernest A. Wiltsee MSS, Roy P. McLaughlin, *The Tenderfoot Comes West,* pp. 42–43.

65. Durbrow interview, p. 14. See also Ralph M. Ingersoll, *Point of Departure,* p. 83.

66. *M&SP, 123* (September 10, 1921), 354, *124* (February 25, 1922), 253; interview with Francis Seeley Foote, Berkeley, California, May 8, 1964.

67. See *EMJ, 29* (February 7, 1880), 95, *36* (September 1, 1883), 125, *82* (August 18, 1906), 319; Louis Janin to John Hays Hammond, San Francisco, July 30, 1896, John Hays Hammond MSS, Box 3.

> science, wider learning in the humanities with the higher ethics of a profession ranking with law, medicine and the clergy.[68]

On the eve of World War I, engineer-journalist Thomas A. Rickard noted the difference in prestige the profession carried in the United States and in Britain when he wrote that the young American woman regarded the mining engineer "as an energetic explorer on the frontier, as a resourceful technician that finds the metals required in civilized life," whereas the young Englishwoman had "a vague idea that the mining engineer is a somewhat nomadic person connected with queer doings on the stock exchange." [69] In the same vein, Herbert Hoover tells of being with a charming female companion, an Englishwoman, on a trans-Atlantic voyage. When the trip was nearly over, in response to her query, he told the lady he was a mining engineer and she was shocked. "Oh, I thought you were a gentleman," she said! [70]

In America, prestige and a reasonable standard of living went with the profession, but undoubtedly, for many, certain intangibles were among the benefits. As one engineer-editor wrote in 1910: "When we consider what such a life means—isolation, rough surroundings, poor food, danger and the thousand and one inconveniences which one has to endure, we feel at a loss to offer a sensible explanation." [71] Prospects for adventure, excitement, and friends in international circles were positive attractions, and engineers might derive a certain satisfaction from constructive accomplishment. Eben Olcott, a young New York engineer, confided in 1878, "Sometimes I have wished that I had selected a different occupation, but then look what I would have missed. There are more wonder-

68. Herbert Hoover, *The Memoirs of Herbert Hoover,* 3 vols., *1,* 131.

69. Rickard, *Retrospect,* p. 138. In the United States, said J. H. Curle, an English engineer, the mining engineer rates "high in the social scale; he ranks with the best." But in Britain, to many people, "he figures as a superior mechanic" or "the superintendent of a coal pit"; "to women. . . he is a sort of a stoker." J. H. Curle, *The Shadow-Show,* p. 56.

70. Hoover, *Memoirs, 1,* 131–32.

71. R. Stuart Browne in *The Pacific Miner* (Portland), *16* (May 1910), 163.

ful satisfactions from being right than there ever are disappointments in making the wrong decision." [72] Years later, another, then in the twilight of his career, remarked that he had not made a great deal of money as a mining engineer, but that he did not consider this a major consideration:

> To be a spectator at the pageant of humanity, to observe the marts and peoples of the world, to be sufficiently in touch with affairs, to avoid becoming the theoretical dry as dust, and to have enough of earth-science to read with keen enjoyment of the parched desert or the snowy sierra, these are the privileges of the engineer who elects to take the world for his oyster, and to open it.[73]

72. Eben Olcott to Katharine Olcott, November 20, 1878, Ebenezer E. Olcott MSS (Wyo.), Box 1.

73. Chester W. Purington, quoted in *M&M, 4* (November 1923), 579.

CHAPTER 2

"& They Call That a Mining School!"

With the summer sun and sunshine come green and greener trees;
Come the college boys to work the while to earn tuition fees,
With a chance to hatch an egg or two of theories in mind,
In the old-time incubator of experience and grind.

—"Summer Jobs for College Students"[1]

The mining engineer, it was said, achieved his status by doing the work of a mining engineer. This, of course, oversimplifies the matter. Once a young man had decided on the pursuit of such a career, he did not attain his goal merely by proclaiming his intent and going to work. He must enter by one of two doors—experience or technical training. Increasingly, as the nineteenth century wore on, he found that these were double doors, to be used in conjunction with each other. During much of the nineteenth century, most mining engineers were of the practical variety—men who through circumstance, ability, hard work, and experience in the different aspects of mining and milling had won their positions without special education. But from the Civil War on, the trend moved in the opposite direction, at first slowly, then more rapidly, and in 1921 it was estimated that of the major mining engineers in the country, six out of seven were college-trained.[2]

Although time saw him decrease in number and in importance, the self-taught engineer who learned the "dips, spurs, and angles of mining" on the job was never one to be ignored. The Comstock Lode was developed (albeit wastefully) by men of this ilk, though it has been pointed out that many of

1. From *Safety Bulletin* of the Nevada Consolidated Copper Company, quoted in *EMJ, 100* (August 28, 1915), 366.

2. Thomas A. Rickard, "The Education of a Mining Engineer," *M&SP, 123* (December 10, 1921), 815.

the old Comstockers were more business managers than mining experts. But they were a colorful lot, equally at home in selecting fast horses or good subordinates, who brought to bear much experience, some talent, and varying degrees of success. Among the prominent Comstock superintendents was showy Bob Graves of the Empire; genial Thomas Taylor of the Yellow Jacket; handsome Fred Tritle of the Alpha. Isaac Requa of the Chollar was considered "the most industrious"; James G. Fair of the Consolidated Virginia "the best miner"; and Sam Curtis of the Ophir, "Glorious old boy that he was," "the oldest, biggest, and ugliest." [3] Their experience was not merely in mining; rather it embraced many areas—agriculture, merchandising, banking, publishing, and transportation—and their focus was on handling men, rather than techniques.

Apart from the old Comstockers was another variety of engineer, more visionary, more adaptable, more theoretical, yet at the same time self-taught and self-made. In the early period, these broad-gauge men were invariably classed as "professors," a term reserved not only for learned geologists and engineers, but also for bootblacks and red light piano players. Thomas Davis was typical. A world traveler, veteran of the Texas revolution and the Mexican War, this "frontier mineralogist," as described by a contemporary in 1877, already had a quarter of a century in both practical and theoretical mining and was "hardy, studious, observant, self-opinionated, but wise in his specialties, honest and temperate in character and habits and almost as open and simple as a child in his ways." [4] Another, "Professor" Frank Stewart, a California Argonaut and filibusterer with Walker in Nicaragua, "might have passed for a twin brother of Abraham Lincoln." Self-tutored, widely read, and highly respected (even for his curious earthquake theories), Stewart was considered "a wonderful geologist, botanist and all around scientist," and more than once clashed

3. Roswell K. Colcord, "Early Comstock Days," *M&SP, 124* (March 11, 1922), 329; *Virginia Evening Chronicle* (Virginia City, Nevada), August 3 & 8, 1877; Thomas Donaldson, "Idaho of Yesterday," typescript, 1941, p. 381.

4. "R. J. H" Hinton to editor, Tucson, May 15, 1877, San Francisco *Post,* May 26, 1877.

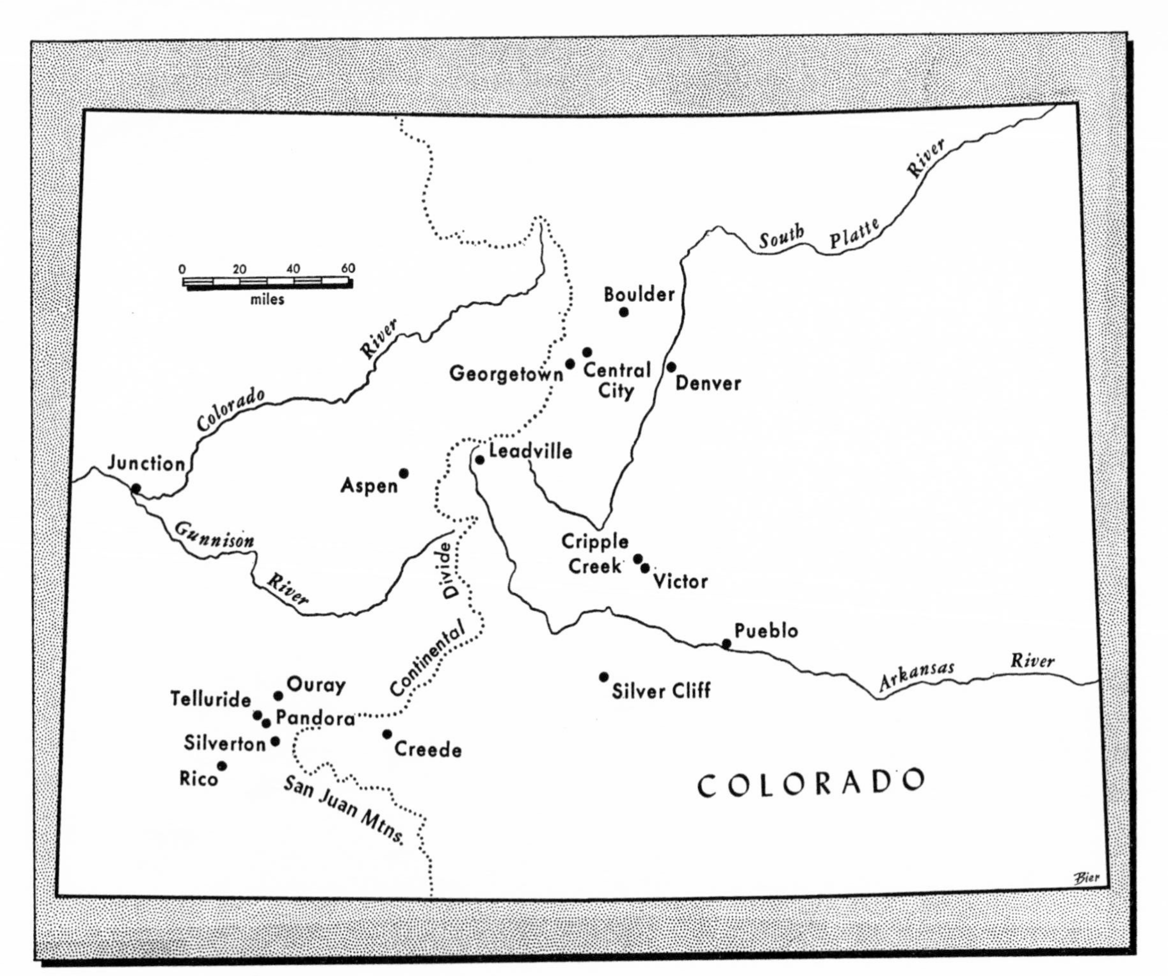

Mining centers of Colorado

with such stalwarts as Benjamin Silliman, Jr., of Yale, over questions of vein formation and came off best.[5] Even more widely known was Joshua E. Clayton, who had been born in squalor and poverty in the hills of Georgia. A mining man even before he came to California at the time of the rush, with only a rudimentary education, Clayton was an "omnivorous reader, a lifelong student of nature," who, according to one admirer, "would have been a boon companion of John Muir." Although for forty years he made "but a doubtful living" as an engineer in the West, Clayton attained the top rung of recognition in the field and at the time of his death in a stagecoach accident in 1889 was "without a peer in his profession." [6]

The career of William Kett illustrates how a self-made engineer learned his trade. Born in Massachusetts in 1870, young Kett grew up in Chicago, and his mother hoped he would go east to MIT. But for financial reasons, Kett forwent college and, at the age of eighteen, took a job as draftsman and machinist at $100 a month for a mining company at Elkhorn, Montana, ostensibly managed by G. C. Vawter of Helena, but actually under the charge of superintendent J. W. Pender, who was to give the young Chicagoan a liberal education in the various branches of mining and milling. As Pender's private secretary, Kett became familiar with many typical problems; as assistant to the assayer and chemist, a young Missouri School of Mines graduate, he learned much; at the same time, an old Comstocker palsied by mercury poisoning taught him amalgamating and retorting. Kett would follow Pender to England and to Central America as a kind of right-hand man, learning all the while and becoming an engineer (sometimes in spite of his mentor's slipshod practices) by the simple process of doing the work of an engineer.[7]

5. Charles C. Goodwin, *As I Remember Them,* pp. 192–93.

6. Ibid., pp. 236, 238–39; *EMJ, 48* (August 3, 1889), 90–91; *Helena Herald,* July 5, 1889.

7. Kett, *Autobiography,* pp. 19–21, 23–25. Amalgamating was based on the proclivity of mercury to unite readily with gold or even silver under the proper circumstances. Quicksilver combined with the metal from pulverized ore to

This experience was repeated many times over, with variations, by others who achieved eminence in the profession. William H. Storms, who for a time served as California state mineralogist (1911–13), had drifted West as a teenager, drifted in and out of journalism, and then settled into mining, first as a day laborer, then as manager, and finally as consulting engineer.[8] William A. Farish, one of an illustrious pair of mine-engineering brothers who advised George Hearst and other major investors, came up the ladder of practical experience.[9] So did Albert Burch, son of a Methodist minister, who, after a single year of college, turned to construction work on the Burlington Railroad. From there, Burch proceeded to railroad surveying, then mine surveying, and finally, after study on his own and the leasing of a small mine, he amassed the experience that ultimately brought him the superintendency of the great Bunker Hill & Sullivan.[10]

No survey of mining engineering in the West would be complete without mention of a highly important triumvirate—Hamilton Smith, Henry C. Perkins, and Thomas Mein—all disciplined in the hard school of experience. Born in Kentucky and educated in New Hampshire, Smith early went to work in his father's coal mines and cotton factory in Indiana. There and in Kentucky, he developed several collieries, but was drawn to California in 1869 and soon became manager of the North Bloomfield mine in Nevada County, where he was instrumental in the establishment of low-cost powder works and where he became the recognized authority on hydraulic mining techniques. When hydraulicking was drastically curtailed after the debris controversy, Smith turned to quartz mining in Alaska and in Venezuela and then, with Edmund DeCrano, formed the Rothschild-backed Exploration Company, Ltd., in

produce an amalgam, and the mercury was then removed by volatization in a heated iron retort. Albert H. Fay, "A Glossary of the Mining and Mineral Industry," U.S. Bureau of Mines *Bulletin,* no. 95, 30, 567.

8. Storms was dismissed as state mineralogist by Governor Hiram Johnson in a patronage squabble and was subsequently shot and killed by a miner trying to interest him in a prospect. *M&SP, 115* (August 25, 1917), 261.

9. *EMJ, 103* (June 2, 1917), 981.

10. Rickard, *Interviews,* pp. 97–98; *M&SP, 121* (August 28, 1920), 295–96.

London, a pioneer among exploration and finance companies in mining.[11] Henry Perkins had migrated to California at the age of seventeen to do office work for Frederick Law Olmsted, who in the 1850s was trying to develop the mineral resources of John C. Frémont's Mariposa Estate. When these efforts failed in 1865, Perkins moved to the New Almaden Quicksilver Company, using this as a springboard to the North Bloomfield, where he became Hamilton Smith's assistant in 1871.[12] Thomas Mein came to California in 1854, and for thirty years built a mining reputation in Nevada County, when Smith persuaded him to take charge of the El Callao mine in Venezuela. Perkins likewise spent time at the El Callao, and, ultimately, he and Mein joined the exodus of American engineers to South Africa, where both their reputations and fortunes reached new heights.[13]

A few of the early engineers in the British tradition came up through the formal apprenticeship system. John Callow, who later made real contributions in the field of mining and milling machinery, was articled to a firm of engineers in Norwich, then did formal course work at a local school, traveled to Australia, and completed the apprenticeship before coming to Colorado in 1888.[14] Some were "privately educated," which usually meant that they learned the profession under the watchful eyes of their engineer-fathers. George Attwood, whose career spanned the Atlantic, studied under his father, Melville Attwood, and began active work for him at the age of sixteen; two years later, he held the responsible position of assayer, metallurgist, and chemist at the Ophir Company on the Comstock.[15] Henry P. Lowe, prominent Colorado engi-

11. *EMJ, 70* (July 14, 1900), 34; *DAB, 17,* 273; see also clippings from the London *Mining World,* July 14, 1900, Samuel Benedict Christy MSS. Hydraulic mining, or hydraulicking, was a technique developed in California. A powerful jet of water washed gold-bearing earth or gravel, carrying it through sluices, where the gold was separated by specific gravity or by the use of mercury. Fay, "Glossary," p. 352.

12. Rickard, ed., *Interviews,* pp. 413–14.

13. *EMJ, 69* (May 12, 1900), 554.

14. *M&SP, 117* (August 3, 1918), 142.

15. *Who's Who in Mining,* p. 3.

neer, was educated in the profession by his father, Theodore H. Lowe, a Kentucky-trained civil engineer who also turned to mining.[16] The four sons of Joseph H. Collins were "privately" educated at home,[17] and Louis Janin's sons, Charles and Louis, Jr., were educated primarily in the offices of their father, though both had some formal course work.[18] Young Clifford Dennis spent his boyhood around the mines in New Almaden, or Nevada City, where his father was manager, took a brief special course at Berkeley, and built up a vast store of experience as a migratory "ten-day miner" in California, Nevada, New Mexico, and Arizona before coming into his own as an engineer.[19]

Editors of the technical journals sometimes urged practical mining men to take advantage of correspondence courses,[20] and occasionally an engineer did receive such training. Frank Johnesse of Boise, for example, completed a full four-year mining course from the International Correspondence School in Scranton.[21] But the need and desirability of formal training rapidly became apparent, and even mining men who made no pretense of following the engineering career often took time to learn the rudiments of assaying, elementary geology, or mineralogy.[22] And completion of a technical course in an American

16. *M&M, 1* (April 1920), 24; Marshall Sprague, *Money Mountain*, p. 13.

17. *Who's Who in Mining*, p. 19; *Who's Who in Engineering, 1922–23*, p. 284.

18. *EMJ, 97* (March 21, 1914), 633; *109* (May 29, 1920), 1215; Handwritten note by Louis Janin (July 1893), James Duncan Hague MSS, Box 12.

19. *M&M, 10* (March 1929), 168.

20. It was pointed out in 1899 that the United Correspondence Schools of New York offered excellent courses in mining and metallurgy. *EMJ, 68* (July 22, 1899), 98. Apparently, Richard Rothwell had endorsed this school and had influenced a number of his fellow engineers to subscribe to stock, but by 1901 it was in trouble. Daniel M. Barringer to James Douglas, July 25, 1901, Letterbook E, Daniel M. Barringer MSS.

21. *Who's Who on the Pacific Coast*, p. 300. See also *Pacific Miner, 10* (September 1906), 5; George Sutcliffe, ed., *Who's Who in Berkeley, 1917*, p. 18. (n.p., n.d.)

22. Examples are Winfield Scott Stratton of Colorado and Charles D. McLure and William A. Clark of Montana. Harry J. Newton, *Yellow Gold of Cripple Creek*, p. 23; unidentified clipping, Scrapbook of Montana clippings, Western American Collection, Yale University; William D. Mangam, *The Clarks: An American Phenomenon*, p. 45.

or foreign university was regarded as highly desirable, even in the 1860s and 1870s, and increasingly as a sine qua non in the years that followed. Prior to about 1870, probably most of the trained mining engineers were the products of European schools; after that time, American institutions took over the role of providing skilled personnel.

On the Continent, undoubtedly the most important school, so far as the American West was concerned, was the Königliche Sächsische Bergakademie at Freiberg in the silver-lead country of Saxony, just up the valley of the Mulde from Dresden. Freiberg in its day was to mining what Heidelberg and the Sorbonne were to arts and letters and numbered among its students some of the brightest lights in the field of geology and engineering. Academicians George Brush of Yale, Thomas Drown of Lafayette and MIT, and Waldemar Lindgren of the U.S. Geological Survey and MIT all studied there.[23] So did Raphael Pumpelly, "the American Humboldt" —engineer, geologist, archaeologist, and explorer—and Rossiter W. Raymond, the most versatile of all, a founder and guiding light of the American Institute of Mining Engineers for more than forty years, "sailor, soldier, engineer, lawyer, orator, novelist, storyteller, poet, biblical critic, theologian, teacher, chess-player."[24] It was at Freiberg that Pumpelly first formed what was to be an enduring friendship with "a tall and strikingly handsome American," James Duncan Hague, and with the Janin brothers, Louis and Henry.[25] Here studied Franklin Guiterman, later executive director of the great American Smelting & Refining Company; James B. Grant, subsequently governor of Colorado; A. J. Seligman, a financial giant in Montana and New York; and John Hays Hammond, that "wizard of modern mining," who ultimately

23. *Festschrift zum hundertjährigen Jubiläum der Königl. Sächs. Bergakademie zu Freiberg,* pp. 277, 287, 291; Frederick G. Corning, *A Student Reverie,* p. 35.

24. Ibid., p. 35; *Festschrift,* pp. 279, 286; Thomas A. Rickard, ed., *Rossiter Worthington Raymond,* pp. 4–5.

25. Raphael Pumpelly, *My Reminiscences, 1,* 117, 138.

would become the best known of all mining engineers except Herbert Hoover.[26]

Only a handful of American students registered at Freiberg before 1850, and the number fluctuated from one to six each year during that decade, with a sharp upswing during the early 1860s to twenty-five.[27] When William P. Blake visited the school as California commissioner to the Paris exposition in 1867, he was welcomed enthusiastically at the Hôtel de Saxe by about forty American students, including six or seven from California.[28] In the late 1860s, probably about half the Bergakademie's enrollment of slightly more than a hundred came from the United States,[29] and after that it began to decline, so that in 1876 only eighteen out of 139 were American.[30] Such statistics, however, take no account of the fact that a substantial number of European students studying at Freiberg eventually migrated to the Rockies and the Pacific Coast.[31]

American students came with varying backgrounds, but until the early 1870s most had college work before they entered. Almon D. Hodges had served in the Civil War and had completed work at Harvard;[32] James D. Hague had a Harvard degree and had attended the universities of Göttingen and Hanover;[33] John Hays Hammond had graduated from Yale;[34] Charles C. Rueger had graduated from the Institute at Benicia, California, had undertaken additional training in a

26. Corning, *Student Reverie,* p. 35; John H. Hammond, *The Autobiography of John Hays Hammond, 1,* 63–73; Edwin Wildman, *Famous Leaders of Industry,* p. 123.

27. William F. Faber had enrolled in 1846, but records list his address as simply North America. Henry Rohdewald of Baltimore and Franz Lenning of Philadelphia entered in 1849. *Festschrift,* pp. 270, 272, 272–290.

28. *M&SP, 15* (September 28, 1867), 194.

29. Cited in *Opinions of the Press and of Eminent Public Men on the Importance of Our Mineral Resources and the Advantages To be Derived from the Establishment of a National School of Mines,* p. 11.

30. *EMJ, 15* (June 10, 1873), 360.

31. See pages 238–40, this book.

32. *EMJ, 52* (July 18, 1891), 71.

33. *DAB, 6,* 87.

34. Hammond, *Autobiography, 1,* 63; Hammond to S. L. M. Barlow, New York, July 8, 1884, S. L. M. Barlow MSS, Box 165.

San Francisco chemical laboratory and assay office, and had worked briefly in the mines of the Comstock.[35] A. F. Wuensch, on the other hand, had no such academic background, but had been in the newspaper business in Leadville before enrolling at Freiberg.[36]

College work was not essential for admission, but applicants were required to pass examinations roughly equivalent to those for admission to Harvard, to be followed by a practical four months' preparatory course in the mines and smelters. Most American students at Freiberg were not degree candidates, although many remained for three years and some even longer.[37] If they were not working toward degrees, they were not required to take examinations on their work, and because the Bergakademie used the lecture system, with lax discipline, an American student could do very little, if he so desired. A correspondent of the Chicago *Tribune* who visited the school in 1866 noted that about half the Americans there were industrious and studious, a quarter were working but not profiting much, and the remainder were simply taking life easy.[38] Other observers agreed that a few Americans did not exert themselves, but that they could not match the German students for dissipation.[39]

As individuals, American students reacted differently to the Freiberg environment. George Brush, later professor at Yale, thought the town itself "very stupid"; "were it not that one has plenty of study, I think I would prefer—for enjoyment—

35. *M&SP, 18* (June 5, 1869), 353; *Progressive Men of the State of Montana,* pp. 395–96.

36. *M&SP, 100* (February 5, 1910), 242.

37. Louis Janin to Louis Janin, Sr., Freiberg, April 23, 1860, Janin family MSS, Box 10. Between 1872 and 1886, thirteen Americans graduated from Freiberg, but a total of 249 attended during that period. R. H. Richards, "American Mining Schools," AIME *Trans., 15* (1886–87), 329.

38. "A college at home is bad enough," said the *Tribune,* "but an institution of that kind abroad, and in a community where little moral restraint is thrown around the young men, there is no limit to the recklessness which may be indulged in by those who have plenty of money and but little self-restraint." Quoted in *M&SP, 13* (July 28, 1866), 53.

39. J. C. Bartlett, "American Students of Mining in Germany," AIME *Trans., 5* (1877), 441.

some portion of the interior of Africa." [40] Nor was Brush impressed with his studies in metallurgy and geology, which he termed "miserable." "I really believe it would be possible for a man who had never seen the inside of a smelting works, to give a far better course than is given here," he complained.[41] Brush also found fault with his fellow students, although their cosmopolitanism impressed him. Their "beer-drinking & 'kneiping' habits," caused him to shun the German students; the Spaniards and Latins he considered "lazy"; whereas the English were "very narrow minded & selfish." [42]

On the other hand, young Alexis Janin, following in the footsteps of his two older brothers at Freiberg, held much more charitable views of the town, the Bergakademie, and its personnel, professors and students alike. Despite the hard work of the practical side of his course, Janin in 1865 enjoyed himself, although he admitted that there were too many Americans at the school and that one was no longer "forced to study in self defense and to pass away the time." [43] Many of the Americans were spoiled, he said, and came with $80 or $100 a month for expenses, whereas the $50 a month he received from his father was perfectly adequate.[44] And although Alexis helped arrange a Fourth of July dinner for the American students that year and "as a souvenir of old times" hung a Confederate flag out of his window, he tended to ignore most of his countrymen and to associate instead with the German students, even switching meals from the Hôtel de

40. George Brush to Benjamin Silliman, Sr., Freiberg, November 10, 1854, Brush family MSS.

41. Brush did admit that the course in engineering mathematics was excellent. Ibid., Freiberg, March 5, 1855.

42. In 1855, the student body included one from Siberia and two from elsewhere in Russia, two from England, and one each from Italy, France, Spain, Sweden, Denmark, Chile, and Bolivia. Ibid.

43. Alexis Janin to Juliet C. Janin, Freiberg, November 3, 1865, Janin family MSS, Box 5.

44. His father also bought his books, paid the Bergakademie expenses, and in addition was billed for a blowpipe apparatus purchased by his son. Alexis Janin to Juliet C. Janin, Freiberg, May 11, 1865, and Alexis Janin to Louis Janin, Sr., Freiberg, May 11, 1866, ibid.

Saxe, monopolized by the Americans, to "a regular German hotel." [45] Both Janin and Augustus Bowie of California, who gave him piano lessons, fought duels, which by then were mere social occasions.[46] Edwin Hatfield Garthwaite, who attended Freiberg a dozen years later, noted that dueling still went on, but that the wounds "in most cases do not amount to much more than glorified razor cuts." [47]

Like many other students, Garthwaite was intrigued with the cobblestone streets and tile roofs of Freiberg and with the internationalism of its student body. Like other students, too, Garthwaite quickly bought the conventional miner's outfit, including the traditional brimless hat with lamp and exceptionally thick felt top, a leather apron on which to sit in damp places, and the inevitable tall boots—the insignia of every engineering student.[48]

Garthwaite fitted into the little American colony more readily than Alexis Janin, and a noticeable esprit de corps held together most of the English-speaking students at Freiberg. At a disadvantage because of the language, they tended to flock together. In the 1860s and 1870s, they maintained their own reading room, containing many journals and magazines from home.[49] Edward Dyer Peters, subsequently a leading metallurgist and professor at Harvard, was president of the American colony in 1869, and even as late as 1885 the Anglo-American student club at Freiberg collected maps, reports, and papers relating to American mining.[50] In the twentieth century, nostalgic engineers in the United States organized the Old Freibergers in America, a group that met periodically at the

45. Alexis Janin to Juliet C. Janin, Freiberg, July 9, 1865, and Alexis Janin to Louis Janin, Sr., Freiberg, May 11, 1866, ibid.

46. "I fought my first duel on the other day," Alexis wrote his mother, "and had the good luck to cut my adversary and to come off without a scratch. Bowie fought at the same time. Both he and his adversary got an insignificant scratch." Alexis Janin to Juliet C. Janin, Freiberg, November 3, 1865, ibid.

47. Edwin H. Garthwaite, "Reminiscences of a Mining Engineer," typescript, p. 20, ca. 1936.

48. Ibid., p. 17.

49. *Opinions of the Press,* p. 11.

50. *EMJ, 103* (March 3, 1917), 380 *40* (July 18, 1885), 37.

Hofbrau-Haus in New York to quaff their lager and reminisce.[51]

The Bergakademie combined theory and practice. The practical aspect, covering both mining and smelting, was a spring and summer program designed to familiarize students with mining and milling practices before the regular fall classes convened. "The practical course was very fatiguing," wrote young Alexis Janin in 1865, as he described his normal routine of arising at 5:30 A.M. and walking two miles to the smelting works, where he remained four hours, observing and taking notes.[52] There, Edwin Garthwaite, with two years of the University of California behind him, had his first taste of underground drilling, timbering, and track-laying, and learned to climb 1,000 feet to the surface with dispatch when the half-day's work was done.[53]

By the 1870s, Freiberg had four principal courses: mining engineering, metallurgical engineering, mine surveying, and iron mine engineering and metallurgy. Most American students combined courses in mining engineering and metallurgical engineering, probably on something of a "pick and choose" basis. Except for the preparatory course and work in drawing, surveying, determinative mineralogy and petrography, physics and chemical analysis, blowpipe and assaying, all teaching was done by the lecture system, with a minimum emphasis on the laboratory, an emphasis that American schools would soon shift. The course in blowpipe reflected the influence of Carl Plattner, who perfected blowpiping as a science, and students learned the technique under Plattner or his successor, Theodore Richter. Mathematics, physics, and mechanics they received from Julius Weissbach, and they often dozed through the dull mineralogy lectures of his son, Albin Weissbach. Geology they learned at the knee of the great Bernhard Von Cotta or his popular protégé, Alfred Stelzner. Both Von Cotta and

51. *EMJ, 95* (January 4, 1913), 38.

52. Alexis Janin to Juliet C. Janin, Freiberg, August 22, 1865, Janin family MSS, Box 5.

53. Garthwaite, "Reminiscences," p. 18.

Stelzner, as well as Élie de Beaumont of the École des Mines at Paris, subscribed to a compromise theory on the formation of ore deposits and attempted to reconcile the Wernerian neptunistic idea that veins were formed by deposition from surface waters and the Huttonian plutonistic concept that they were the result of igneous actions. Both aqueous and igneous agencies were important, insisted the Von Cotta–Beaumont school, but the latter were the essential features.[54]

After leaving Freiberg, many American students spent time at other mining schools on the Continent and in addition made a tour of the mines and smelters in eastern Europe.[55] No exception, the Janin brothers capped off their education with a stay at the École des Mines in Paris, a sojourn that was quite brief for the two oldest ones, but quite lengthy for Alexis, the youngest, who was not at all anxious to leave Europe and who spent twelve years there before completing his instruction.[56] Charles Rueger climaxed his study at Freiberg with additional work at schools in Clausthal and Berlin, made a seven months' circuit of the mines, then returned to his native California amidst considerable fanfare from the press.[57]

If Freiberg was the most popular and the most important, American students also traveled to other European schools for mining engineering training. Clausthal, in the Hartz mountains, "one of the most celebrated institutions of its class in Europe," attracted some. Under the patronage of the Prussian government, Clausthal was open to all nationalities, and

54. Corning, *Student Reverie,* pp. 19–20, 28, 30; *EMJ, 63* (March 6, 1897), 232; Bartlett, "American Students of Mining in Germany," AIME *Trans.,* V (1877), p. 447; John A. Church, *Mining Schools in the United States,* pp. 11–14; Benjamin Smith Lyman, "The Freiberg School of Mines," in Rossiter W. Raymond, *Mineral Resources of the States and Territories West of the Rocky Mountains,* (1869), pp. 230, 234–35. The blowpipe uses a forced jet of air to direct and intensify the heat of a flame against a mineral compound to produce an identifiable reaction and is thus important in analyzing ore content. Fay, "Glossary," p. 89.

55. Louis Janin to Louis Janin, Sr., Schemnitz, Hungary, August 13, 1860, Janin family MSS, Box 10.

56. Louis Janin to Juliet C. Janin, Paris, November 5, 1860, ibid., Box 9; *EMJ, 63* (January 23, 1897), 95.

57. *M&SP, 18* (June 5, 1869), 353.

its curriculum resembled that of Freiberg, but with more emphasis on the practical side of mining and smelting.[58] A few studied at the Bergakadamie in Berlin, or as one student put it in 1881, referring to himself, were "alternating between the Bergakadamie, acquiring the language and enjoying life in the German metropolis."[59] A few, such as the youngest son of Adolph Sutro, builder of the Sutro Tunnel, were trained at Frankfurt am Main, or as George W. Maynard and John C. F. Randolph, spent time at Göttingen after attending Columbia.[60] Schemnitz in Hungary was, like Freiberg, under government direction and was located in a highly developed mining area, so that its course of study incorporated four years of general study, much of it theoretical, with two final years of practical experience in the mines.[61]

A handful of Americans attended the École des Mines at Paris, either for brief terms, as did the two older Janins, or for extended periods, as was the case with William Ashburner and Francis Vinton. But the École was under the auspices of the French government, and foreign students might be admitted only on recommendation of their government's ambassador in Paris and only if vacancies existed in laboratories and drafting rooms. Consequently, the number of Americans was small: only nineteen enrolled during the period from 1850 to 1890, all but four of them in the 1850s and 1860s. They studied metallurgy with Gruner, mineral chemistry with Rivot, mining with Callon, minerology with Daubrée, geology with Élie de Beaumont and De Chancourtois, and learned surveying from Fuchs, both in the streets of Paris and in the catacombs below.[62]

58. *M&SP,* 12 (May 19, 1866), 313, *14* (March 23, 1867), 181; *AJM, 2* (March 23, 1867), 409.

59. W. R. Sherwood to Samuel B. Christy, Berlin, March 1, 1881, Christy MSS.

60. Wolfe, *Men of California,* p. 21; *EMJ, 91* (February 18, 1911), 363, *95* (February 22, 1913), 431; Richard P. Rothwell, ed., *The Mineral Industry, 1893, 2,* xxxv.

61. *The Miner: A Monthly Magazine* (San Francisco), *1* (March 1866), 5.

62. *AJM, 6* (August 8, 1869), 90; Thomas T. Read, *The Development of Mineral Industry Education in the United States,* p. 28; Raymond, *Mineral Resources (1869),* p. 242.

An American engineer, John A. Church, wrote in 1871 that there were four first-class mining schools in Europe—Freiberg, Berlin, Paris, and the Royal Mining Academy at St. Petersburg. There is no indication that any Americans were trained at St. Petersburg, although a few Russian graduates did find their way to the West and were entitled to wear the huge gold or silver Imperial Mining Badge, the symbol of the highest scientific knowledge in mining bestowed in the czar's domains.[63]

Church included no British schools. A few Oxford and Cambridge graduates did become engineers and practice in the West,[64] but the technical training at these universities was limited. An outstanding English engineer spoke disparagingly of the course of study he had taken at Cambridge to help prepare him for a mining career and insisted that "a course in Chinese or Hebrew, with the same object in view, would have borne just the same results, and would have been less mentally confusing." [65]

Even the Royal School of Mines, first located on Jermyn Street and later in South Kensington, lacked the prestige and influence of the major Continental and American institutions, although a number of Americans, including Charles C. Broadwater of a prominent Montana family, studied there, and such British graduates as Thomas and Forbes Rickard would be intimately associated with western mining for half a century.[66] But even Thomas Rickard agreed with the editor of a Pacific Coast journal who thought that, in the mineral field, "The British system of education is in but a small degree better than none." [67] If British technical men held their own, said Rickard in 1904, "it is rather through inherited ability than the aid afforded by the miserably financed institutions which do duty

63. Church, *Mining Schools,* p. 11; "G.W.M." (George W. Maynard?) to editor, Kargalinsky, Russia, July 6, 1876, *EMJ, 22* (September 23, 1876), 204.

64. For examples, see *EMJ, 69* (March 30, 1900), 266; *Who's Who in Engineering, 1922–23,* p. 666.

65. J. H. Curle, *The Gold Mines of the World,* p. 18.

66. *Royal School of Mines Register,* pp. 24, 161; Rickard, *Retrospect,* pp. 17, 22.

67. *The Engineer of the Pacific* (San Francisco), *3* (March 1880), 7.

for schools of mines in the various states that fly the British Flag."[68] If English technical academies, including the Royal School, were "quite behind the times," as another observer noted,[69] and never achieved the distinction of their counterparts elsewhere, it was in part because the M.E. degree was less highly valued and the profession itself carried less social and intellectual prestige in England than in other areas.

As time passed, a few Canadian universities would make contributions in providing trained engineers for the West. "What a splendid group of metallurgists has been given to this continent by McGill University," exclaimed T. A. Rickard in 1916, referring to the Montreal school, which since 1879 had been training technical men, including C. A. Molson, prominent in Montana mine management, and E. P. Mathewson, of the Pueblo Smelting and Refining Company.[70] David W. Brunton, one of the most versatile and distinguished of engineers, had studied his profession at Toronto, before finishing up at Michigan.[71]

The heyday of American study in European mining schools came in the 1860s and gradually tapered off as specialized American technical curricula developed. And at least some of the agitation for the creation of such courses of study in the United States came from critics of the graduates of Freiberg, Clausthal, or Paris. Some westerners in the 1860s believed foreign-educated engineers superior, and it was argued that Europeans were more advanced in assaying, analytical chemistry, and underground engineering, though not in the evolution of such machinery as the revolving stamp mill, the American pan amalgamation process, and better hoisting works.[72] On the other hand, some westerners tended to identify "bogus representatives of science" almost exclusively with Europe and to attribute the common distrust of trained engineers and

68. *EMJ, 78* (September 22, 1904), 459.

69. George J. Young to George D. Louderback, London, April 17, 1910, George D. Louderback MSS.

70. *M&SP, 113* (July 22, 1916), 111; *EMJ, 57* (March 31, 1894), 291.

71. *EMJ, 87* (February 27, 1909), 458.

72. *M&SP, 12* (January 27, 1866), 56, *18* (March 27, 1869), 194.

geologists to "the scientific mining expert armed with spectacles, magnifying glass, and blow-pipe. . . . This individual is usually of foreign birth, and has studied mineralogy at the opera houses of European capitals, although we have the native American article." [73]

Many Americans remarked on the excellent theoretical training of men from foreign schools, but had reservations on the practical application of that theory. Josiah Whitney, of the California Geological Survey, in 1866 thought it typical of most Freiberg men that his new assistant seemed "pretty smart & well posted theoretically in surveying, but practically as green as grass." [74] Rossiter W. Raymond pointed out that theory remained constant the world over, but that the economics of different processes and techniques varied with the cost of labor, fuel, transportation, or other factors. Young engineers, fresh from countries with lower working costs, might naturally be tempted to consider ore deposits "very rich" or "highly promising" by foreign standards, when on the western scale they might not be profitable at all. Raymond continued:

> It is a severe requisition that mines make upon a young graduate of a foreign school of mines, when we ask him to adapt his acquired science at once to our widely different circumstances. Every such graduate has to reconstruct, alone and for himself, the whole art which he has learned—a work requiring genius as well as intelligent perseverance, and one in which many men fail, who, *if they had been educated in the region where they were to practice,* would have been respectably and deservedly successful all their lives.[75]

73. San Francisco *Examiner,* April 28, 1866(?), clipping, Hayes Scrapbook, *3,* Vol. *3.*

74. Josiah D. Whitney to William Brewer, San Francisco, September 2, 1866, Brewer-Whitney MSS, Box 2.

75. Raymond, *Mineral Resources (1869),* p. 226. For similar comments, see Henry Janin to Louis Janin, Sr., Freiberg, January 25, 1859, Janin family MSS, Box 6; Church, *Mining Schools,* p. 16; *M&SP,* 12 (January 27, 1866), 56.

Thomas Drown, a Freiberg graduate himself, agreed with Raymond, and from the vantage point of the early 1880s described the young American who returned after several years of study abroad.

> He brought back with him an elaborate case of blowpipe instruments, a collection of minerals and hammers, a well-assorted library, practically rich in the German classics, an intimate knowledge of foreign social customs, a vocabularly literally enriched in foreign technical terms, and a look and an air that seemed to say, "I am ready to show you how it is done."

A few were successful, said Drown, but many "made lamentable failures at the start, causing loss to mine and mill owners, and bringing discredit on systematic instruction." But Drown was careful to point out "that the early graduates of our American mining schools were also often unsuccessful at first." [76]

These American technical schools developed slowly. Graduates of the military academy at West Point sometimes entered the field as mining engineers, although ill-equipped for this specialization without additional training or experience. Henry W. Halleck, Lincoln's chief of staff, directed engineering at the New Almaden Quicksilver mine in California in the 1850s, and William S. Rosecrans, another notable Civil War general and a product of "The Point," was regarded as "a talented engineer and practical and scientific mineralogist and metallurgist" in the post-Appomattox years.[77] An occasional Annapolis graduate drifted into the profession,[78] but the Naval

76. Thomas M. Drown, "Technical Training," *Journal of the Franklin Institute, 116* (November 1883), 334–35.

77. "Burgher," *The New Almaden Mine,* p. 21; *M&SP, 12* (May 19, 1866), 305, *17* (July 25, 1868), 49. Hiero B. Herr, another West Point graduate, taught at Lehigh and was for many years active in western mine engineering and promotion work. *EMJ, 18* (September 26, 1874), 203.

78. Charles W. Haskell, Annapolis, 1874, remained in the navy until 1881, then worked as a surveyor in the West, took work at the Colorado School of Mines, and practiced as a mining engineer, before giving it up for newspaper editing. Charles W. Haskell dictation, Grand Junction, Col., January 18, 1887.

Academy in general played no significant role in training men for the mineral industry.

In addition to West Point, Rensselaer Polytechnic Institute of Troy, New York, was by 1835 producing civil engineers, some of whom engaged in mining, but probably the first institution to offer a degree in mining, was the Polytechnic College of the State of Pennsylvania, chartered in 1853. Of some 369 graduates prior to 1890, perhaps 50 went into the mineral industries, including J. P. Wetherill and William P. Miller, both well known in western engineering circles.[79] Yet, in the long run, the Philadelphia school was not successful in attracting either funds or good faculty and students, and the first successful school of mines in the country was that established at Columbia College in New York in 1864.

When it first opened its doors that autumn, the new Columbia School of Mines enrolled twenty-nine students—twice as many as anticipated—and before the year was over, the number had nearly doubled.[80] The first lectures were given in the cellars of the Academic Institute and in makeshift quarters that included "the old buildings east of the College (heretofore leased as a paperhanging factory)." [81] Of seventy-nine students enrolled in 1865–66, seventeen already had a bachelor's degree, only nine were not from New York or New Jersey, and none came from west of Ohio.[82] With a new building constructed in 1866, Columbia Mines was well on its way to becoming the outstanding institution of its kind in the United States.

In the beginning, it offered a three-year program. Its admissions standards required a student to be at least sixteen years of age and to pass examinations in algebra, geometry, and trigonometry. First-year courses included drawing, stochiometry, mathematics (analytic geometry and differential and inte-

79. Read, *Development of Mineral Industry Education, pp.* 36–39; *M&SP, 106* (June 28, 1913), 1008.

80. Entry for November 25, 1864, Allan Nevins and Milton H. Thomas, eds., *The Diary of George Templeton Strong, 3,* 502.

81. Entry for May 23, 1865, ibid., *3,* 600; *AJM, 3* (April 13, 1867), 51.

82. *School of Mines, Columbia College, 1865–66,* pp. 9–10.

gral calculus), physics (heat, the steam engine), electricity and magnetism, inorganic chemistry, quantitative analysis, mineralogy (including blowpipe use and analysis), French and German. In the second year, the student continued mineralogy and quantitative analysis and added metallurgy, mechanics, geology, botany, and mining engineering (machines). In the final year, metallurgy and quantitative analysis were continued, and courses added included theory of veins, "exploitation of mines," assaying, "conservation of force," and an attempt to connect the various sciences.[83]

Although a Yale professor in 1866 thought that the course was not "very thorough or complete," he had to admit that the new technical school was "creating a great sensation."[84] Editors lauded the institution appreciatively, looking to it to "supply a real want," and visitors were impressed with the rigor of the curriculum. "A graduate of the School of Mines," said one in 1867, "will be well worthy of his degree."[85] To John A. Church, an expert on the subject, Columbia Mines in 1871 was already "one of the best schools in the world—more scientific than Freiberg, more practical than Paris."[86]

By this time, the school of mines had been reorganized as a general school of technology, offering in 1870 five regular courses of instruction: mining engineering, metallurgy, civil engineering, geology and natural history, and analytical and applied chemistry.[87] This broadening of curriculum was primarily the work of Charles F. Chandler, who came from Union College as professor of Chemistry, but it was to two other early faculty members, Thomas Egleston and Francis Vinton, classmates at the École des Mines in Paris, that much of the success was due. Egleston, a protégé of the elder Silliman at Yale, had conceived the idea of a school of mines and pushed it to fruition, volunteering to stay on to teach both mining and

83. Ibid., pp. 15–16. See also advertisement, *AJM, 1* (March 31, 1866), 16.

84. William Brewer to Josiah D. Whitney, New Haven, February 28, 1866, Brewer-Whitney MSS, Box 2.

85. *AJM, 1* (April 7, 1866), 24, *3* (April 13, 1867), 51; *Journal of Commerce* (November 24, 1864), clipping, Hayes Scrapbooks, *2*.

86. Church, *Mining Schools,* 19, 21–22.

87. *EMJ, 10* (July 26, 1870), 56.

metallurgy. Vinton, a West Point graduate before completing the Paris course, had retired from the Union forces after having been wounded at Fredericksburg and was named the first professor of mining engineering at Columbia, where he taught until 1877, when he opened a consulting practice in Colorado.[88]

Through such men, the European impact lingered for some time. Until at least 1884, Kerl's *Metallurgy,* Plattner's *Blowpipe Analysis,* and several of Von Cotta's treatises were standard texts. Beudant's *Géologie* was used, both for its value in its discipline and as a source of French instruction, and Burat's *Géologie appliqué* and his *Exploitation des mines* were also assigned. However, from the beginning, Dana's *Manual of Geology* was a basic tool, even down to the turn of the century, and students were exposed to a growing number of American points of view as the European influence gradually waned.[89]

During the 1870s, because of the superior facilities in metallurgy, students tended to gravitate toward this field, rather than toward mining itself. As early as 1872, a preparatory year had been added at the School of Mines and the minimum age for admission to the regular program set at eighteen. Tuition had risen, but in 1873 the enrollment of 140 students surpassed that of Columbia College.[90] Of the thirty-one M.E. degrees granted in the entire United States in 1879, twenty were awarded by Columbia Mines, a school that by the end of

88. Read, *Development of Mineral Industry Education,* p. 51; *EMJ, 28* (October 11, 1879), 257; *Mining Record* (New York), *6* (October 18, 1879), 303; *DAB, 3,* 613–14, *6,* 56, *19,* 282.

89. *School of Mines, Columbia College, 1865–66,* p. 16; *School of Mines, Columbia College, 1872–73,* pp. 49–50; Columbia College, School of Mines, *Circular of Information, 1883–1884* pp. 18, 19, 74–75. However American viewpoints on the theory of vein formation, for example, often followed those of the European masters. Joseph Le Conte taught his Berkeley classes a slightly modified version of the Von Cotta position, and Le Conte's *Geology* was widely used. So were the books of J. F. Kemp and Waldemar Lindgren, both of whom espoused essentially the Von Cotta approach. *Register of the University of California, 1894–95,* p. 111; *Catalogue of the Colorado School of Mines, Golden, Colorado, 1898–99,* p. 88. Thomas Crook, *History of the Theory of Ore Deposits,* pp. 96–98, 108–12.

90. *School of Mines, Columbia College, 1872–73,* pp. 19–20; entry for November 3, 1873, Nevins and Thomas, eds., *Diary, 4,* 499; *M&SP, 18* (May 22, 1869), 322; *EMJ, 16* (October 21, 1873), 265.

1881 had bestowed 198 such degrees. Of this number, nearly 60 percent were engaged in mining or kindred fields: 4 percent were geologists or mineralogists; slightly over 6 percent were metallurgists; nearly 29 percent were chemists or assayers; and some 17 percent were listed as mining engineers. Of the fifty-two mining engineers, twenty-two were consultants, ten were fieldworkers, sixteen were mine managers, and four were instructors.[91] Compilations of Samuel B. Christy, at the University of California in 1893, indicated that of 871 mining engineering graduates in the country prior to 1892, 402 were from Columbia. Thus, to this time, Columbia Mines had given

Table 1. Total Number of Graduates from Mining Schools in the United States, to 1892

DATE OF FIRST GRADUATION	INSTITUTION	NO. GRADS TO 1892	AVERAGE PER YEAR
1867	Columbia School of Mines	402	15.46
1867	University of Michigan	41	1.57
1868	Massachusetts Institute of Technology	126	5.04
1869	Washington and Lee (none since 1875)	8	.03
1871	Lehigh University	48	2.28
1871	Lafayette College	40	1.71
1874	Missouri University	26	1.31
1874	Washington University, St. Louis	43	2.26
1877	University of California	55	3.44
1878	University of Illinois	6	.40
1879	University of Wisconsin	12	.92
1882	Colorado School of Mines	26	2.60
1888	Michigan School of Mines, Houghton	27	5.40
1890	Alabama Polytechnic	4	1.33
1891	Montana School of Mines	6	3.00
1892	Pennsylvania University	1	1.00
	Total in 26 years	871	33.05

SOURCE: Samuel B. Christy, "The Growth of American Mining Schools and Their Relation to the Mining Industry," *AIME Trans.*, *23* (1893), 445.

91. *SMQ*, *3* (November 1881, March 1882), 57, 241–42.

about as many mining degrees as the total given by all other American schools offering work in this field. (See Table 1.)

Columbia graduates quickly scattered throughout the West. They were so numerous in Colorado by 1884 that the Mines Alumni Association voted to establish a branch in Denver.[92] In early 1892, at least twenty-five were in Colorado (twelve in Leadville alone), twelve in Mexico, seven in Utah, six each in California and Arizona, and four in Nevada.[93] Such figures ignore countless others based in eastern or midwestern cities who ranged in and out of the West on consulting trips.

Meanwhile, other institutions were also establishing programs of technical mining education. Rensselaer Polytechnic in 1866 advertised its "Civil and Mining Engineering" curriculum, and Yale announced that its "mining department" was offering instruction, including "practical training in Civil and Mining Engineering, Metallurgy, Analytical Chemistry, Assaying, Mineralogy, Geology, and the French and German languages." [94] But Yale's approach was more a series of lectures on mining, rather than a degree program, and most of the sons of Eli who became mining engineers in the West were either trained in civil engineering, as was Arthur D. Foote, or, like John Hays Hammond, had taken additional work at Columbia or abroad.[95]

At Harvard, where Abbott Lawrence in 1847 had endowed the technical fields, including mining and metallurgy, the situation was roughly analogous until the 1890s.[96] But Nathaniel Southgate Shaler came to have an influence on many a budding engineer, and between 1897 and 1905 inclusive, Harvard's Lawrence Scientific School graduated thirty-nine men in the mining fields. Five never entered the industry, and of the thirty-four who responded to a survey in 1906 or 1907, eleven were mine superintendents, nine were consulting or examining

92. *SMQ, 5* (May 1884), 394.
93. *SMQ, 3* (November 1881), 62–64.
94. *AJM, 1* (July 28 & September 8, 1866), 285, 384.
95. Rickard, ed., *Interviews*, p. 171; *Pacific Miner, 10* (August 1906), 3; Hammond, *Autobiography, 1*, 39–66.
96. *DAB, 10*, 55; *M&M, 1* (May 1920), 19.

engineers, two were mining geologists, two were metallurgists, one was a teacher of metallurgy, and one was in the metallurgical supply business.[97]

In 1894, Richard P. Rothwell listed twenty-six mining schools in the United States and Canada, including several—Texas A. & M., Armour Institute, and the University of Wyoming, among them—that were peripheral, to say the least.[98] One was the Massachusetts Institute of Technology, where technical training had been offered, especially in metallurgy, since the late 1860s. During the late 1890s, with a great upswing in mining owing to new mineral strikes and a new appreciation of engineers throughout the world, the old buildings on Huntington Avenue were jammed with aspiring M.E.'s, some of whom, such as young Robert Livermore, were in deadly earnest and anxious "to get a mining education and get into the field as quickly as possible."[99]

Washington University of St. Louis had a department of mining and metallurgy as early as 1871; at Rolla, the Missouri School of Mines conferred its first degree in mining engineering in 1874, had awarded 22 by 1888, and a total of 377 by 1920, numbering among the recipients its most distinguished alumnus, Daniel C. Jackling, class of 1892.[100] The Michigan School of Mines opened in 1886 in the basement and attic of the fire station at Houghton. It produced a number of western mining engineers, and in 1894 was regarded as "the only school in the country devoted exclusively to mining engineering."[101]

Herbert Hoover, of course, was a graduate of Stanford, an institution highly recommended by other prominent engineers

97. George Packard to editor, Boston, July 18, 1907, *M&SP, 115* (August 10, 1907), 173.

98. Rothwell, ed., *Mineral Industry, 2,* 812–24.

99. Gressley, ed., *Bostonians and Bullion* pp. 75, 76; Read, *Development of Mineral Industry Education,* pp. 173–74; Richards, *Robert Hallowell Richards* pp. 35, 99.

100. Rickard, ed., *Interviews,* pp. 193–94; Clarence N. Roberts, *History of the University of Missouri School of Mines and Metallurgy, 1871–1946,* pp. 26, 34, 42, 57, 75.

101. Rothwell, ed., *Mineral Industry, 2,* 819.

and geologists.[102] A number of other schools developed in the West also—South Dakota School of Mines (1887), University of Arizona School of Mines (1891), University of Nevada School of Mines (1892), Montana Mines (1891), and New Mexico School of Mines at Socorro (1893)—but only two would prove real contenders to Columbia by the time of World War I—the Colorado School of Mines and the University of California.[103]

At Golden, the Colorado School of Mines had opened its doors in 1872, had commenced mining instruction the following year, but granted no degree until 1882.[104] Although local editors urged the state to support the school adequately "and make it equal to the Columbia School of New York," [105] the institution did not prosper immediately. Trouble among students, faculty, and trustees was common, and more than one president resigned in protest. Moreover, the school had the reputation of admitting large numbers of freshmen in order to gain larger appropriations, then weeding them out drastically. But between 1883 and 1902, it graduated 224 mining engineers, and with the presidency of Victor Alderson, beginning in 1903, commenced an era of vigorous expansion and the development of an international reputation.[106] Even so, one professor in 1917 found the school "a dreadfully lonesome and stupid place," with a "fair attendance of the roughest specimens of youthful humanity that can be found anywhere." [107]

Meanwhile, the school of mines at the University of California had already emerged as the leading institution of its kind in the West. In the early 1860s, Pacific Coast editors had

102. Daniel M. Barringer to Guy P. Bennett, July 29, 1899, Letterbook C, Barringer MSS.

103. Rothwell, *Mineral Industry, 2,* 812–24.

104. Read says that E. C. Van Deist was granted the E.M. degree in 1882, but that the first formal commencement was in 1883. Read, *Development of Mineral Industry Education,* p. 90. The school's historian says that the first degree was awarded to Milton Moss in 1882. Morgan, *A World School,* p. 66.

105. *Georgetown Courier* (Georgetown, Colorado), April 26, 1883.

106. *EMJ, 75* (May 16, 1903), 739.

107. George J. Young to George D. Louderback, Golden, April 6, 1917, Louderback MSS.

called for the establishment of a school in California "whereby the miner, engineer and prospector may be drilled in the art to prosecute mining in all its bearings" to offset the waste and inefficiency of methods that seemed not much more advanced than those of antiquity.[108] In 1864, the College of California, a private institution, added a mining and agricultural college—a course of instruction to be taught by William P. Blake, just returned from helping organize a scientific program in Japan, and Sherman Day, manager of the New Almaden Quicksilver mine.[109] When the college petitioned the legislature for the state university funds, representing itself as being in full operation, it became the object of scorn for such critics as Josiah Whitney, who described the school as largely "an asylum for decayed clergymen"; as for mining, said Whitney, "one of the professors was in the Colorado mines and another at Washoe." [110] Of the same enterprise, Whitney could comment two years later: "W.P.B. is now lecturing at Oakland to the College boys & they call that a Mining School! This is the era of Mining Schools in the U.S.! Heaven save the mark!" [111]

But from these inauspicious beginnings, once the denominational college had been merged into the new land grant University of California later in the decade, would arise a major source of technically trained mining men. Although the university lists Gardner F. Williams, class of 1865, as its first mining engineer, no mining degrees were given until 1873 or 1874.[112] During its early years, the faculty was unstable: such prominent engineers or geologists as William Ashburner and George F. Becker were appointed, but were too busy with private practice to teach. Only when one of its own graduates,

108. *M&SP, 4* (January 11, 1862), 4.

109. *M&SP, 8* (January 30, 1864), 72.

110. Josiah D. Whitney to William Brewer, Placerville, April 8, 1864, Brewer-Whitney MSS, Box 1.

111. Whitney to Brewer, San Francisco, March 7, 1866, ibid., Box 2.

112. A WPA survey cites Williams as the first mining engineer, Frank Otis, class of 1873, second, and Harry Webb, class of 1875, third. Bunje et al., *Careers,* 4. Read says the first mining graduate was in 1874. Read, *Development of Mineral Industry Education,* p. 84.

Samuel Benedict Christy, became professor of mining and metallurgy in 1885, was real stability achieved. Even then, the appointment of a "book miner" with little practical experience was greeted with disdain by some, including a correspondent of the *Tuolumne Independent:*

> Scientific mining has nearly ruined this coast, and we have had enough of it; and now, we want *science and practice combined.* We have been flooded with whole torrents of Freiberg and Heidelberg, Columbia and Yale College Scientific Mining Experts, or Engineers, and want no more of them; but now, we suppose we are to have a "regular green-goggled-mining mill" at Berkeley. Well, we can only say, we have an overplus of mining experts now, and we want to stop this Berkeley mill before it grinds out too many of the same dangerous kind.[113]

Despite such criticism, surely minor, Berkeley's "regular green-goggled-mining mill" turned out a limited number of graduate engineers, averaging slightly more than three per year between 1880 and 1898. Then, thanks to increased demand in Australia and South Africa, where spectacular achievements and salaries went hand in hand, the number rose sharply, averaging twenty-nine per year for the period from 1899 to the beginning of World War I. (See Table 2.) By 1917, 536 alumni of the California School of Mines were scattered through the mineral world. One had been president of Mexico; another would soon rule Columbia.[114] In mining—in the West or elsewhere—few ranked higher than Harry Webb, Charles Butters, Charles W. Merrill, or Stanley Easton, and the reputation of these and countless other California graduates left no doubt of the school's importance in the profession.

If Columbia and California were the two most important technical schools, followed by Colorado Mines, MIT, Michigan Mines, Missouri Mines, and a host of others, prominent

113. Undated clipping, Christy MSS.

114. Francisco I. Madero, 1895, became President of Mexico; Pedro Nel Ospina became President of Colombia in 1922. Ibid., p. 16; *EMJ, 113* (April 22, 1922), 687, (November 15, 1924), 764.

mining engineers came from numerous other institutions, ranging from the University of Pennsylvania to Virginia Military Institute, William Jewell College, and Blue Mountain University in Oregon. How much specialized training these men may have had is conjectural. Certainly some engineers entered through the side door, with degrees in totally unrelated fields: Alvah G. Briggs, who managed California mines for at least a dozen years, was a graduate of the classical course at Northwestern University; Robert Brewster Stanton, who covered much of the mining West, had studied a similar curriculum at Miami of Ohio, where his father was president.[115]

Table 2. Mining Degrees Awarded Annually at the University of California, 1877–1936

YEAR	DEGREES AWARDED	YEAR	DEGREES AWARDED	YEAR	DEGREES AWARDED
1877	7	1897	5	1917	14
1878	3	1898	8	1918	6
1879	9	1899	17	1919	8
1880	4	1900	27	1920	15
1881	2	1901	25	1921	15
1882	2	1902	26	1922	25
1883	0	1903	30	1923	41
1884	1	1904	30	1924	43
1885	3	1905	21	1925	30
1886	3	1906	41	1926	25
1887	7	1907	43	1927	26
1888	2	1908	43	1928	35
1889	3	1909	26	1929	13
1890	0	1910	28	1930	23
1891	6	1911	44	1931	12
1892	4	1912	26	1932	22
1893	5	1913	25	1933	26
1894	2	1914	25	1934	10
1895	4	1915	15	1935	32
1896	1	1916	13	1936	24

SOURCE: E. T. H. Bunje, F. J. Schmitz, and D. I. Wainwright, *Careers of University of California Mining Engineers, 1865–1936* (Berkeley, 1936).

115. Sutcliffe, ed., *Who's Who in Berkeley,* p. 18; Robert B. Stanton, *The Hoskaninni Papers: Mining in Glen Canyon, 1897–1902,* ed. C. Gregory Crampton and Dwight L. Smith, x.

And a few of the most successful engineers never completed their college training, for it is estimated that at least 40, and perhaps as much as 60, percent of the entering freshman class never graduated.[116] Thus, the mineral industry lost the services of Douglas Fairbanks, Sr., who once matriculated at the Colorado School of Mines.[117] But others who dropped out made their mark as engineers. John B. Farish, one of the true veterans, left college for lack of funds, just as Frederick W. Bradley, who was rated "one of the foremost mining engineers of the world" in 1907, had dropped out of the University of California at the end of two years to support his family. "I have always intended to go back," he once wrote Samuel Christy, "but have not yet been in a position to afford the time."[118] Karl Krug, a great baseball player on the Berkeley campus, had left school after his junior year because of trouble with mathematics, but later achieved real success as a dredge engineer.[119]

In general, mining engineering was based on a broader training curriculum than other professional engineering programs. The nature of his work required the mining engineer to have a smattering of knowledge in a number of different areas. "No profession requires so broad and complete a knowledge of scientific and practical subjects," remarked an editor in 1905:

> In addition to having acquired a comprehensive knowledge of practical mining in all its branches, he must have had a thorough grounding in mathematics and have training in civil engineering, and in chemistry and metallurgy. Besides this, he must have had training in geology, physics, architecture, natural and social sciences, practical business experience and above all, he must be endowed

116. Allen H. Rogers, "How Should a Mining Engineer Be Trained?" *EMJ, 113* (January 21, 1922), 121, 123; Rothwell, ed., *Mineral Industry, 2,* 813.

117. *Who's Who in America, 1928–1929,* p. 739.

118. *M&SP, 44* (August 31, 1907), 275; Frederick W. Bradley to Samuel Christy, Wardner, February 22, 1892, Christy MSS; *EMJ, 112* (December 17, 1921), 980.

119. Durbrow Interview, 31–32.

with a substantial amount of what is usually spoken of as common sense.[120]

Some felt that the mining engineer was at a disadvantage because of the very breadth, and thus the shallowness, of his education. In the words of a critic of 1892,

> He is neither fish, flesh, nor red herring. He is not so good an engineer as those who graduate from the other engineering courses, because he has not been so thoroughly grounded in engineering. He is not so good a chemist as those who graduate from a chemical course for the same reason. Other things being equal the civil or mechanical graduate will beat him on his own ground, while he is not nearly so well qualified to compete with them on theirs.[121]

On the other hand, it was also argued that, because the mining engineer's work was basically destructive, rather than constructive, he needed to know less about fundamental engineering theory than his colleagues in other fields.[122]

Some critics sought an even broader educational base. Many mining engineers look on themselves as apostles from a higher civilization to a lower one; through a broad and even liberal education, they must be prepared to provide an example of the cultured man to others.[123] As James D. Hague told his son, who was in college taking some economics preparatory to an engineering career, "In Economics you deal with everlasting questions and principles, while Boilers are temporary arrange-

120. *M&SP, 90* (April 18, 1905), 214.

121. *Engineering News and American Railway Journal, 28* (October 6, 1892), 328.

122. Ibid., *28* (November 3, 1892), 414–15.

123. "Personally I think the man goes further in the end who has broad cultural training joined with thorough training in the fundamentals of engineering than the man who neglects English and languages, history and philosophy and economics, however skilled an assayer or surveyor he may become." Christopher M. Weld to Sara Weld Blake, New York, November 16, 1922, Letters of Christopher Minot Weld. See also *Mining Science* (Denver), *61* (March 10, 1910), 217–18.

ments that endure only for a season and sometimes not so long." [124]

Herbert Hoover believed that commercial, as well as technical training was vital, for increasingly the engineer was called on to be both manager and organizer. And all knowledge must be up to date and pertinent, for, said Hoover, "We have not the time nor the inclination to knock out of the heads of these men misimpressions which they have gained by so-called practical training in the technical school." [125] Some schools, notably Michigan Mines, established departments of technical writing, so that future engineers might be trained in the writing of precise reports.[126] Most students were exposed to French and German, and language was a useful tool. It was not unusual for an engineer's card to read "familiar with all the modern languages" or "si habla española," [127] and men in the profession were constantly urged to increase their language facility, for, said Thomas A. Rickard in 1915, "When opportunity knocks at the door, she does not wait for an interpreter." [128]

The Nevada School of Mines instituted a course in cooking, and at least one editor believed such training should be universal for mining students, because engineers in the wilderness were often forced to rely on their own culinary talents.[129] Others, including Richard Rothwell, thought the engineer should receive as part of his education some medical knowledge, including "at least a limited knowledge of operative surgery." He should "certainly know enough of the art to be able to tie an artery, stanch the flow of blood, or give temporary relief to those who are suffering from burns or scalds." [130] Nor was this idle dreaming; engineers in the field were called on to perform medical chores—to tend the sick, set broken

124. James D. Hague to William Hague, New York, February 11, 1902, Hague MSS, Box 2.

125. Quoted in *Science,* n.s., *20* (November 25, 1904), 718.

126. *EMJ, 93* (June 1, 1912), 1066.

127. *M&SP, 12* (March 3, 1866), 136, 144, *46* (March 17, 1883), 194.

128. *M&SP, 110* (June 19, 1915), 938.

129. *Mining Reporter* (Denver), *55* (February 28, 1907), 195.

130. *EMJ, 35* (January 13, 1883), 13.

limbs, sometimes even to amputate or to serve as obstetricians.[131]

It was agreed that the graduate of a mining school was not a mining engineer and that only practical experience and common sense could convert him into one. Most mining schools, by one means or another, sought to take this into consideration, and to provide some kind of exposure to the realities of mining apart from what was taught in their own classrooms and laboratories. In its catalogue for 1865–66, Columbia Mines stipulated that during vacation "each student is expected to visit mining and metallurgical establishments, and to hand in, on his return, a journal of his travels, and a memoir on some subject assigned him." [132]

Soon, however, Columbia had a more formal summer visitation and work program as an integral part of its curriculum. Between his third and fourth years, each student was to spend six weeks in a mining region at a "Summer School of Practical Mining," under the supervision of one of his professors. This involved a detailed on-the-spot study of geology, mine plant and methods of working, surveying, shafting, tunneling, and the other aspects of day-to-day mining.[133] Sometimes, the professor was able to combine his direction of summer students with professional work of his own.[134]

Other schools followed the same procedure. Robert H. Richards, professor at MIT, made his first summer school trip in 1871 to the mines of Missouri, Colorado, and Utah, with an additional visit to the mine machinery factories in San Francisco.[135] The program of Washington University in St. Louis required students to visit and study mines and mills in

131. See Frank E. Johnesse to Mayme Patten, Lewiston, Idaho, December 21, 1899, Johnesse family MSS; John Hays Hammond, "The Human Side of Engineering," typescript (November 20, 1930), Hammond MSS, Box 16.

132. *School of Mines, Columbia College, 1865–1866,* p. 18.

133. Rothwell, ed., *Mineral Industry, 2* 814–15; EMJ, *24* (August 11, 1877), 102, *58* (October 27 1894), 388; interview with Francis Seeley Foote, Berkeley, California, May 8, 1964.

134. *EMJ, 35* (June 30, 1883), 375.

135. Richards, *Robert Hallowell Richards,* pp. 65–66.

the area each Saturday during the term, and in addition to attend a six-week summer school in the Lake Superior copper and iron regions, the Colorado gold and silver fields, or the Missouri lead districts.[136] In 1904, a joint endeavor, the George Crocker Summer Mining School, in charge of Henry Munroe of Columbia Mines, attracted eighty-five students from Columbia, MIT, Yale, and Harvard to the area around Silver Plume, Colorado.[137]

These field expeditions were not pretentious. Students provided their own transportation, often traveling by "Side-Door Pullman." As a group, they often traveled horseback in the mountains, with a pack train of supplies and a herder and cook hired for the occasion, and for many, students and teachers alike, these might well be enjoyable outings.[138] Still, one University of Minnesota professor could complain in 1914 about the need to take a class to Arizona for field work: "It makes me groan to think of being on the wheels again but one cannot lie in the lap of luxury & expect things to come by automatically." [139] Nor did all students show up well on such trips. One Yale undergraduate lasted only about two weeks on a field trip to Arizona. "He didn't do anything but keep a chair warm and eat grub," and turned out "to be a loafer and a quitter," according to his disgusted professor.[140]

Because of their proximity to the mines, western schools often did not have formalized practical summer programs, but their students were urged to find vacation jobs in the mines, even at substandard wages, merely for the experience to be gained.[141] Samuel Christy of the University of California

136. Rothwell, ed., *Mineral Industry, 2,* 823.

137. *EMJ, 78* (July 21, 1904), 113; *M&SP, 89* (December 24, 1904), 419.

138. *The Black Hills Engineer* (Rapid City), *13* (1925), 200–01.

139. George J. Young to George Louderback, Minneapolis, April 17, 1914, Louderback MSS.

140. Joseph Barrell to William S. Bayley, New Haven, September 10, 1911, William S. Bayley MSS.

141. William Read to Thomas Price, Capitola, California, March 15, 1903, and Read to L. W. Jefferson, Capitola, March 15, 1903, copies, Letterbook 1900–07, William M. Read MSS.

frequently found cooperative mine superintendents (some of them former students) who would put a number of students to work at $1.50 a day pushing wheelbarrows, timbering, surveying, or drilling under supervision.[142] But managers and foremen could not be expected to take any real interest in summer transients; indeed, in some instances, because of the miner's skepticism of "them college fellers," the student might prefer to conceal his educational background and take his chance with other day laborers.

By the turn of the century, some mining schools sought to reach a compromise by leasing, or even buying, mining property for summer operation by their students. Washington State leased a copper mine for this purpose in 1905, and in the previous year, the New Mexico School of Mines had evoked much favorable comment when it purchased a mine in the Socorro Mountains for the practical training of its students.[143] Other western schools earlier had demonstration mines; now the movement was toward operating ones.

But a mining engineering course was limited in the extent to which it could emphasize practicality. Experience the young engineer could gain on the job. Nor could the education be too generalized. As the body of technical information expanded, specialization was also a consideration, with Columbia Mines conferring its first degree in metallurgy in 1873 and in metallurgical engineering in 1887.[144] In the twentieth century, a few curricula were still three years, most were four, and a handful, including Stanford and Columbia, required five or six years beyond the high school level. A survey of the programs of twenty-four American mining schools immediately following World War I, indicated considerable disparity in the amount of time devoted to professional and nonprofessional subjects. By this time, more than half required no foreign language and over 60 percent permitted no electives.

142. Charles C. Derby to Samuel Christy, New Almaden, July 1, 1897, Christy MSS; Mary H. Foote to Helena Gilder, Grass Valley, June 1, 1906, Mary Hallock Foote MSS.

143. *M&SP, 90* (July 23, 1904), 51; *EMJ, 89* (December 24, 1905), 419.

144. *EMJ, 92* (November 4, 1911), 877.

Thus, increasingly, technical education was becoming more specialized. Schools were assuming a more liberal training before the student entered into an engineering program. Many deplored this trend, but the profession demanded it.[145]

Table 3. Percentage of Time, Maximum and Minimum, Devoted to Different Subjects in Mining Curricula, 1919–20

SUBJECT	MAXIMUM PERCENT	MINIMUM PERCENT
Mathematics	16.8	6.1
Physics & mechanics	13.2	4.
Mechanical drawing	8.6	2.8
Chemistry	15.7	7.3
Geology & mineral	17.7	4.3
Engineering, economics, & history	18.8	2.
Foreign language	7.9	3.8
Surveying	7.4	1.
Civil, chemistry, & electrical engineering	17.9	7.1
Mining engineering	19.1	4.4
Metallurgy & ore dressing	18.4	4.4
Thesis	4.3	1.
Electives	20.4	1.3

SOURCE: Allen H. Rogers, "How Should a Mining Engineer Be Trained," *EMJ, 113* (January 21, 1922), 120.

145. See *M&SP, 96* (June 13, 1908), 787; *EMJ, 99* (June 5, 1915), 1000.

CHAPTER 3

"One Who Has the Theory and Needs the Practice"

There's no tellin' how big that lode, nor tellin' how much she will run;
Comstock was located by men with sagebrush in their hair:
Look at Butler, Stratton, or Frenchy that lit on the Flin Flon—
Yet the puttee-legged experts had damned 'em all fer fair!

—Anonymous [1]

When a young man had completed his prescribed course of studies and emerged, sheepskin in hand, a graduate "Mining Engineer," or "Engineer of Mines," what then? What kind of job could he expect? What kind of reception could he anticipate in the mining world? Although answers are not clear-cut, it seems safe to say that the word "mixed" might be used to describe both the initial position he might take and the reaction he might receive in Western America.

To begin with, a number who left college with M.E. degrees never practiced as mining engineers and, indeed, maintained little contact with the mineral industries. Edwin Lefevre, Lehigh (1890), for example, immediately began a highly successful career in journalism, whereas Rube Goldberg, California Mines (1904), promptly went to work as a cartoonist for the San Francisco *Chronicle*.[2] Records indicate that graduate M.E.'s turned to a variety of nonmining professions, including banking, law, farming, teaching, construction, retailing, real estate, advertising, park work, laundering, the ministry, and numerous others.[3] Undoubtedly, however, the

1. "The Lone Prospector," *EMJ, 110* (November 27, 1920), 1045.
2. *Who's Who in America, 1928–1929,* pp. 877, 1278.
3. Josiah Spurr, "Successful Engineers Need Not Apply," *EMJ, 111* (June 18, 1921), 1014; Bunje, et al., *Careers,* pp. 6, 29, 35, 38, 39, 48, 56, 68, 71, 91; see also undated list of graduates in Christy MSS.

majority sooner or later did go into some aspect of mining or metallurgy, and a substantial number made it their life work.

But many delayed entering the profession and went momentarily into something else. George Parsons, Columbia, 1868, spent ten years in the nursery business before engineering and investment work attracted him to Colorado.[4] One of his classmates, William Allen Smith, became private secretary to George Bancroft, U.S. Minister to Germany, and became so interested in Bancroft's historical research that he remained in Europe until 1872, when he returned to take a position with Phelps Dodge & Company.[5] William W. Walker, a graduate of Rennselaer Polytechnic Institute, paused long enough to study medicine and to complete his M.D. before coming to Colorado as an engineer; Franklin R. Carpenter, an alumnus of Rector's College in West Virginia, also studied medicine and taught school, then opened an engineering office in Colorado and eventually became dean of the Dakota School of Mines.[6] Another went into the Swedenborgian ministry out of college, but ultimately came back to engineering in the West and Latin America.[7] Swiss-born Alfred Wartenweiler, educated at the École des Mines in Paris, came to California in 1869 to work for the Giant Powder Company, but in a few years turned to mining proper.[8] Axel O. Ihlseng, Columbia, 1877, spent four years as a chemist for American Sugar Refining before taking up mine engineering in the Rockies.[9] Gerald F. G. Sherman, a Columbia product of a decade and a half later, cut his professional teeth on an irrigation project in Ohio prior to moving on to California to make his reputation in James D. Hague's Grass Valley mines.[10]

A surprising number of fledgling engineers were attracted to first jobs with newly developing railroad lines, both at home

4. *Lists of the Alumni of Columbia College School of Mines, 1892,* p. 49.
5. *SMQ, 20* (April 1899), 279–80.
6. *DAB, 3,* 511; *EMJ, 51* (May 23, 1891), 612.
7. *EMJ, 72* (August 3, 1901), 145.
8. *M&SP, 105* (December 21, 1912), 812.
9. *Who's Who in Engineering, 1922–23,* p. 649.
10. *Who's Who in Arizona* (n.p., 1913), p. 471.

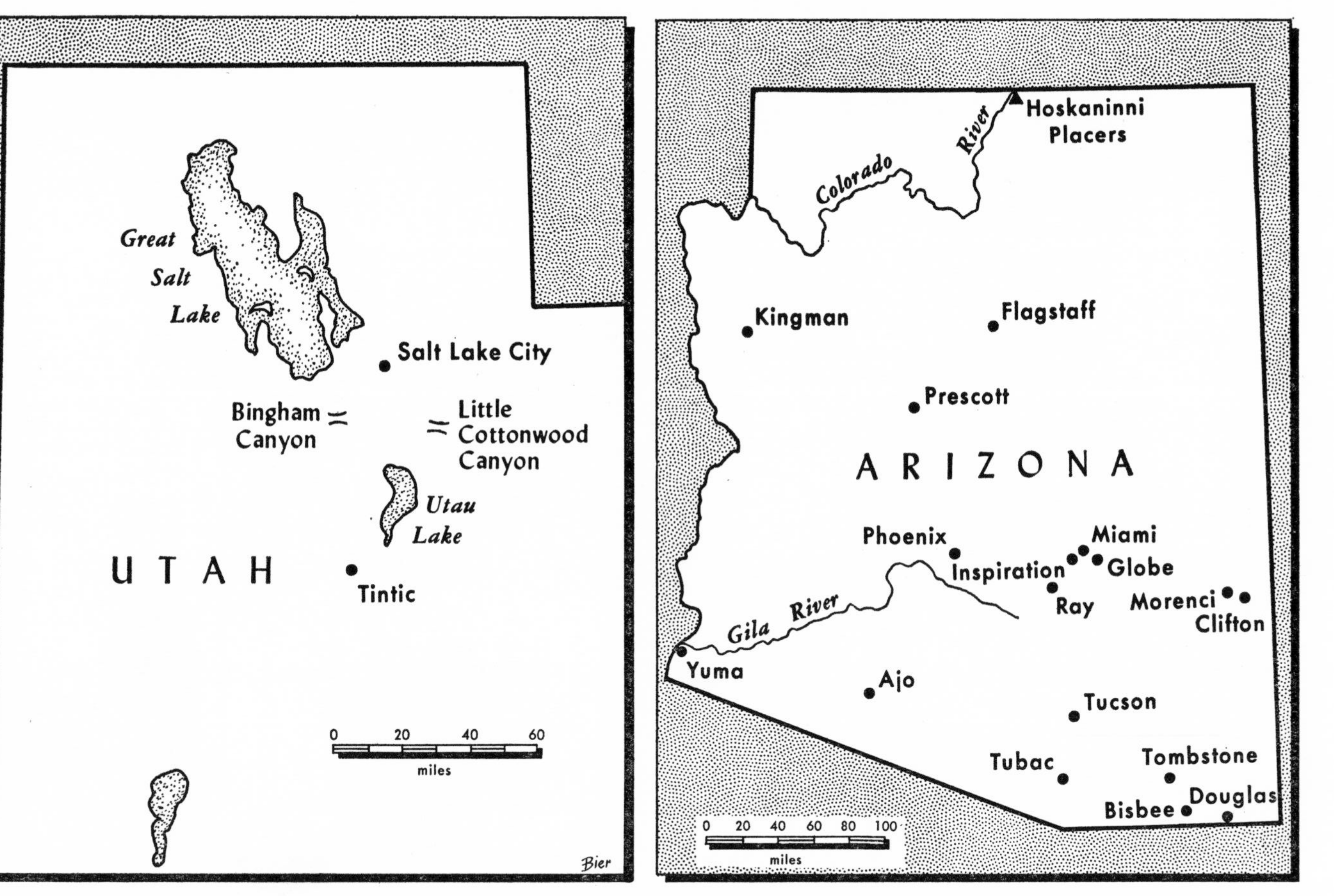

Mining centers of Utah

Mining centers of Arizona

and abroad. Some were civil engineers by training, and would come into mining by the side door, as it were;[11] but many had gone through the conventional M.E. education, and adapted themselves temporarily to the needs of an expanding transportation network. Eugene N. Riotte, for example, one of the real old-timers, a graduate of Frieberg in 1864, spent a year in Costa Rica conducting explorations for the Panama Railroad Company before coming to the Pacific Slope.[12] Arthur Macy, Columbia, 1875, began as engineer for the Ontario Southern Railway Company, constructing harbor facilities until the line was wrecked on financial rocks; then he began his mining career.[13] Charles E. Graff finished up at Columbia ten years after Macy, and his first job was with a railroad company in Argentina. After two years, he returned as engineer for an Arizona mining firm, resigned after a few months to do manufacturing work in reapers and binding twine, but by 1890 was back in mining engineering.[14]

A substantial number of newly minted M.E.'s found employment temporarily with public or semipublic agencies. One of Columbia's first mining graduates, John C. F. Randolph spent two years with the Federal Lighthouse Board at the start of his career; James T. Beard, also a Columbia man, started as assistant engineer on the Brooklyn Bridge (1877–79), turned to engineering work for the C.B. & Q. Railroad, before being attracted to the Colorado mines.[15] Max Boehmer, born in Luxemburg and educated at Hanover Polytechnic Institution, spent his initial half dozen years in the United States as a federally employed engineer making improvements on the Mississippi and Missouri rivers, then showed up in Leadville in 1879 to commence a distinguished

11. For examples, note the careers of Isaac E. James, Luther Wagoner, A. A. Blow, and George H. Robinson. *EMJ, 52* (August 1, 1891), 121, *82* (July 21, 1906), 119; *M&M, 3* (November 1922), 60; *Northwest Mining Journal* (Seattle), *1* (August 1906), 36–37; John Debo Galloway, *Early Engineering* Works Contributory to the Comstock, pp. 51–52.

12. *EMJ, 51* (May 16, 1891), 583.

13. *EMJ, 51* (May 9, 1891), 561.

14. *Lists of the Alumni of Columbia, 1892,* p. 30.

15. *EMJ, 91* (February 18, 1911), 363; *Who's Who in Mining,* p. 6.

career of thirty-four years as a western mining engineer, losing his life in a fall at the Bunker Hill & Sullivan in 1913.[16] One of those working with Boehmer on the rivers, Thomas Weir, a young civil engineering graduate of Union College, would also be associated with him at Leadville, and would ultimately manage some of the most prominent mines in the West for leading industrialists, including the Guggenheims and the Rockefellers.[17]

Neophyte engineers sometimes accompanied the reconnaissances of the army corps of engineers. Theodore B. Comstock, later an important figure in western mining, was with the expedition headed by Captain W. A. Jones ranging north from Fort Bridger to the Yellowstone region in 1873 and prepared an elaborate and important geological report and map of the country traversed.[18] One member of the scientific group accompanying Colonel D. S. Stanley's exploration of the Yellowstone country that same year was Lionel R. Nettre, who would settle and practice engineering in Montana.[19] And when the federal government dispatched a scientific expedition in the summer of 1875 to check on rumors of gold discoveries in the Black Hills, the party included at least eleven professional mining men.[20]

James D. Hague and Robert Goering, both German-trained, accompanied naval expeditions to investigate guano deposits in the Pacific, before embarking on distinguished mining careers.[21] A few M.E.'s made their debuts with the U. S. Coastal Survey,[22] but it was the state and federal geological surveys in particular that came to be regarded as the most tra-

16. Ibid., p. 9; *EMJ, 95* (March 1, 1913), 490.

17. *EMJ, 51* (March 21, 1891), 451; James A. MacKnight, *The Mines of Montana,* pp. 86–87. See also Rickard, *Interviews* pp. 256–57; Wiltsee, "Reminiscences," pp. 5–7.

18. Theodore B. Comstock, "Geological Report," in William A. Jones, "Report upon the Reconnaisance of Northwestern Wyoming, Made in the Summer of 1873," *House Executive Document,* no. 285, 43d Cong., 1st Sess. pp. 85–184.

19. William H. Goetzmann, *Exploration and Empire* p. 415.

20. Ibid., p. 422.

21. *DAB, 6,* 87; *Who's Who in Mining,* pp. 34, 37–38.

22. Ibid., p. 31; *Lists of the Alumni of Columbia, 1892,* p. 26.

ditional and legitimate training ground for youthful engineers and geologists fresh from college.

With the California State Geological Survey of the 1860s were a number of men who would eventually be among the elite of the mining engineering profession—William Ashburner, Charles Hoffmann, Watson Goodyear, and Clarence King.[23] Countless others, including George C. Swallow, Herbert Hoover, Frank L. Nason, and H. Foster Bain, received their introduction to their life's work through the apparatus of one or the other of the state surveys.[24]

The list of those who entered their profession through the U.S. Geological Survey would be even more impressive. Among those associated with the Hayden Survey in the late 1860s and early 1870s were Persifor Frazer, Jr., and Frederick M. Endlich, both trained as mining geologists. Frazer would pursue an academic career with mine engineering as a sideline, whereas Endlich, with an impressive array of European degrees, would be connected with the Smithsonian Institution until 1880, when he turned to western mine management and smelter construction.[25] John A. Church, one of the pioneering graduates of Columbia Mines, was also a part of the early surveys west of the 100th meridian, and made a careful geological study of the Comstock Lode, which was accepted as a graduate thesis by his alma mater.[26] Early in his career, John Hays Hammond was employed by the survey to help collect mining statistics, and Herbert Hoover found his summer employment with Waldemer Lindgren of the survey instrumental in forming the contacts that would send him on his way professionally.[27] Such stalwarts as Walter H. Weed,

23. William Brewer, *Up and Down California, in 1800–1864,* ed. Francis P. Farquhar. pp. 18, 63–64, 451.

24. George P. Merrill, *The First Hundred Years of American Geology,* p. 425; *Who's Who in Mining,* p. 42; *Who's Who in America, 1928–1929,* p. 1549; *M&SP, 124* (January 13, 1922), 50; Winfield S. Downs, ed., *Who's Who in Engineering, 1941,* p. 72.

25. *EMJ, 68* (July 29, 1899), 126; LeRoy & Ann W. Hafen, eds., *The Diaries of William Henry Jackson* p. 215; Goetzmann, *Exploration and Empire,* p. 497.

26. *Mining Magazine* (New York), *11* (May 1905), 399–400.

27. Hammond, *Autobiography, 1,* 85; Hammond to S. L. M. Barlow, New

Chester W. Purington, and Howland S. Bancroft, all eminent engineers, began as green college lads on the payroll of the U.S. Geological Survey.[28]

The survey was regarded as a steppingstone—as a place to learn and gain experience at government expense—and nearly every position connected with it—even as cooks and teamsters—seemed to have commercial value. So much was this so, that the survey itself issued a warning in 1909, cautioning the public to look carefully into the standing of any mining engineer who advertised his association with it. Only 10 percent of the survey staff was in geology, it was noted, and no member was permitted to do private professional work in the United States while employed by the survey.[29] Nonetheless, the importance of the survey and its strong hold "on the imaginations of the young graduates" was recognized by the editor of *Mining and Scientific Press* in 1912:

> Good men go but others come, and the Geological Survey is fast becoming, as an incident to its main work, a great graduate school of instruction. To call the role of ex-employees is almost to list the successful men of the mining profession. So long as the spirit remains right and the work is honest and thorough, the Survey can continue, like a university, to send its men out into other branches of professional service.[30]

To be sure, many young engineers went directly into the mineral industries. As in most fields, the influence of friends, relatives, and teachers was instrumental in placing the beginner on his first job. Young Alexis Janin, finishing his study in 1865, was correct in his assumption that his brother, Louis, already established in San Francisco mining circles, would "be of

---

York, July 8, 1884, Barlow MSS, Box 165; *Who's Who in Mining*, p. 42; Hoover, *Memoirs*, *1*, 26.

28. *Who's Who in Mining*, pp. 77, 102; *Who's Who in Engineering, 1922–23*, p. 95; *EMJ*, *112* (November 5, 1921), 737.

29. *EMJ*, *87* (January 23, 1909), 216.

30. *M&SP*, *104* (June 22, 1912), 850–51.

great advantage." [31] John B. Farish received his start at the Sierra Butte mine in California through his brother, who managed the property.[32] Young Tom Rickard graduated from the Royal School of Mines, London, in 1885, was presented by his father with a ticket, steamer rug, and $100 in cash and was on his way to become assayer for one of the British mining companies in Colorado, then supervised by his uncle, who promptly returned to England leaving "T. A." in charge.[33] James D. Hague made several diligent and successful efforts to place his brother-in-law, Arthur D. Foote, and Adolph Sutro was reminded in 1873 by John C. Churchill, former representative from New York, that he—Churchill—had supported the Sutro tunnel scheme in Congress and that his son, just graduated from Lafayette College in mining engineering, was seeking a position.[34]

Undoubtedly, as correspondence of such men as Samuel Christy of the University of California indicates, some of the most important clearing houses were the faculties of the technical schools, with their wide range of acquaintances. Young engineers often sought an association with older, more experienced members of the profession, and the veteran often felt some responsibility to be as helpful as possible during the post-college "period of adjustment." [35] In this respect, established engineers were sometimes known for their role in sending newcomers on their way up the ladder, finding experience-building positions for them and broadening their world of professional contacts. Louis Janin earned a reputation of this sort, and the North Star Mine at Grass Valley, California, of

31. Alexis Janin to Juliet C. Janin, Freiberg, August 22, 1865, Janin family MSS, Box 5.

32. *M&M, 10* (December 1929), 580.

33. Rickard, *Retrospect*, pp. 28–29.

34. James D. Hague to Clarence King, San Francisco, November 14, 1873, Hague to Arthur D. Foote, San Francisco, September 18, 1873, and Hague to B. B. Thayer, San Francisco, June 30, 1875, copies, Letterbook 6, Hague MSS; John C. Churchill to Adolph Sutro, Oswego, July 13, 1837, Sutro MSS, Box 3. For another example, see G. H. Robinson to Wilbur F. Sanders, Marysville, December 5, 1891, and Ferdinand Van Zandt to Sanders, Butte, December 7, 1891, Wilbur Edgerton Sanders MSS.

35. *The Mining World* (Chicago), *33* (December 24, 1910), 1192.

which James D. Hague was an important shareholder, was known as a training ground for young engineers. Mary Hallock Foote commented in 1897:

> We laugh at Mr. Hague a little about his mining kindergarten. It is one of his practical ways of doing good. Many a young man will date his prosperity from the day James D. took him in hand. He doesn't give them much to start with, and promotion is slow; but steady work counts; and he sifts them and sifts them, and the few who stand the sifting process are fixed for life.[36]

Perhaps in the 1860s and 1870s, when trained mining engineers were comparatively rare, men fresh from college could step directly into engineering positions,[37] but the general tendency was for a man to come up "via the muckstick route," and to serve in a number of subordinate positions while learning the practical side of the profession. Samuel Christy was well aware of the need for the new M.E. to begin at the bottom, to achieve a combination of "brain and hand." To turn the young graduate loose on "an unprotected public," he said, was "often to their mutual injury." "Everyone recognizes that no matter how brilliant the student, or how exceptional his training, or the advantages he has had in the laboratories and workshops of his college he's not yet an engineer." [38] "The college is but the preparatory school—the beginning of knowledge—not its ending," wrote another. "It does not intentionally propose to turn out fully equipped mining engineers, any more than a naval academy undertakes to turn out efficient

36. Mary H. Foote to Helena Gilder, Grass Valley, June 30, 1897, M. H. Foote MSS.

37. Young Louis Janin came fresh from Freiberg in 1861 to become manager and engineer for the nearly defunct Enriqueta Mining Company in California, but his brother Henry started as bookkeeper and time keeper. Louis Janin to Juliet C. Janin, Enriqueta Mine, July 16, August 25 & October 18, 1861, Janin family MSS, Box 10.

38. Samuel B. Christy to Rossiter W. Raymond, Berkeley, May 13, 1886, copy, Christy MSS.

steamship captains, for that is manifestly impossible." [39] In order to direct mining work, an engineer must first know mining work—and this came through experience. The English expression, "it takes a million pounds to educate a mining engineer," contained more than a grain of truth, considering that experience was too often obtained at the expense of the employer.[40]

Thus, although some argued to the contrary,[41] the opinion generally prevailed that only one who knew "the business end of a shovel" could truly appreciate and understand the basic problems and the human element of mining engineering. Berton Braley caught this mood in his "The Gentleman Mucker," depicting the neophyte engineer set on acquiring practical experience:

The gentleman mucker comes out from the east
In his niftiest college suit.
With the legs of his trousers nicely creased
He lands on the hills at Butte.
But he changes his tweeds for overalls
His coat for a flannel shirt
And down on a level he works like the devil,
Shoveling copper dirt.

. . .

The gentleman mucker learns the ropes
And he talks in the miners' slang,
Of "chutes" and "raises" and "sumps" and "stopes,"

39. *M&SP, 91* (September 2, 1905), 150; *EMJ, 80* (December 2, 1905), 1029. See also National Mining & Industrial Exposition Association, *The Mining Industry,* p. 37; *M&SP, 85* (November 22, 1902), 293; *Mining and Engineering World* (Chicago), 40 (May 23, 1914), 958.

40. "Mining School Graduates and the Cost of Education," *Mining Reporter, 53* (May 31, 1906), 529.

41. Some argued that it was useless for college men to attempt day work in the mines, for they could never become as proficient as the old-timers. Others insisted that manual labor would tire the young engineer so much that he would gain little sympathy for or understanding of the miner's problems. *EMJ, 95* (January 4, 1913), 32; *M&SP, 112* (May 27, 1916), 776.

Of mining he gets the hang,
And the first thing you know he's a man of fame
And boss of a mine or two,
And works no more where he did before—
With the mucking job he's through![42]

Many of the "new school" of college-bred engineers made deliberate attempts to broaden their base of practical knowledge as quickly as possible. Some roamed the West as "ten-day miners," working at a variety of jobs, spending a few months in each camp, then moving on as they rounded out their fund of experience.[43] Bartlett L. Thane, former quarterback on the University of California football team, focused his career in Alaska, where he was successively laborer, miner, shift boss, foreman, assistant superintendent, and finally engineer.[44] When William Hague completed his formal technical training early in the twentieth century, he stepped into a round of miscellaneous jobs in mines and mills throughout the Southwest, and wrote his father, engineer-financier James D. Hague, "I am as anxious as anyone to step into partnership as it were with you, only I don't want to do it until I am on enough in the game."[45] Even H. Foster Bain, the first director of the Geological Survey of Illinois, editor of *Mining and Scientific Press* for several years, aide of Herbert Hoover in administering Belgian relief, and ultimately director of the U.S. Bureau of Mines, came out of college, Ph.D. in hand, and after a brief fling with the Iowa Geological Survey, took a job as shift boss in a Colorado mine, simply to learn the business. There he had

42. *The Butte Inter Mountain,* December 30, 1905. The word "mucker" applied to the men who loaded mine cars and in most mines it also included those—sometimes called "trammers"—who pushed the cars to the shaft or the mouth of the tunnel. Fay, "Glossary," p. 452.

43. For example, note the careers of Stanley Easton and Robert Tally. *EMJ, 112* (September 25, 1921), 503; *M&M, 18* (February 1937), 133.

44. *EMJ, 110* (July 3, 1920), 21.

45. William Hague to James D. Hague, El Paso, November 2, 1907, Hague MSS, Box 2. The sons of some engineers obtained early experience at home. The son of Arthur D. Foote graduated from MIT and went to work temporarily at the North Star stamp mill to learn what he could, but not for pay, "so he can sleep late of mornings," explained his mother. Mary H. Foote to Helena Gilder, Grass Valley, June 28, 1896, M. H. Foote MSS.

to stop a feud, teach Italian miners to use the new hammer drills, and in the process helped pass pajamas off to a group of cowboys as Hindu garb worn to combat hot weather![46]

Two special avenues open to young mining school graduates were assaying and surveying, both important jobs and natural points of departure for beginners. To be sure, there were plenty of self-appointed experts in both fields, but these left much to be desired. Observers of the 1880s described some of these "pseudo-assayers" as men who made "cheek take the place of brains," trained by "two weeks spent in sweeping out the laboratory of some alleged assayer," and able to get a yield of from forty to ninety ounces of silver per ton from a "pulverized jug-handle, or grindstone."[47] Others took aim at the self-styled surveyors that infested the boom camps in the same period, Leadville in particular:

> Then there is our army of surveyors with their corduroy suits, high-top boots with cotton-batting calves and leather leggings. . . . No two of them have ever been known to take a bearing or draw a horizontal line the same. An old woman with one eye and a tape-line, can, with less trouble and more accuracy, take the bearings of a piece of ground than two-thirds of the surveyors in Colorado.[48]

Thus, men with some real training were welcomed, as one contemporary wrote in 1883, with regard to the assayer,

> among the first to hasten to every new camp is an enterprising graduate from Freiburg [*sic*] or some American school of mines, eager to put his newly acquired learning

46. *EMJ, 111* (January 8, 1921), 73.
47. Frank Triplett, *Conquering the Wilderness,* p. 353.
48. James D. Murray to editor, Leadville, June 24, 1880, *Virginia Evening Chronicle,* July 1, 1880. The work of the surveyor was both above ground and below. On the surface, he was concerned with the conventional business of determining the position and limits of a mill site, a tract of placer ground, or a quartz claim. Or he might be engaged in determining the route of an aerial tramway, a road, or even a rail line. Underground, he laid the guidelines for cutting the various subterranean passageways and also measured the ore body. Fay, "Glossary," p. 665.

> to practical use. He is a mere boy, perhaps. His hands are soft, his tongue unused to all the rough phrases and quaint slang of the diggings, his frame so light that one of those brawny pick-swingers could hurl him over a cliff with a single hand; but they are glad to see him, and, however much they may laugh at his greenness in mountain manners, hold in high respect his scientific ability, and wait with ill-suppressed eagerness for his report upon the samples they have brought to him for analysis.[49]

No doubt fledgling engineers were regarded as a cheap source of labor in assaying. Luther Wagoner, manager of a Washington reduction company, wrote Samuel Christy in 1892, asking for an assayer: "What I need is about a 75 dollars per month man, one that has the theory and needs the practice." [50] Despite this, countless engineers began their professional careers in the "poison shop" of some mine or mill and found it rewarding. Among them was George Attwood, who at eighteen became assayer, metallurgist, and chemist for the Ophir Company on the Comstock, having learned the trade at the knee of his engineer-father, Melville Attwood.[51] John M. Adams, Columbia's first graduate mining engineer, made his first trip West via Panama to take a position as assayer for a concern in Silver City, Idaho.[52] Edward Dyer Peters, a Freiberg man, began his active career at the Caribou mine in Colorado in 1869, first as assayer and millman, then as metallurgist and superintendent. In 1872, he became territorial assayer for southern Colorado, but returned east to take a medical degree and to practice medicine between 1877 and 1880, when he returned to mining and was in great demand as

49. Ernest Ingersoll, *Knocking Round the Rockies,* pp. 116–17. The function of the assayer, of course, was to test ores by various processes—chemical, blowpipe, or cupeling—to ascertain mineral content. He might work in a commercial capacity for the public, but he might also be employed by a specific mining company, for an efficient concern kept an accurate day-by-day check on the value of the ore being extracted from its property. Fay, "Glossary," p. 47.

50. Christy had a man to suggest, but also proposed that one-way travel fare be included. Luther Wagoner to Samuel Christy, Ruby, Washington, June 25, 1892, and Christy to Wagoner, Berkeley, July 7, 1892, copy, Christy MSS.

51. *Who's Who in Mining,* p. 3.

52. *SMQ, 14* (April 1893), 187, 189.

a smelting consultant ranging from Montana to Spain and Tasmania.[53] Daniel Jackling taught a year at the Missouri School of Mines after his graduation in 1892, and eventually arrived in Cripple Creek with $3 in his pocket and took a job in an assay office, bucking samples. Soon he was assayer for the group installing the first barrel-chlorination plant in Colorado, and he remained as metallurgist at the works until it burned down in 1895.[54] Theodore J. Hoover, brother of Herbert Hoover, and a capable mining engineer in his own right who ultimately became dean of engineering at Stanford, began as assayer at the Keystone mine in Amador County, California.[55]

Mine surveying was another important area where beginners could augment their theoretical knowledge with practical experience and an opportunity to observe. As early as 1869, Rossiter W. Raymond had commented on surveying as a desirable first job, and Thomas A. Rickard echoed that sentiment in 1921. "The surveying department is a good ante-room for mining engineering," said Rickard, "if the young man be not too content to remain there, because it brings him in contact with operations underground and gives him an opportunity to study the geologic relations of ore deposits."[56] Louis D. Ricketts passed up a teaching position at Princeton in 1883 in favor of a surveyor's job at the Morning & Evening Star mines at Leadville. For the tall, soft-spoken easterner later dubbed by the Mexicans "el polo seco" ("the dry pole"), this was the beginning of a strenuous two-year apprenticeship preparatory to his taking charge of a group of small claims near Silverton.[57]

Arthur D. Foote, a civil engineer by training who had once

53. *DAB, 14,* 504; *EMJ, 103* (March 3, 1917), 380.

54. Rickard, ed., *Interviews,* pp. 194–96; Wolfe, ed., *Men of California,* 101. A barrel chlorination plant extracted gold or silver through treatment of ore in a revolving barrel or drum, using chlorine or another reagent. Fay, "Glossary," p. 64.

55. *EMJ, 74* (August 9, 1902), 192.

56. Raymond, *Mineral Resources (1869),* p. 225; *M&SP, 123* (September 10, 1921), 353.

57. Rickard, ed., *Interviews,* pp. 431–33; Walter R. Bimson, *Louis D. Ricketts,* pp. 10–12. For another example, see Harry M. Gorham, *My Memories of the Comstock,* pp. 192–93.

dropped out of Yale because of eye trouble, was given his first real chance in mining as underground surveyor at the New Almaden Quicksilver mine in California, but soon the position opened new vistas. His wife wrote a friend in 1877:

> The "Old Man" apparently thought his experience included nothing but surveying. He has been put in charge of the "Construction" at the mine and added $300 to his salary. The money is not very much, though, at this time particularly, it is very acceptable—but A. says the experience he will get is worth three times the advance in his wages. He has been restless because his work was getting too easy—now he has something to think about—"Construction" in A's business means the creative—Surveying is a matter of study and work but Construction is designing.[58]

That assayers' and surveyors' jobs often involved much more than the titles of the positions implied is clear. When Samuel Christy placed one of his students as a surveyor at the New Almaden mine in 1887 at a salary of $50 a month, he admonished him that the experience would be far more valuable than the pay. Christy also gave some gratuitous advice:

> Spare no pains to be perfectly certain that your work is correct—make yourself useful, indispensible if you can—and above all mind your own business—interfere with no one else—and keep your opinions and criticisms of others to yourself. . . . Remember you go to learn and not to teach.[59]

One of Christy's former students wrote back from Nevada that his "assaying" duties also included sampling and weighing, making up the ore bids and the daily reports, and doing

58. Mary H. Foote to Helena Gilder, Almaden, April 26, 1877, M. H. Foote MSS.

59. Samuel Christy to J. F. Wilkinson, Berkeley, August 5, 1887, copy, Christy MSS.

all the purchasing for his company.[60] Another, who went to the Kootenay District of British Columbia, found himself occupied with "assaying, mining law, surveying, mining, tool sharpening &c." "When not otherwise engaged," he wrote, "I worked with miners to learn what I could about 'breaking ground.' "[61] Twenty-two-year-old F. Sommer Schmidt finished up at Columbia Mines in 1903 and went out to do surveying and assaying for Mark Requa in White Pine County, Nevada, at $100 a month. Wearing "a derby hat, and kid gloves, and what were known at that time, as college-cut clothes," he stepped off the stage at Ely to find himself a jack-of-all-trades. Schmidt later recounted,

> I quickly discarded that derby hat, and fitted into my position of roustabout, scientific and otherwise. I helped the carpenter build houses using a single jack to drive nails, because there was only one claw hammer on the job. I unloaded and checked freight, made out pay rolls, bucked samples, surveyed the underground workings, and made underground connections between workings.

He did assaying in an unheated building, working in subzero weather in overcoat and overshoes, stamping his feet to keep warm, and he helped build the first real laboratory there. Schmidt studied in his spare time and learned much from such established engineers and geologists as Oscar Hershey, Andrew Lawson, and John B. Fleming. Moreover, for him it was a desirable, fruitful experience.

> I was playing on an industrial team. I was taking a post graduate course. I did a job then like patenting claims, or laying out the preliminary lines for a railroad, or making plans for a pilot mill for the first time. What a wonderful thing it was to get paid for learning all this.[62]

60. Thomas Rickard to Samuel Christy, Eureka, August 26, 1887, ibid.

61. John G. Sutton to Samuel Christy, Gold Ridge, Oregon, February 20, 1886, ibid.

62. F. Sommer Schmidt, "Early Days at the Nevada Consolidated Copper Company, Ely, Nevada," September 23, 1949, pp. 4–12.

Whatever his education and his background, and whatever aspect of mining he entered initially, the trained engineer had to overcome an early prejudice of the practical miner against "them damned eddicated fellers." A strong segment of public opinion in the West often supported the practical man, although enlightened sources rightfully held that only theory and experience combined could successfully and efficiently exploit mineral resources.

In the first place, it was charged that mining engineers and geologists—"the lace-boot, tack-hammer brigade"—were worthless in discovering minerals. Again and again, this theme was reiterated—"no wise man from the east—no scientist, no geologist—has ever found a valuable mine."[63] In his own special brand of pungent prose, Wyoming's Bill Nye summed up much of the sentiment of the ordinary westerner in 1882, when he castigated one particular "mining expert and geological toadstool," "a graduate in metallurgy and a keen student of true fissure pie," who like others of his ilk, had made "a galvanized ass of himslf."

> The only man in the West at present who knows the whole secret of creation and how the geological bowels of the earth were put in place is the mining expert. . . . The hoary-headed rock sharps who were sent to Leadville took out their pocket glasses, looked critically at the rich carbonates lying in deposit, shook their heads sadly, got drunk and went back to 'Frisco.

Then, said Nye, a few men such as H. A. W. Tabor, "whose knowledge had previously been in the great field of family mackerel and the study of rock salt and codfish float," plunged in and came out millionaires.

> It remained for the train dispatcher of a six-mule team to make money, while the scientific fungi found nothing but

63. Quoted in *Mining World* (Las Vegas, New Mexico), *2* (September 1, 1881), 12. See also San Francisco *Examiner,* April 28, 1866; "Amador" to editor, San Francisco, January 23, 1893, *M&SP, 66* (January 28, 1893), 52; Richard Harding Davis, *The West from a Car Window,* pp. 70–71.

> porphyry and pneumonia. The talented jackass with gold-bowed spectacles walked away and took the home-bound stage, while the man whose pantaloons were held in place with a strap and who didn't know whether the gneissic schist ought to overlie penurious bromide or not, became extremely wealthy and was prominently mentioned as a possible senator.[64]

A standard witicism that went the rounds in the next decade reflected the same disdain: "How are things in Cripple Creek?" "Oh, they are about the same as usual. The tender-feet are taking the ore out where they find it, and the mining engineers are hunting for it where it ought to be." [65] Even in the twentieth century, home-grown poets put the same sentiments into verse,[66] and Senator Albert B. Fall of New Mexico could deny on the floor of Congress that any geologist "ever discovered or developed a lode of minerals." [67]

Not only were technically trained experts unable to find metals, it was charged, they failed to make mining pay. As early as 1849, a California gold-seeker complained, "These mere finding of gold mines here is nothing, but the labor of getting it out is the grand item, therefore scientific geologists are worthless and laughed at." [68] Scientific experts were "at a great discount" on the Comstock Lode in 1865, said Samuel Bowles, "as all the rules of science with which they come equipped, are outraged and defied by the location and combi-

64. From the Laramie *Boomerang,* quoted in *Mining World, 2* (August 15, 1882), 292.

65. *M&SP, 73* (August 1, 1896), 86; *MI&R, 18* (July 23, 1896), 27.

66. "I have seed them science fellers with the glasses in their eyes
Jammin' round these Rocky Mountains lookin' ruther overwise,
An' explainin' the formation in a lot of hefty words
'Bout as meanin' to a miner as the chirpin' of the birds.
I have seed them walk so clus to payin' leads if they'd a been
Rattlesnakes a layin' fur 'em they'd a got it in the shin—
Payin' leads 'twas with a million an' that arterwards was struck
By an ol' prospector bankin' on his durn fool luck."

From James Barton Adams, "Just Durn Fool Luck," *The Double Jack* (Dillon, Wyoming), May 13, 1905.

67. *Congressional Record,* 65th Cong., 2d Sess. (1917–18), *56,* Pt. 10, 10184.

68. Quoted in Walker Wyman, ed., *California Emigrant Letters,* p. 77.

nation of ores, rocks, oils and soils on this side of the Rocky Mountains." [69] College-trained engineers were quoted as saying, "I have been very lucky; I generally guess right!" Educated men might fill mine reports with " 'ites, 'ides, 'ics and 'ers—words foreign to the masses of the people," but this did not demonstrate that their opinions were preferable "to a report of an honest, intelligent miner." [70] Even as late as 1913, a University of California graduate could write his old professor from a mine at Camp Seco about a former classmate: "Ollie Wyllie was working here for a month, and did some good work in the mine. The Shift-Boss said that he was the first college fellow that he had ever run into that was any good." [71]

Part of the hostility stemmed from the fact that trained engineers, like their practical brethren, made mistakes. "Scientific men, so called, have made stupendous mistakes of judgment; but they have been surpassed by the blunders of the practical men, so called," observed Rossiter W. Raymond in 1869.[72] Sometimes, "theoretical miners" were even more optimistic than the practical, and their enthusiasm did more harm than good.[73] Practical men referred caustically to the errors of the "school-bred miner" and the financial wreckage left in his wake. "From the days of the Frémont estate in the early fifties to the Bears' Nest fiasco in 1890," wrote one, "the track of the professionally, scientifically educated mining engineer has been one of ruin." [74]

Part of the antagonism undoubtedly came from a superior, "I-am-ready-to-show-you-how-to-do-it" attitude frequently displayed by fledgling engineers. Even veteran engineers on occasion damned the young "metallurgical wiseacres" who were quick to produce processes that were "utterly absurd and im-

69. Samuel Bowles, *Across the Continent,* p. 155.

70. "T.S." to editor, *M&SP, 26* (June 21, 1873), 386.

71. E. H. Clausen to George Louderback, Camp Seco, California, April 19, 1913, Louderback MSS.

72. *EMJ, 8* (July 13, 1869), 21.

73. *Among the Silver Seams,* p. 20.

74. J. H. Morton to editor, Redding, California, January 21, 1893, *M&SP, 66* (January 28, 1893), 53.

practical, even if theoretically correct." "Sweet innocence! What refreshing coolness of one just hatched, and of one not so much as introduced into the business of making quicksilver!" was the response of Eugene Riotte to a proposal made by a very stuffy young neophyte in the profession.[75] Residents of Leadville could only react negatively to the "big batch" of green engineers, many of them "theoretical frauds with diplomas," that descended on the camp in 1880, "most of them born, reared, educated and disciplined to mining pursuits in the city of New York," and who came with unbecoming arrogance and condescension.[76] At Ouray, the *Solid Muldoon* spoofed a little when it announced that, under local mining laws, "mining expert scalps will be taken at $4 each in payment of taxes,"[77] but the meaning was clear. So was the moral of the story about the newly minted "M.E." who arrived in the Lake of the Woods district much later:

> A certain individual dressed in all the equipment common to his type—khaki suit, knee-boots, gaiters, a brace of six-shooters prominently displayed in his carved leather belt—was accosted by a little urchin selling the local newspaper, with the cry, "Miner, sir, Miner!" With a look of the utmost contempt, and a voice choking with suppressed anger, the man replied: "D— you, boy, I'm not a miner; I'm a mining expert."[78]

At least in the early years, skepticism of technically trained men made it more difficult for young engineers to obtain jobs in some quarters. When John Hays Hammond returned from Freiberg in 1879, he sought employment with George Hearst, a family friend and one of the most important mine owners in

75. Quoted in *M&SP, 27* (October 18, 1873), 242; Charles H. Aaron to editor, Montgomery, Nevada, November 14, 1869, *M&SP, 19* (December 4, 1869), 354.

76. San Francisco *Bulletin,* March 15, 1880; *Mining World, 1* (November 1880), 6; *Mining Record* (New York), *6* (May 29, 1880), 509.

77. Cited in *M&SP, 40* (May 1, 1880), 278. See also *Virginia Evening Chronicle,* July 1, 1880.

78. "Engineer" to editor, Los Angeles, October 3, 1904, *EMJ, 78* (October 20, 1904), 621.

the West. The vinegary Hearst at first turned him down. "The fact of the matter is, Jack, you've been to Freiberg and have learned a lot of damn geological theories and big names for little rocks. That don't go in this country." "Anything else?" asked Hammond. "No," said Hearst, "Freiberg is enough." "Well," said Hammond, "I'll make a confession to you if you won't tell my father. I *didn't* learn anything of importance." He got the job—at $50 a month.[79] Even in the 1890s, a college degree brought snorts of disapproval from mine owners,[80] and an engineer in charge of undergound work at the Anaconda Mining Company in Montana explained why:

> A situation that would be acceptable to a graduate is often very hard to obtain, with a mining Co. for the reason that those who have the positions to give are often ignorant persons whom luck has placed them where they are and they discourage the employment of educated persons except in positions which absolutely require college bred men.[81]

Such attitudes were falling more and more into the minority. Mine operators everywhere became increasingly aware of the value of expert geologists and engineers, though some drew a line of demarcation beyond which they would not go. William A. Clark, for example, probably piqued by the adverse testimony of experts in the Butte litigation at the turn of the century, subsequently refused to permit a geologist to examine his United Verde Extension in Arizona, thus missing the great riches later opened up by James Douglas, Jr.[82] Others regularly consulted engineers, but proceeded warily on their recommendations. Marcus Daly, one of the most practical of men, was quoted as saying, "I listen to the reports of my engineers and then I lock myself in my room lest they influence my judgment." As a mine owner, he had to consider questions

79. Hammond, *Autobiography, I,* 83–84.
80. H. J. Jory to Samuel Christy, San Francisco, May 3, 1891, Christy MSS.
81. M. K. Rodgers to Samuel Christy, Butte, June 6, 1892, ibid.
82. *M&SP, 116* (January 12, 1918), 41–42.

other than purely engineering ones. John D. Ryan, another outstanding mining man, made a similar point:

> If the Anaconda company should do all the good things its engineers recommend, it would never pay a dividend. They are good engineers, and most of the projects they urge are good, but if we carried out all of them our capital would be perpetually tied up.[83]

From the beginning, fair-minded commentators decried the squabble between practical miners and trained engineers, and saw cooperation as the solution. "Mining being a science as well as an art, requires an educated head as well as an educated hand," wrote a San Francisco editor in 1868. "Either can do but little singly; conjoined they can accomplish almost everything." [84] But the quarrel died slowly. "Like the wrestle of the ministers with the devil," it threatened to continue indefinitely, unless a spirit of understanding could be achieved.[85] Yet die it did, and gradually, with the passage of time and an increasing number of experienced engineers producing tangible results, the technically trained man came into his own.

Despite all the criticism, that the college-bred engineer commanded much respect and that his services were increasingly in demand was indicated by his many imitators—the "local butchers, bakers and candlestick-makers, who put M.E. to their names" and posed as mining experts throughout the West.[86] At Leadville, during the boom, "there as nowhere else before the crop of self-made 'professors' and 'mining engineers' and 'experts,' seemed to spring out of the ground or to fall from the clouds. Their name was legion, and their performances marvelous beyond credulity." [87] In the Black Hills in

83. Quoted by Walter R. Ingalls, "The Business of Mining," *M&SP, 113* (August 19, 1916), 277.

84. *M&SP, 17* (July 18, 1868), 40.

85. Denver *Tribune-Republican,* quoted in *M&SP, 53* (July 17, 1886), 34. See also *The American Mining Gazette, and Geological Magazine* (New York), *5* (August 1864), 312; *AJM, 3* (April 20, 1867), 71; *EMJ, 36* (December 15, 1883), 366; *MI&R, 19* (January 7, 1897), 2.

86. *Anglo-Colorado Mining and Milling Guide* (London, *1* (October 29, 1898), 141.

87. "The Plague of Experts," *M&SP, 41* (August 21, 1880), 120.

the 1890s, these self-styled "engineers" and "professors" were "thicker than flies around the bunghole of an empty beer barrel in summer time." [88] "It doesn't take much to make a mining engineer and metallurgist professor now-a-days," said a Denver editor in 1898:

> A few undigested facts, not even distantly related, and a few long words, are enough stock-in-trade to get the title. Then he plays scientist among the "mob," until someone who really knows something comes along—and then he hides. He bewilders real workers with a professed long "practical" experience, and wandering easterners with his jargon about sulphides, sulphates, oxides and tellurides. He isn't fit to be even a collector of facts, for he don't know them when he meets them. He is no more qualified to speculate upon the sciences than a bricklayer is to rival Paderewski.[89]

Because the public often lumped together indiscriminately the bona fide engineer and the pseudoscientist, some of the opprobrium directed at the latter undoubtedly fell on the shoulders of the former and caused concern. One engineer believed in 1881 that his profession had "been more than any other abused by the pretensions of a class of untrained men." [90] No other profession, insisted T. E. Schwarz, a Freiberg man, was so poorly defined in the public mind and so invaded by "quacks and shysters." [91] Others urged mining engineers to organize in self-defense, to educate the public and restore prestige lost by "the boldness of charlatans." [92] Technical editors flailed away at the unauthorized use of the suffix "M.E.," whereas others deprecated the cheapening of the term "engineer" to include, by 1911, anyone who "has an idea that they can sell

88. *Black Hills Pioneer,* quoted in *M&SP, 60* (February 15, 1890), 110.
89. *MI&R, 20* (April 14, 1898), 449.
90. Theodore Comstock to editor, Durango, Colorado, November 23, 1881, *EMJ, 32* (December 10, 1881), 390.
91. T. E. Schwarz to editor, Denver, *EMJ, 36* (December 15, 1883), 366.
92. *EMJ, 34* (December 30, 1882), 351.

cheap advice or run anything from a 'skin game' to a Hupmobile." [93]

One suggestion frequently made was to register or license mining engineers to screen out the unqualified, and by the turn of the century this movement had gained some support. Some believed that engineers, like doctors and lawyers, should be examined and licensed by the state; others felt that the American Institute of Mining Engineers should regulate its own membership. By 1921, nineteen states did have laws governing the practice of engineers, but those of Arizona, Minnesota, Virginia, and West Virginia made registration optional, and those of Idaho, Illinois, Louisiana, and Wyoming specifically exempted mining engineers. In general, many mining engineers opposed such legislation and used their influence against their enactment in Colorado, Nevada, and California, arguing that any system of state registration would hinder the engineer's normal need to move freely across state lines in conducting his work.[94]

So the mining engineer was forced to live with his imitators, but even before the turn of the century it was clear that he had won his battle, not only with the pseudoengineer, but with the mining public as well. By the 1920s, the old-time prospector, lacking both capital and technical knowledge, had largely disappeared. "He is not yet as extinct as the dodo," said one of the new breed, "but his glory, like that of Tyre and Nineveh, has departed, and he has become shrouded with the mist and mystery of a hoary and semi-historical antiquity." [95] The "Cousin Jack," with a touch of Cornwall in his speech, and an able but narrow miner, would have trouble in breaking away

93. *M&SP, 45* (August 5, 1882), 88; *EMJ, 91* (April 8, 1911), 700; *Mining and Engineering World, 42* (February 13, 1915), 333.

94. For the question of licensing, see *MI&R, 20* (February 3, 1898), 337; *Mining Reporter, 54* (October 18 & November 25, 1906), 384, 486; *Northwest Mining News, 1* (August 1907), 6; *Mining Science, 62* (July 14, 1910), 33; *Mining American* (Denver), *72* (November 6, 1915), 5; *M&SP, 119* (July 5, 1919), 1, *120* (February 14, 1920), 218, 232; *M&M, 1* (January 1920), 17, 19, *2* (May 1921), 8, *3* (February 1922), 35.

95. Horace W. Winchell, "The Future of Mining," *M&M, 4* (January 1923), 12.

from the methods of his fathers and might be inclined to continue looking askance at newfangled notions.

But in the end, it was the precise methods and scientific knowledge brought from the technical schools and tempered with experience that made the western mineral industry a commercial success. The failure of self-made men to adapt their processes to new ores or to stem the loss in tailings or to handle low grade ores in bulk left the way open for the technician, who combined theory and practice and who moved about the globe with impunity unencumbered by the inertia of his forefathers. As the editor of *Mining and Scientific Press* noted in 1903,

> Within the past twenty years this old-time practical miner has been slowly forced into the background and in his place is found the technically educated man, who with a world-wide knowledge of his business is equal to almost any emergency. . . . The Western United States, Australia and South Africa have furnished a practical school, where unusual conditions were found and had to be met —novel situations in mine and mill. The old-timer would have undertaken these varied problems in the old-time manner and he would have failed, or at best would have made but moderate success.[96]

96. "Advantages of Technical Education," *M&SP, 87* (November 21, 1903), 332.

CHAPTER 4

"Be as Cold Blooded and as Unenthusiastic as a Clam"

Who is the man who views the mines and promptly turns them down?
Who is the one that thinks this is the short cut to renown?
Who is it gives the bum advice to the innocent financier?
The knowledge-feigning, theory-straining mining engineer.

—Anonymous [1]

The mining engineer was a man of several hats. Often his prime function was to serve as manager of a mine or mill property, supervising part or all of the operation. Sometimes he acted as a professional witness, giving technical testimony before the courts in controversies over title. But perhaps his most important role was as a consultant, dispensing advice to whomever was willing to pay—mine owner, vendor, promoter, or investor, potential or actual. In this respect, the mining engineer served as a confidential counselor of the capitalist, the middleman, the operating company, and the mine manager.[2]

He might be employed by a lawyer interested in assessing an estate or by a county or state concerned with mine valuation for tax purposes.[3] He might be called in as a troubleshooter to handle any one of a number of general or specific problems. Sewell Thomas once had the job of investigating "high grading" for his Goldfield, Nevada, employer—of discovering how and by whom ore was being stolen and shipped out of the state, although more commonly this task was assigned to a professional infiltrator, such as Charles Siringo.[4]

1. From "The Engineers," *EMJ, 105* (April 13, 1918), 697.
2. "The Mining Engineer, His Functions and Importance," *Northwest Mining Journal, 3* (June 1907), 87.
3. See pages 197–98, 250, this book.
4. Sewell Thomas, *Silhouettes of Charles S. Thomas*, pp. 94, 95–96; Charles A. Siringo, *A Cowboy Detective*, pp. 231–32.

He might be called on when an owner wished to determine whether continued operation and outlay of capital were justified[5] or to formulate working plans for opening newly acquired property, as in 1896, when E. H. Harriman engaged James D. Hague to examine his mines at Deadwood and devise operating procedures.[6]

Likely he would be brought in when production fell off or when a technical question arose that could not be solved by those on the scene. Thus, Daniel M. Barringer, himself an engineer, sought a cyanide expert to improve the rate of extraction at a mill in Mexico, and he solicited a dredge-engineering specialist when a dredging concern in which he was interested ran into difficulties in California.[7] Often a simple recommendation, such as a change in the type of flux, might spell the difference between success and failure.[8]

Not infrequently it was individual or minority shareholders who hired the consulting engineer to see why no returns were forthcoming. Edward T. McCarthy was sent to the Drumlummon mine in Montana in 1884 by an irate stockholder in India. McCarthy's report touched off an investigation by two company directors and an independent engineer, with resulting improvements in working efficiency and eventual prosperity.[9] Less fruitful were the results of an examination of copper property in Washington made in 1899 by an engineer sent out by a handful of unhappy shareholders: their mill could not operate, he reported, because their property contained no paying

5. James D. Hague to Directors, Ferguson Gold Mining Company, San Francisco, November 17, 1873, copy, Letterbook 5, Hague MSS; O. M. Stafford to Eben Olcott, Cleveland, August 4, 1894, Olcott MSS (N.Y.).

6. James D. Hague to Ellsworth Daggett, May 18, 1896, copy, Letterbook 21, Hague MSS; T. B. Ladlum to L. Bradford Prince, Denver, March 12, 1895, L. Bradford Prince MSS.

7. Daniel M. Barringer to T. Gordon Janney, Meteor, Arizona, May 23, 1908, copy, Letterbook N, and Barringer to F. H. Minard, December 16, 1912, Letterbook R. Barringer MSS.

8. *EMJ, 34* (October 21, 1882), 209.

9. Edward T. McCarthy, *Incidents in the Life of a Mining Engineer,* pp. 158–59; *EMJ, 38* (December 4, 1884), 373; Montana Company, Ltd., *Report of Messrs. N. Story-Maskelyne and J. R. Armitage to the Board of Directors,* November 12, 1884, pp. 8–10.

Mining centers of Idaho and western Montana

ore. But at least they were deterred from sending good money after bad.[10]

Critics sometimes deprecated the consulting engineer as "a positive parasite on the mining industry," one who "tended to diminish the efficiency of the superintendent," and who built "his reputation on bluff and the work of other abler men," [11] but this may have been merely the natural reaction of the manager who resented someone looking over his shoulder in critical fashion. As early as 1877, the editor of the *Alta California* urged every incorporated mining company expending more than $10,000 a year to have its property examined annually by a reputable engineer. Without such scrutiny, managers proceeded as they and their ill-informed directors chose; thus "their blunders are concealed; the true condition of their mine may be hidden from the stockholders, and year after year of bad management runs by with no correction of the evil." [12]

In several ways, the outside consultant, less hedged in by restrictions, was better able to accomplish change than the manager or directors. In court, it was established that a consulting engineer was "one who manages managers," and hence his opinion was worth more.[13] Alexis Janin, one of the great metallurgists of his era, once wrote a friend:

> Young men, as we old men know, are generally fools, but it is quite permissible in the young to be fools, provided that they recognize the fact before it is too late. . . . There are sometimes *old* fools. Directors of a Company are generally old and also fools. I have more than once put things over merely because, as expert, I had the authority to bring about changes which were denied to a superintendent, although he was more capable than myself.[14]

10. George E. Vigouroux, ed., *The Diary of a Mining Investor,* p. 15.

11. Carl E. Morris to editor, New York, May 12, 1913, *EMJ, 95* (May 24, 1913), 1062.

12. *Alta California,* February 23, 1877, clipping, Hubert Howe Bancroft Scrapbooks, *51,* no. 4, 1362.

13. *EMJ, 87* (May 29, 1909), 1101.

14. Alexis Janin to Charles W. Merrill, Santa Barbara, February 14, 1893, quoted in David W. Ryder, *The Merrill Story,* p. 23.

Because of the expense, small companies or individual mine owners could not usually afford to maintain a consulting engineer on a permanent basis. When an emergency arose, this meant a special study of each case by the consultant, who, because such jobs were irregular and uncertain, was forced to charge a high fee. One solution was for several companies to make joint use of an engineer, retaining him on an annual basis. This permitted him to become familiar with plant and property, yet to be free of all the distractions of the superintendent in charge. At the time he was appointed California state mineralogist in 1901, for example, Lewis E. Aubury was developing some gold mines of his own in Mariposa county and was acting as consulting engineer for three companies with property in California and Arizona.[15] On the other hand, this arrangement sometimes prompted a concern to overextend itself in acquiring as much property as it believed its consultant could oversee.[16]

His employment varied, but undoubtedly the most important work of the consulting engineer was the examining and reporting on mines when a change of ownership was contemplated. This "experting" might be done on behalf of any of a number of interested parties—the owner, an intermediary vendor or promoter, a prospective buyer or investor. As an arbiter between the investor and the prospector or between the investor and the vendor or promoter, the responsible engineer did not risk his reputation lightly, for if a transaction were completed, hundreds of thousands of dollars might be invested on his opinion. The reputable engineer laid his good name on the line only for fees that were worthwhile.

Thus, it was generally agreed that this was no field for the neophyte, that a broad and varied experience as well as native common sense were prerequisites to success, especially where undeveloped property in a new district was involved. "It requires the basis of a thorough education, coupled with years of

15. *EMJ, 50* (July 19, 1890), 68, *85* (February 8, 1908), 313–14; *M&SP, 82* (April 13, 1901), 175.
16. *M&SP, 73* (September 19, 1896), 234.

field experience, and a familiar acquaintance with the locality examined," concluded Rossiter Raymond in 1869.[17] "To ask a man who has no experience in the business and management of mines to appraise the value of a prospect situated in a new region is to court disaster," wrote Thomas A. Rickard, thirty-five years later.

> In sizing up the situation it is necessary that a man should know what are likely to be the costs of stoping, timbering, road-making, erection of machinery equipment, etc., and these things he can only know through actual underground experience and personal participation in the administration of mines.[18]

One editor in 1871 believed that not 5 percent of even the trained mining engineers in the country were then experienced enough to judge a new enterprise safely.[19]

During the 1860s and 1870s, when technical experts were in short supply, many a new graduate of Freiberg or Columbia went almost immediately into experting, but he did so at his own risk. Most followed the more conventional pattern of working up the ladder as assayer or surveyor, assistant superintendent of mine or mill, then superintendent, and finally manager in charge of all operations, before moving into the circle of examining consultants.

Both trained and untrained mining men performed this work. For many, it was a full-time occupation, but for others it was a sideline activity. More than one engineer employed as a mine or mill manager reserved to himself the right to take on additional consulting and examining work.[20] Walter McDermott, a well-known engineer who was for years affiliated with Fraser & Chalmers in the sale of mining machinery,

17. Rossiter W. Raymond, *Mineral Resources* (1869), p. 225.

18. Thomas A. Rickard, *The Sampling and Estimation of Ore in a Mine*, p. 14. See also G. W. Miller, "Mining Engineering in the Valuation of Mines," *M&SP, 86* (April 11, 1903), 228. "Stoping" means to excavate ore on a vein by means of a series of horizontal workings. Fay, "Glossary," p. 652.

19. "The Examination of Mines," *EMJ, 12* (November 14, 1871), 313.

20. See pages 191–92, this book.

often inspected mines as he traveled about the West on company business.[21] Sometimes a man temporarily out of work could turn his hand to "experting" in a pinch. One engineer wrote his mentor at Berkeley in 1914 that his mine had shut down for the winter and that, meanwhile, he had done some examining, "but there was not much in it." [22]

Professors in colleges across the land found "experting" a welcome (and profitable) break from the routine of campus life. Charles F. Chandler, first at Union College, then at Columbia Mines; George W. Maynard, professor at Rensselaer Polytechnic Institute; and the younger Silliman at Yale were typical examples.[23] When Theodore B. Comstock became Director of the University of Arizona School of Mines in 1893, it was with the express understanding that he be permitted to continue mine reporting as a private undertaking.[24] Probably the majority of mining, mineralogy, and metallurgy professors did some examining of western property,[25] although occasionally they came under fire as "unfair competition." [26]

Trained engineers and geologists in journalistic positions or in certain stations of government employ might also pursue mine inspection in their free time.[27] On the other hand, tradi-

21. *EMJ, 113* (April 29, 1922), 733.

22. Erich J. Schrader to George Louderback, Yerington, Nevada, January 5, 1914), Louderback MSS.

23. *American Mining Gazette, 5* (1864), unnumbered pages at end of volume; *EMJ, 8* (July 13, 1869), 29; *The Mining Magazine* (New York), *1* September 1853), 228; William Brewer to Josiah D. Whitney, New Haven, April 2, 1874, Brewer-Whitney MSS, Box 4.

24. *EMJ, 55* (April 1, 1893), 300.

25. See Robert Peele to Eben Olcott, New York, 1894, Olcott MSS (N.Y.); *EMJ, 68* (July 29, 1899), 134; *M&SP, 92* (May 13, 1916), 727, *114* (January 6, 1917), 33.

26. S. A. Crandall to editor, Tacoma, June 23, 1915, *EMJ, 100* (July 10, 1915), 68. On the other hand, many deemed this a desirable practice, for, given the low rate of university pay, it helped retain men in the teaching profession who could have done better financially in mining, and it undoubtedly helped university professors keep abreast of progress in the mineral industry and give a touch of realism to the theoretical side of mining in the classroom. Augustus Locke to D. M. Riordan, San Francisco, November 6, 1919, copy, Louderback MSS.

27. For examples of "moonlighting" editors, see *Mining World, 1* (June 1881), 4; *EMJ, 49* (May 24, 1890), 593. For examples involving state geologists

tion and express edict often barred members of the state or national geological surveys. Thus, Josiah Whitney, head of the California Geological Survey, was precluded from making an appraisal of the Comstock Lode in 1864, and Clarence King, of the U.S. Geological Survey, had to write in 1880 that his official appointment "positively prohibits me" from making a private report on the Sutro tunnel.[28]

In practice, advertising was limited in the profession, and the examining engineer depended on "the publicity of success, the scholarly distinction of the pen, and the plain advertising of a directory" to bring him employment.[29] The engineer who had "arrived" had little occasion to worry,[30] but often circumspect inquiry behind the scenes helped drum up business in dull times or before a reputation had been fully established. Early in his career, John Hays Hammond gently sent his card and a brief résumé of his experience to prospective clients; once his name became eminent, he was flooded with business from unsolicited quarters.[31] Daniel Barringer and Richard Penrose of Philadelphia sometimes promoted themselves by letter, and John Farish, during the depression of the early

---

or officials of the Assay Office, see *Mining Record* (New York), *6* (September 13, 1879), 205, *7* (June 5, 1880), 531; *MI&R, 17* (May 7, 1896), 520.

28. Josiah D. Whitney to William Brewer, Virginia City, Nevada, April 18, 1864, Brewer-Whitney MSS, Box 1; Clarence King to Joseph Aron, (April 8, 1880), copy, Letterbook 9, Clarence King MSS. See also Henry Janin to S. L. M. Barlow, San Francisco, September 29, 1879, Barlow MSS, Box 127.

29. "To blow a horn or to hire the town-crier is to provoke a derision; . . . to advertise the individual ability after the manner of a soap manufacturer is undignified and therefore ineffective; to provoke constant notice is tiresome." *M&SP, 92* (October 10, 1908), 474. See also *Mining & Metallurgical Journal* (San Francisco), *19* (September 1911), 2; Herbert C. Hoover, *Principles of Mining,* p. 193.

30. James Hague declined to insert his card in a leading professional journal in 1897, commenting that "having for so many years neglected that excellent method of promoting business, I am unwilling to begin now and only hope for a nice obituary notice by and by." Hague to Sophia Braeunlick, New York, January 22, 1897, copy, Letterbook 21, Hague MSS.

31. Hammond to S. L. M. Barlow, New York, July 8, 1884, Barlow MSS, Box 165; Daniel M. Barringer to W. M. Morgan, July 21, 1905, copy, Letterbook K, Barringer MSS.

1890s, solicited examining work through other engineers.[32] In flush times, however, Farish referred clients to his cousin, also a reputable expert, and Barringer passed along opportunities to Pope Yeatman, pleading overwork: "I am about worn out with travelling and must have a rest, and my own mining interests occupy all of my time." [33]

Once employed, the consulting engineer was charged with evaluating a mine and determining whether or not, given the set of circumstances that prevailed, it could be a profitable undertaking. Whether he was examining a prospect, a partially developed property, or a large operating concern, he had to take into account a host of considerations and he had to rely to some extent on what one editor called "scientific guesswork, guided by 'hoss sense'—the rarest quality on earth." [34]

Sometimes he was called on to make a preliminary examination, especially of a highly speculative prospect or partially opened mine. If the property looked promising, perhaps a potential buyer took an option, then had the engineer make a more exhaustive study or even do limited development work as part of a complete study of the ground. After the turn of the century, despite criticism, it was not uncommon to send a young, relatively cheap engineer to make the preliminary investigation, followed by a more experienced one for the thorough examination.[35]

No examining engineer wished to be hurried. As Louis Janin wrote in 1891, "no experting trip is worth the expenses unless one can be thorough. . . . After the Mulatos row and the Alaska row, no Expert of standing wants to do anything hastily." [36] At the same time, explicit instructions were desir-

32. Daniel Barringer and Richard Penrose to Messrs. Picands, Mather & Co., December 15, 1894, copy, Letterbook A, Barringer MSS; John B. Farish to Eben Olcott, Denver, February 14, 1894, Olcott MSS (N.Y.).

33. John B. Farish to Isaac L. Ellwood, Minas Prietas, June 25, 1903, Isaac L. Ellwood MSS, Box 102; Daniel M. Barringer to W. M. Morgan, July 11, 1905, copy, Letterbook K, Barringer MSS.

34. *MI&R, 20* (April 21, 1898), 459.

35. *SMQ, 11* (April 1890), 193–94; Hoover, *Principles,* pp. 53–54.

36. Louis Janin to James D. Hague, February 27, 1891, Hague MSS, Box 12.

able and necessary. Sometimes the client's orders were brief and to the point. Full instructions to James D. Hague in 1879 were: "Go at once to Leadville, see B. F. Allen & telegraph me whether you can advise buying the Breece Iron mine for sixty thousand dollars. Want reply friday night." [37] Most instructions were more elaborate, especially if from a prospective buyer, in which case the engineer might also receive copies of the option and previous engineering reports.[38]

But no engineer wished to be restricted by instructions that were unrealistic. Albert Ledoux once inspected a mine in the Southwest for a British client, who sent him detailed directions for work estimated to take two weeks. Ledoux and his companions found the property vastly overpriced, completed the job in two days, and started back in heavy snow on snowshoes. En route, they met "a vigorous Englishman in yellow leggings," stuck in a drift, who had been cabled to join them to make sure they followed the prescribed procedures. As Ledoux recalled it,

> Standing up to his waist in snow he catechized us frankly as to our method of examination, occasionally shaking his head in dissent, and at last said, "Well, I must go on anyhow, for you have neglected to take any specimens of the rock for petrographic slides." [39]

Sometimes the examining engineer was dispatched by his client under orders for secrecy,[40] and frequently, once in the field, he communicated with his employer in code, relying perhaps on John C. Bloomer's *Pacific Cryptograph,* or working

37. Thomas Nickerson to James D. Hague, Boston, April 1, 1879, telegram, ibid., Box 12.

38. James D. Hague to Hamilton Smith, Victor, Colorado, June 14, 1895, copy, Letterbook 19, and Smith to Hague, London, May 8, 1895, ibid., Box 12.

39. Albert R. Ledoux, "The American Mining Engineer," *EMJ, 77* (February 25, 1904), 310.

40. See Eben Olcott's private secretary to editor, *EMJ* (October 27, 1891), copy, Letterbook 18, Olcott MSS (N.Y.). Daniel M. Barringer to T. Gordon Janney, Meteor, Arizona, May 23, 1908, copy, Letterbook N, Barringer MSS; McCarthy, *Incidents,* pp. 158–59.

out a private cipher of his own.[41] One Californian devised an elaborate ten-page code covering most eventualities, including "Abigail"—"Not much pleased with the mine"—and "Faith"—"Don't say anything to anyone." He even added a few words with personal meaning—"Bogus"—"Am suffering for someone to pet me"—presumably for use with his (or someone else's) wife.[42]

Whether sworn to secrecy or not, the inspecting engineer had a reputation for being close-mouthed. New York engineer Arthur M. Wendt was described in 1881 as "a very clever kind of fellow," one who "doesn't say much. Keeps his counsel to himself as he ought." [43] Daniel M. Barringer once advised a young man who aspired to the profession on the need of being noncommittal on the job and to

> keep a perfectly cool head and to believe nothing until it is proven beyond all shadow of doubt and to at all times keep your mouth shut as to your own opinion or my opinion or as to anything concerning which your judgment will tell you is not wise to talk. Ask plenty of questions and get all the information you can, *but impart none*. If you have the least bit of enthusiasm in mining or even in mining expert work you can never hope to succeed. Be as cold blooded and as unenthusiastic as a clam.[44]

Once he arrived at the mine, the examining engineer had to keep in mind three all-important basic questions: How much ore was in the property? What was its average value? What

41. See John C. Bloomer, *Pacific Cryptograph, for the Use of Operators in Mining Stocks, Mining Superintendents, Bankers and Brokers;* John B. Farish to Messrs. Ellwood & Fleming, St. Louis, January 1, 1890, Ellwood MSS, Box 102; Barringer & Penrose to E. H. Harriman, December 17, 1894, copy, Letterbook A, Barringer MSS; Compagnie Française de Mines d'Or & d'Exploration Société Anonyme to James D. Hague, Paris, May 15, 1895, Hague MSS, Box 11.

42. Cipher copy, in T. J. Lamoureaux MSS, Box 1.

43. Entries for October 24 & 25, 1881, Parsons, *Journal,* pp. 269, 270.

44. Daniel M. Barringer to Guy P. Bennett, January 14, 1899, copy, Letterbook C, Barringer MSS. Of examining engineers, a veteran mining man commented in 1904: "some of them are very close and do not say anything for or against, and it is very hard for you to guess their opinion in regard to it." William Read to Alex Muir, Capitola, California, February 12, 1904, copy, Letterbook (1900–07), Read MSS.

would be the cost of extracting and marketing it? Or as Daniel Barringer liked to express it, he had to consider mine examining as a three-legged stool, the legs of which were quantity, average quality, and availability. All three had to be firmly established before a report could be favorable.[45]

To establish the amount of ore and its average value, the engineer had to measure what was obvious, sample the ore, and make educated estimates. On an undeveloped prospect, especially, he was forced to rely more heavily on guesswork, which in turn was based on his own ability, prior experience, and familiarity with local conditions. Even in the case of an established producing mine, it would be impossible to assign a wholly accurate value; estimates above the minimum always carried varying degrees of risk, depending on the individual engineer's qualifications and outlook. Each mine was different and had to be judged on its own merits, but "Experience, silvered with age, is the presiding judge, and will decide whether this or that piece of evidence is relevant or not." [46]

Nineteenth-century inspecting engineers understandably were more primitive in their approach than their twentieth-century counterparts. The earlier engineer looked over the records and maps, broke samples here and there, and hopefully and unsystematically "put his finger on the vital spot." [47] But by the turn of the century, mine examinations were much more thorough than before, a fact one engineer attributed mainly to accomplishments on the Rand in South Africa.

> It is probable that the very atmosphere of thought breathed by the earnest men of the profession has been influenced by the great developments in the Transvaal and the consequent introduction of a degree of system previously rare in metal mining,

wrote T. A. Rickard in 1904.[48]

Graduates of modern mining schools were taught how to

45. Daniel M. Barringer to Tatsuzo Kosugi, August 8, 1906, copy, Letterbook L, Barringer MSS.
46. Rickard, *Sampling,* p. 54.
47. *M&SP, 121* (August 14, 1920), 223.
48. Rickard, *Sampling,* p. 10.

sample and survey; they knew how to map and estimate ore; they were familiar with amortization and the use of Inwood's tables. In examining a mine, they often made a topographical survey, and if the property were developed, they might also run an underground survey or hire a surveyor for the purpose. If the property were a large one, the inspecting engineer employed not only a surveyor, but often a draftsman and a sampling crew as well.

On small properties, data could be obtained relatively simply, sometimes in a matter of a few days; on large mines, with from ten to a hundred miles of workings, the task might involve taking from ten to fifty tons of samples and require several months. One of the most elaborate examinations, that made for the Guggenheim Exploration Company of Utah Copper Company property in 1903, under A. Chester Beatty, Henry Krumb, and Seeley Mudd, required sixteen junior engineers as assistants, involved 3,500 samples, took seven months and more than $150,000 to complete.[49]

After a brief period to familiarize himself with the mine, the inspecting engineer began sampling or brought in his sampling crews. Thomas Rickard believed that the mark of a successful engineer was to know how to sample accurately and estimate tonnage and that weaknesses in this area ruined more engineering reputations than any other cause.[50] It was agreed that sampling should be closely supervised and organized, without interference from the mine owner, for the validity of sampling rested on the premise that mineral was distributed more or less regularly through an ore body and that samples taken at a number of points would, when averaged, give a reasonable idea of the unit value of the ore. But it was up to the expert to determine how frequently samples were to be taken and where without coaching from the management. In regular ore, perhaps every ten feet was adequate, but in case of fluctuation, every three feet might be desirable. On certain types of ore, notably the common sulphides, variations might be smaller and sampling could be done at larger intervals. The

49. Arthur B. Parsons, *The Porphyry Coppers*, pp. 73, 75.
50. Rickard, *Sampling*, p. 1.

examining engineer or his chief assistant marked in chalk or whitewash where samples were to be taken on the face of the ore and across its course, for these were important decisions. "A few cuts placed by an engineer of experience are worth a raft of samples poorly placed," insisted one expert.[51] As sampling proceeded, variations in vein, the number and position of each sample cut, and other information were plotted on a master plan, which ultimately would show the assay value at each point. "Given a good sampling and a proper assay plan," said Herbert Hoover, "the valuation of a mine is two-thirds accomplished." [52]

The sample itself, a trench-like cut, perhaps four inches wide and two inches deep, was usually taken with a moil, a short steel-cutting tool struck by a four-pound sledge and preferred by most engineers, because unlike a pick or geologist's hammer, it did not seek out crevices or soft places in the vein and produced a truer sample.[53] Sampling with a moil was "hard, uncongenial, manual labor," and experts recommended that the assistants who did it be given other tasks occasionally to break the monotony:

> the dirt, the wet, the strained positions, the splinters that hit the face and hands, the obstinacy of rock and circumstance, the weary iteration of it—these require something better than mere mule-like persistence to overcome them.[54]

Each sample was marked with a tag of wood, metal, or paper, and might be reduced to manageable size by "quartering." Using a rock breaker or a cobbing hammer, the expert crushed his samples on a square sheet of canvas or oil cloth, mixing well and discarding alternate quarters repeatedly to produce a small sample representative in value to the original. All the while, precautions had to be taken against fraud, and

51. Morton Webber, "Systematizing Large Mine Examinations," *M&SP, 121* (August 14, 1920), 234–35.

52. Hoover, *Principles,* p. 5.

53. Rickard, *Sampling,* p. 17.

54. Ibid., p. 26; Hoover, *Principles,* p. 5.

most engineers preferred a large leather sample bag, metal cases or tubes with locks.[55] As a further check, they often introduced a few worthless samples of "country rock" and utilized a system of duplicate or even triplicate samples, each set assayed by a different assayer.[56]

Most did not do their own assaying, but included an assayer in their crew or hauled samples to a professional.[57] When Penrose and Barringer examined an Arizona property for E. H. Harriman early in the 1890s, they sent the samples to Burlingame in Denver for assaying; likewise, when Robert B. Stanton inspected a placer mine north of Cheyenne in the spring of 1903, he shipped the samples to Denver, then had to borrow money to obtain the results.[58]

But samples alone were not the only standard by which to judge. As colorful old John Gashwiler, a California mine operator of considerable success, once put it when shown some rich ore specimens and asked his opinion on the value of the property, "You might as well show me the hair from the tail of a horse, and then ask me how fast the horse can trot." [59] In addition to sampling and estimating the average ore value, the examining engineer also had to calculate the amount of ore in the mine, and in so doing, he employed various terms to classify the ore according to the degree of risk of returns. "Ore in sight" was an expression commonly used and often damned for its elasticity, although reputable men men insisted that it be applied only to ore exposed on at least three sides. "Ore being developed," or "probable ore," might legitimately

55. Ibid., p. 7; O. H. Packer, "How Mines Should Be Examined," *EMJ, 70* (October 20, 1900), 457; Rickard, *Sampling,* pp. 23–25.

56. Hoover, *Principles,* p. 7; Josiah Spurr, "Salting," *EMJ, 110* (September 11, 1920), 506.

57. Arthur B. Foote to J. B. Furguson, October 4, 1907, copy, A.D. Foote MSS.

58. Daniel M. Barringer to Richard A. F. Penrose, March 5, 1895, copy, Letterbook A, Barringer MSS; entries for April 15, 23, 24, 25, 27 & May 2, 1903, field notes 20, Robert Brewster Stanton MSS.

59. Quoted in John Hays Hammond, "Suggestions Regarding Mining Investments," *EMJ, 89* (January 1, 1910), 10. In 1880, Jessie Benton Frémont made the same point: "selling mines on 'specimens' turns out as badly often as the old business of marrying royal brides from a portrait miniature." Frémont to William K. Rogers, New York, April 3, 1880, William King Rogers MSS.

refer to ore exposed on two sides, whereas "ore expectant," or "possible ore," was suggested as descriptive of what might exist beyond or below the last visible ore.[60] But any responsible expert had to be careful of his use of terminology, lest his client or subsequent shareholders interpret his comments literally, disregarding qualifying adjectives.

Even the process of sampling, when "scientifically" done, assumed a sameness of ore value and volume that seldom existed; it was a basis for an estimate, but not an undisputed fact. In general, sampling in place gave a higher average value than actual reduction might give, and the engineer had to keep this in mind. What might actually exist beyond the sampled face was uncertain. The character of the ground might change unexpectedly; hard rock might give way to soft or vice versa. One portion might require timbering, another not; veins might pinch out or contract; water might be struck. Any unpredicted change would disrupt cost estimates. The engineer, therefore, had to fall back on geological evidence and take into account such factors as the general character of adjoining mines, the size, origin, and structural nature of the deposit, the position of the mine openings with regard to the secondary alteration, and the depth to which the mine had already been exhausted.[61]

But ore quantity and average value were but two legs of Barringer's three-legged stool. The third had to do with the cost of extraction and marketing. The examining engineer had to put the property into proper perspective with economic reality. He was expected by his client to "give an opinion, not a mere description." [62] Of what use was a report made by one engineer who stated that there was $4 million in sight, when that of a second engineer sent to verify his work estimated that $5 million would be required to get the ore out? [63]

60. For comments of various engineers on the use and misuse of such terms, see Rickard, *Sampling,* pp. 51–52, 81, 117; Hoover, *Principles,* p. 17; G. W. Miller, "Mining Engineering in the Valuation of Mines," *M&SP, 86* (April 11, 1903), 229.

61. Hoover, *Principles,* pp. 21–23; *M&SP, 89* (November 5, 1904), 304; *90* (March 11, 1905), 149.

62. "The Examination of Mines," *EMJ, 12* (November 14, 1871), 313.

63. "The Valuation of Mines," *M&SP, 86* (February 21, 1903), 114.

In new districts, reliable cost data would be lacking, and it was easy for an inspecting engineer to go astray when trying to answer the basic question, "Will it pay to work this property?" As Rossiter W. Raymond pointed out in 1869, foreign-trained engineers were often tempted to consider western veins rich or highly promising, forgetting that labor costs were much lower in the European mining districts with which they were familiar.[64] In estimating the expense of mining and preparing the mineral for market, the isolation of an area and costs of transportation had to be considered. Smelting fees and even fluctuating metal prices, except for gold, had to be taken into account. The examining engineer would have to determine in his own mind whether the price being asked for the property was reasonable and how much capital would be required to develop and work it. Were there special problems? Was the ore such as could be worked by standard processes? Was it in combination that made it refractory? Gold combined with copper or arsenic would exclude cyanidation, for example. What of the human element? Why did the owners wish to sell? Were adequate timber and water resources available? The engineer had to look at all factors bearing on the working of a mine.[65]

Even if a mine was producing or had behind it a history of production, determination of costs was not always easy. The examining expert, of course, had access to the books, and sometimes he had them audited by an accountant, but he had to be able to read and interpret company accounts himself.[66] Not that such records were always readily available. Robert E. Booraem, a Columbia M.E. conducting an examination of a Colorado mine in 1882, belatedly relayed the information that he had overlooked mine expenditures of more than $12,000 because they had been entered in a second account book. A

64. Rossiter W. Raymond, *Mineral Resources* (1869), pp. 225–26.

65. For general discussions of the many factors to be considered in mine examination, see *SMQ, 11* (April 1890), 193–94; Hoover, *Principles,* p. 55; Rickard, *Sampling,* pp. 10–13; *M&SP, 121* (August 14, 1920), 223.

66. Charles V. Jenkins, "The Auditing of a Mine Company's Accounts," AIME *Trans., 33* (1902), 93, 105.

dozen or so years later, Ellsworth Daggett inspected another Colorado property. It was in one town, he said, and had three sets of books and papers in four different offices in another.[67]

Where accounts were available, the engineer could not be certain of their accuracy. Owners sometimes ignored improvement outlays or depreciation of equipment; they might disregard the difference in transportation costs in winter and in summer; or they might not distinguish between costs of development or "dead work" and actual stoping operations when ore was being produced. For safety, engineers learned to evaluate costs over a long period, if possible, in order to include seasonal and market fluctuations.[68]

Examination techniques varied with the type of mining. Obviously, an elaborate sampling plan could be applied only to a mine developed enough to present a sizable ore body. After the turn of the century, new methods were commonly applied to both dredge and open-pit copper prospects. In the case of dredge ground, the engineer began by dropping simple shafts and working his samples by pan or rocker. If results were promising, he might recommend further testing, probably with a Keystone or Cyclone drill. The churn drill, of whatever make, was a self-contained unit powered by steam, oil, or even electricity and bored a six-inch hole. It was more expensive than shafting, but was not bothered by groundwater, and its same pump recovered samples from the bottom of the hole. These were then washed for recovery and the results computed in terms of the amount of gold per cubic yard of gravel. Normally, the engineer determined where preliminary drill holes were to be dropped in order to gauge the extent of workable ground; then he commenced a pattern of holes—sometimes twenty-five or thirty per acre—to evaluate the whole.[69]

67. Robert E. Booraem to James D. Hague, Sharkill(?), Colorado, August 6, 1882, Hague MSS, Box 16; Ellsworth Daggett to Hague, Salt Lake City, December 14, 1895, ibid., Box 19.

68. Rickard, *Sampling,* pp. 11–12.

69. D'Arcy Weatherbe, *Dredging for Gold in California,* pp. 27–30; Lewis E. Aubury, *Gold Dredging in California,* pp. 20–26, 33–34. Also called the cable drill, or well drill, the churn drill performed its work by means of pulverizing blows. Its drill head was raised by rope or cable and allowed to drop, its heavy force deepening the hole with each impact. Fay, "Glossary," p. 155.

But this was expensive business: drilling costs might average from $1 to $3.75 per foot, exclusive of engineer's fees, and a hole twenty or thirty feet deep was not unusual, and some went down as much as fifty. Thus, when Fred L. Morris finished his preliminary examination of Merced River property in 1911 and recommended the sinking of 100 drill holes as the next step, he was advocating the commitment of considerable capital.[70]

Like his colleague in hard-rock mining, the engineer examining dredging ground had to keep many factors in mind. The value of gold was basic, but he had also to consider the nature and characteristics of the gravel; the size of boulders present; the depth, character, and contour of bedrock; the water level; the availability of water supply; surface configuration and timber growth; normal costs of labor and transportation; and even climate, for in cold areas, especially the frozen North, special equipment and techniques would be required. Moreover, in California, he had to keep antidebris legislation in mind.[71] But, unlike most other types of mining, the life of a dredge proposition could be more closely anticipated. In underground mines, profits in sight could be considered a guarantee for a return of only part of the capital outlay; in dredging, retirement of property and equipment costs was figured on the basis of the value in sight, generally on the basis of recommending machinery large enough to turn over the ground in ten years—the average life of a wooden-hulled dredge.[72]

After its introduction by consulting engineer. W. Y. Westervelt in Arizona, the churn drill also became an integral part of the examining process in the low grade porphyry coppers

70. Fred L. Morris to John Hays Hammond, San Francisco, March 25, 1911, copy, Fred Ludwig Morris MSS, Box 6.

71. Weatherbe, *Dredging for Gold,* p. 27; Oscar C. Perry to Fred L. Morris, August 18, 1912, and Charles Munro to C. K. Lipman August 27, 1912, copy, Morris MSS, Box 2.

72. Aubury, "Gold Dredging," 36. Dredge mining involved a self-contained unit designed to move on water. A power-operated scoop or suction apparatus mounted on a flat-bottomed hull dredged up gravel, which was then screened and the gold separated on the boat itself, with the tailings cast aside. Fay, "Glossary," p. 230.

from about 1908. By 1933, it was estimated that about 90 percent of such ore deposits then developed had been proved by churn drills capable of probing to great depths. Though more expensive to operate, diamond drills were also common in copper examinations, but in either case, it took a skilled engineer to sample accurately and thoroughly and to interpret the results properly.[73]

Generally, unless he had a definite understanding with his employer, the examining engineer of whatever kind of property did not like to give a preliminary report until the work was completed and his data analyzed. Conclusions drawn before the inspection was finished might be only tentative, and their premature announcement misleading and embarrassing. Once the report was concluded, however, the expert might then telegraph an abbreviated version to his client, to be followed by the full report in the mail.[74]

The nature of the expert's report varied. Some were brief and informal, and a few individuals, such as Henry Janin, who detested writing up examinations, occasionally operated on a terse "yes" or "no" basis. Clarence King could be cryptic as well as witty. Once King examined a western property for friends who were alarmed over rumors that the mine had developed a "horse"—a term for a worthless body of rock lying within a vein. King went through the property and shot back a telegram: "The mine is a perfect livery stable." [75]

Most written reports were more elaborate and came to follow a fairly standard pattern, describing the location of the mine, its history, geology, development or lack of development; discussing the samples and the weak points versus the

73. Parsons, *The Porphyry Coppers,* pp. 355–68. Porphyry copper is the name applied to copper-bearing igneous rock with large conspicuous crystals. In common usage, the term "porphyry" came to apply to almost any igneous rock occurring in sheets or dikes. Fay, "Glossary," p. 529.

74. Samuel B. Christy to Henry Perkins, Berkeley, December 14, 1891, copy, Christy MSS; Spencer C. Browne to W. W. Mein, Denver, November 18, 1912, copy, Spencer C. Browne MSS.

75. *EMJ, 91* (May 27, 1911), 1056; James D. Hague, "Memorabilia," in King Memorial Committee of the Century Association, eds., *Clarence King Memoirs* (New York, 1904), pp. 403–04.

strong; and ending with a summary and finally the examining engineer's recommendations.

Much, of course, depended on the purpose for which the report was intended. A prospective buyer wanted as much specific detail and evaluation as possible; so did an owner or shareholder who hired an examining engineer as troubleshooter. A vendor or promoter seeking to raise capital was amenable to detail so long as it was favorable, but invariably preferred a noncommittal report to a negative one. In 1895, Lewin Barringer, brother of an active engineer and himself interested in mining investment, urged greater detail as to geology and possible capital return:

> In other words, make your report such that it could be used *to show parties who think of putting* additional capital into the concern. I do not believe in making reports too short. You can elaborate in the report, *and then summarize in the conclusions*. Of course, if you give a man *too little* paper for his money, he is apt to complain, and besides reports should be made, not only for the party who actually employs you, *but also to be shown to others with whom you have no opportunity to personally interview in order to explain the conclusions arrived at in your report.*[76]

From a professional standpoint, it was agreed that the best report was short and to the point. The British engineer, J. H. Curle, believed that American engineers excelled in this respect, and agreed with the American who said he judged his subordinates "by the way they could condense a report into the limits of a cable; and the shortest cable meant the best man." [77] Perhaps Curle was referring to Herbert Hoover, who consistently emphasized simplicity and brevity. "The essential facts governing the value of a mine," said Hoover, "can be expressed on one sheet of paper." [78]

76. Lewin W. Barringer to Barringer & Penrose, June 14, 1895, copy, Letterbook A, Barringer MSS.

77. J. H. Curle to editor, London, April 25, 1903, *EMJ, 75* (May 9, 1903), 701.

78. Hoover, *Principles*, p. 56. See also *EMJ, 94* (September 7, 1912), 433.

The best examining engineers were decisive and their conclusions clear-cut. Pope Yeatman, one of the "greats" of the profession, was known for his brutal, to-the-point reports. Clarence King minced no words. In his famous report on the Emma mine in 1873, he sorely castigated the management for gutting the property without pushing exploration and without taking precautions against caving and in each case asked "Why?" concluding with the observation that these were "questions which it were highest charity to solve on the theory of stupidity." Or consider the words of Seeley Mudd in his report on the Ord Mountain Group in California (1905): "These notes are so brief because the property is so utterly unpromising that it seemed a waste of time and effort to write or read with regard to it." [79]

A great many reports either displayed too much imagination or were totally neutral. With regard to the first, a reader of *Mining and Scientific Press* in 1893 believed that publishers were overlooking best-seller material. "A collection of mining romances in the shape of reports on mines," he said, "if published, would afford entertainment for readers of fiction unequaled by any modern publication." [80] Technical men and especially some of the moonlighting academicians in the field were sometimes guilty of overusing scientific terms and of "slinging solid chunks of geology around with an appalling looseness." "I have seen a report," said one of the profession, "which started with the nebular hypothesis, and traced the progress of the earth from its pulpy state right down through

79. Edward M. Woolley, "Salary—$100,000 a Year," *McClure's Magazine, 42* (April 1914), 110–11; Clarence King to Secretary, Emma Silver Mining Company, Ltd., Salt Lake City, June 11, 1873, in "Emma Mine Investigation," *House Report* 579, 44th Cong., 1st Sess. (1875–76), p. 634; Seeley W. Mudd, "Report on the Ord Mountain Group of Mines," December 2, 1905, Morris MSS, Box 6. For other examples, see Eugene N. Riotte, *Report of the Havilah Mining Co's Property in Kern County, California* (San Francisco, 1875), pp. 2, 5–6; Ellsworth Daggett, "Report upon the Property of the Atlantic Gold and Silver Consolidated Mines, Ltd.," July, 1897, typescript, Hague MSS, Box 16; Spencer C. Browne, "Report on the Sunnyside Mine, Roosevelt, Idaho," November 1912, typescript, Browne MSS.

80. "Mining Sharp" to editor, San Francisco, January 2, 1893, *M&SP, 66* (January 7, 1893), 4.

its various stages to oxidation of the outcrop of a particular vein in the year of grace in which the report was written." [81] One such report written on an Arizona property had to be returned to the expert for correction because with all its erudite jargon, it failed to mention the quality or quantity of ore in the mine! [82]

Some were simply overly enthusiastic and let their words run away with them. Perhaps few were as intemperate as a legendary expert named "Metalliferous" Murphy, who is supposed to have praised a Nevada property to the skies: "With all Niagara for water power and all hell for a dump, that mine would niver be worked out in tin thousand years!" [83] But Rossiter Raymond could wax near-poetic in describing terrain and setting, though he seldom exaggerated, and Benjamin Silliman, among others, was lavish with superlatives—"Nothing, I am persuaded, since the discovery of California and Australia, is comparable for its measurable reserves of gold, available by the hydraulic process, in these deep placers of the Rio Grande." [84]

The other type—the neutral reports—were common enough, as indicated by a widely circulated burlesque written by a "Professor Noncommittal" in the 1880s:

> There is undoubtedly a mine here if the ore bodies hold out. The gangue rock is favorable to the existence of ore and the over-lapping seams of schistose show an undoubted tendency to productiveness in rock, which may be ore bearing. While I refrain from pronouncing with cer-

81. *M&SP, 55* (February 25, 1888), 120; Walter McDermott, "Mining Reports and Mine Salting," *Transactions of the Institution of Mining & Metallurgy* (London), *3* (1894–95), 118.

82. Reginald W. Petre to editor, Baltimore, May 25, 1903, *EMJ, 75* (June 6, 1903), 849. See also Jethro Jerington, "A Hole-in-the-Ground 'Pome,'" *M&SP, 52* (April 10, 1886), 238.

83. Quoted in Wells Drury, *An Editor on the Comstock Lode,* p. 31.

84. Justus Adelberg & Rossiter W. Raymond, *Manhattan Silver Mining Company of Nevada;* Benjamin Silliman, *Report on the Newly Discovered Auriferous Gravels of the Upper Rio Grande del Norte, in the Counties of Taos and Rio Arriba, New Mexico,* p. 7. See also Herbert Strickland to Stephen B. Elkins, Cerrillos, New Mexico, November 19, 1895, Stephen B. Elkins MSS.

> tainty on the Goosetherumfoodle mine, still I argue that as great expectations regarding the yield of this vein may be maintained as of any ground in the vicinity. The trend of the rock is S.S.E. and the direction of all the dips and angles show this to be a true lead, and as such liable to rich ore. Above all things the ground should be thoroughly prospected. I would advise the sinking of one hundred shafts ten feet apart through the hardest rock which can be found. If water be encountered it should be pumped out. If the rock prove rich the mine will prove valuable. If it prove very rich the mine will prove very valuable. It should be borne in mind that if it is necessary to sink deep on the vein the lode must be penetrated farther than if not.[85]

The noncommittal report was the result of a number of factors. One engineer, William Blake, once wrote a prudent, fence-straddling report under duress. In Baja California to examine a mine, he was forced to compose his report while isolated, except for his employer, the wife of whom was a known murderess, and he dared not write a negative commentary.[86] But this was exceptional. Many neutral reports were the work of engineers who either felt loyalty to a vendor-client with a mediocre property or who wished to be "lukewarm and then blow both hot and cold," thus maintaining a clean reputation for never having made an error.[87] Many were simply the result of an inability of the inspecting engineer to foretell the future. "Mines are like saints," went a western saying, "for many are called and few are chosen." Great mines were rare. To develop a prospect into a paying proposition meant the risk of considerable capital, and even conscientious engineers hesitated to commit themselves when evidence was slim. They qualified their opinions and sought to protect themselves by

85. From the New York *Graphic,* quoted in *The Nevada Monthly* (Virginia City), *1* (July 1880), 247. See also *EMJ, 29* (April 10, 1880), 251; *Helena Herald,* April 21, 1880. For another humorous parody of a mine report, see Oscar Shuck, ed., *The California Scrapbook,* pp. 357–58.

86. Hammond, *Autobiography, 1* 149.

87. *EMJ, 42* (August 13, 1886), 117.

hedging, whereas nonvendor or nonpromoter employers demanded clear-cut answers where there were none. But responsible engineers agreed that as much clarity and unequivocation as possible was a desirable feature of any report.[88]

The path of the examining engineer was strewn with pitfalls, not the least of which was the necessity to deal with overly optimistic or even unscrupulous owners or promoters. Every mine owner, of course, believed that his "was the most workable mine in the world." [89] It was a western truism that more money was made from selling mines than from buying them, just as it was accepted that many a good mine had been spoiled by working it.[90] Promoters especially took on a stereotype, and some engineers were inclined to approach them cautiously. As Robert Peele wrote in 1894, "A mining engineer with but ordinary experience would certainly be justified in losing faith in human nature (as it is manifested among mine promoters) at a very early age!" [91] Some believed that the only safe course was to represent only the prospective purchaser, never the seller; others saw nothing wrong with reporting for the vendor, provided the engineer was of sufficient reputation to take the risk.[92]

Obviously not all mine vendors or promoters were scoundrels, but they were interested in good reports and they con-

88. Frederick G. Corning, *Papers from the Notes of an Engineer,* pp. 92–93; Thomas A. Rickard, "Mine Reports," *EMJ, 79* (June 15, 1905), 1145; *M&SP, 95* (July 20, 1907), 79; *Pacific Miner, 15* (October 1909), 113.

89. William Ashburner to George Brush, Mazatlan, May 15, 1863, Brush family MSS.

90. "Pure gold was hidden in the quartz, they said,
'Twas proved by dreams and signs, and rods divining,
By chemic tests, and spirits of the dead,
In fact by everything—except by mining."

From R. E. White, "The Mining Town," *Alta California,* August 18, 1873, clipping, Bancroft Scrapbooks, *51,* no. 3, 797.

91. Robert Peele to Eben Olcott, New York, May 10, 1894, Olcott MSS (N.Y.). For a graphic description of the classical type promoter—"the Caruso of the mining game, a Corot in his masterful command of colorful profanity, an alchemist in his transmutation of basic quartz into shining coin of the realm"—see Charles T. Hutchinson, "A Pinch of Salt," *M&SP, 121* (July 24, 1920), 123.

92. See David Brunton, quoted in *Northwest Mining News, 6* (December 1909), 10; *M&SP, 97* (August 8, 1908), 181; John Hays Hammond, "Professional Ethics for the Mining Engineer," *EMJ, 86* (October 10, 1908), 717.

sciously sought to employ inspecting engineers known for their optimism. Not that this was easy: the very nature of mining gave engineers a reputation for being pessimists. After all, as Henry Janin wrote his mother in 1875, "the chances are always against a mine."[93] George W. Maynard was once described as "a natural wet-blanket" where mines were concerned, although many colleagues did not share this opinion.[94] Good engineers were cautious, and, in the parlance of the nineteenth century, "would condemn the U.S. Mint." This reputation was earned, for the great majority of mine reports were unfavorable. It was estimated in 1867 and again in 1879 that three quarters of all reports were negative. Many of these "obituary notices" had been written for owners or promoters and suppressed when they threatened to harm the value of a property.[95]

Vendors and promoters, therefore, naturally tried to avoid the pessimists and to engage engineers whom they believed would be helpful. When Adolph Sutro sought capital in England for his proposed tunnel in 1866, Louis Janin suggested that a report on the project by the British engineer J. Arthur Phillips, who was "known to report in favor or against as he is paid for," would carry considerable influence with English investors.[96]

Any good mining man realized that the appearance of a mine could be materially improved in a very short time. Victor

93. Henry Janin to Juliet C. Janin, Denver, July 1875, Janin family MSS, Box 6.

94. Jessie Benton Frémont to William K. Rogers, New York, November 1879, Rogers MSS. A contemporary of Maynard in 1892 insisted that Maynard invariably exaggerated his statements "at least 700%." N. W. Witherell to Eben Olcott, New York, May 24, 1892, Olcott MSS, (N.Y.); see also Charles M. Rolker to Olcott, (New York, July 23, 1894, ibid.

95. *AJM, 4* (August 24, 1867), 120; *EMJ, 28* (October 4, 1879), 235; *Alta California,* September 6, 1869; *Mining Science, 59* (June 10, 1909), 441.

96. Louis Janin to Adolph Sutro, Virginia City, Nevada, February 14, 1866, Sutro MSS, Box 6. See also Henry M. Yerington to Robert Sherwood, Carson City, January 27, 1884, copy, Letterbook 5, Henry M. Yerington MSS; John B. Farish to Isaac Ellwood, San Francisco, May 27, 1902, Ellwood MSS, Box 102; E. S. Van Dyck to D. S. Fletcher, Goldfield, March 3, 1906, copy, Goldfield (Nevada) Mining Companies MSS, Box 1.

Clement, manager of the Bunker Hill, and an honest and able engineer, wrote the owner in 1887:

> Please let me know before hand whenever you expect an expert to examine the mines. The more time I can have for preparation the more advantageous it will be to your interest.—A week's time can set things to look 50% better to the eye.[97]

Stephen Roberts, a San Francisco stock operator interested in the sale of Nevada property in 1880, cautioned those at the mine as to how they should handle the inspecting experts, warning them that one was "a natural Bear on mines" who needed to be shown "all the *bright features*." "He may not measure the water," said Roberts, "if not let him guess."

> Get them to horn all you can and if you can select the rock for them so much the better. . . . They will both be tired when they get there and a little *bracing* in the matter of pointing out the advantages to them, will not be out of order.[98]

Nor were some employers opposed to giving the examining engineer a little additional coaching. Roberts informed the expert who was to look at his Nevada property that a favorable report was required, and that

> There are some features in the mine it will not be necessary to mention in your report, such as the hardness of the rock or the giving out of the ore in the small shaft (as there is but little doubt but what the ledge will soon come in again).[99]

97. Victor Clement to Simeon Reed, Wardner, Idaho, November 7, 1887, Simeon Reed MSS.

98. Stephen Roberts to N. A. Garvin, San Francisco, July 16 & August 4, 1880, copies, Letterbook September 22, 1878–May 19, 1881, Stephen Roberts MSS.

99. Stephen Roberts to M. G. Rhodes, San Francisco, March 1, 1880, ibid.

In 1910, a New York engineer, Robert B. Stanton, was instructed by his client—a mine vendor—that his report should say little about the commercial side of the property and should not discuss transportation rates at all. Moreover, he suggested that Stanton include only the high assays in his report. In his private notes, Stanton recorded: "I told him that would not be honest & refused to consider it. He replied that I, or at least he, knew they should be *high* & why not use what they *should be!*" [100]

On several occasions, at least, vendors or their agents so plied examining engineers with liquor that they were unable to sample ore properly or to take adequate precautions to assure an impartial inspection.[101]

Other types of pressure might also be brought to bear. Charles Janin, a distinguished member of a distinguished family of engineers, reported in 1907: "Our friend in Denver showed the yellow streak when I was out there trying to influence my report first by attempting to bully, his bluff being immediately called, and then by whines without success." [102] Occasionally, an engineer was offered a direct bribe to make a favorable report, the temptation reportedly running as high as $25,000. Thomas A. Rickard was once approached by the manager of the Rico-Aspen mine in Colorado—a man who "lacked a sense of humor, and much else besides"—and offered $1,000 to make an untrue and misleading statement about the property. Rickard suggested that the manager throw his paperweight out the window in hopes of striking someone on the head who might be willing to prostitute himself for less money. A California engineer, Fred L. Morris, once confided that he had been offered "about $200,000" to report favorably on a mine—"that is get a buyer which I take it amounts to the same thing." Morris declined, even though he did "like

100. Entries for November 22 & December 8, 1910, field notes 24, Stanton MSS.

101. Kett, *Autobiography,* pp. 34–35; Henry *v.* Mayer, et al., *Pacific Reporter, 53* (1898), 590–92.

102. Charles Janin to Louis Janin, October 17, 1907, Charles Janin MSS, Box 38. See also entry for April 10, 1889, diary of Frank Leonard Sizer.

money pretty well," concluding that the property might have some value, but "I would want to be in a glass or armored case when I made the examination." [103]

Morris' last reference was to the engineer's need to be free from interference from or dependence on others while making the inspection. Many engineers did call on others for maps or other information as a matter of professional courtesy, and occasionally one was foolish enough to give his report without having seen the property in question for several years or, perhaps without ever having seen it.[104] But most agreed that it was advantageous to gather as much information as possible from those on the scene, provided that due caution was exercised and an independent course maintained.[105] One expert admitted that, when sampling a working mine, he usually asked the superintendent, "in a confidential way, to point out the rich places, trusting to his own observations and judgment to discover the poor ones." Others conceded that a cigar to the shift boss was an excellent investment. "The shift-boss is the boatswain of those who go down to the mine in skips," said Joseph P. Hodgson, consulting engineer for Phelps Dodge. "A hint from a shift-boss may be worth as much as a thousand samples." [106]

On the other hand, the examining expert had to exercise extreme care, for the "knocker" was "a scarce article in a mining

103. Rickard, *Retrospect,* p. 65; Fred L. Morris to W. H. Lanagan, Ross, New Zealand, August 1911, copy, Morris MSS, Box 1. For other references to bribery, see *EMJ, 53* (February 1892), 176; *M&SP, 52* (April 10, 1886), 238, *75* (December 4, 1897), 523; Isaac B. Hammond, *Reminiscences of Frontier Life,* pp. 72–73.

104. See *Alta California,* November 17, 1874; Clarence King to Joshua Clayton, New York, December 2, 1885, isolated letter, Western Americana Collection, Yale University; Benjamin Lawrence to James D. Hague, Denver, June 8, 1895, Hague MSS, Box 12; M. O'Neil to L. Bradford Prince, Cerrillos, June 1907, Prince MSS; *EMJ, 53* (February 6, 1892), 249.

105. *M&SP, 97* (October 17, 1908), 508.

106. *M&SP, 77* (June 3, 1899), 582; *116* (March 2, 1918), 284–85. Ernest Wiltsee was once warned by an old crony who was speculating in worthless claims near the Nevada boom town of Manhattan: "there is nothing here for the likes of you, but plenty here for the likes of me." Wiltsee, "Reminiscences," pp. 224, 226.

camp," and more than one career was blighted or tarnished by too trusting dependence on others.[107] Rossiter Raymond, a man of vast experience who had already examined a host of mines ("It may be five hundred, or it may be a thousand," he testified in 1877),[108] in 1880 reported on the Chrysolite, a Leadville silver property then the center of much stock speculation. Raymond accepted the estimate of ore blocked out and in reserve given by the manager, an old Freiberg classmate, without making more than a cursory examination; wrote up a glowing report the same day; and pocketed his $5,000 fee. Chrysolite shares boomed, but not for long. Soon the blocked-out ore was found to contain a large core of limestone, as the manager already knew, and Raymond found himself the object of a public outcry and at least one lawsuit from an outraged shareholder.[109]

A year later, another veteran engineer, William Ashburner, was similarly embarrassed. Ashburner had been engaged to report on Colorado property on behalf of a large stockholder who was alarmed over plunging share values. Personally conducted through the mine's workings by the manager, he wrote a favorable report showing more than 50,000 tons of workable ore in sight, and his client promptly increased his holdings. But quickly it became evident that Ashburner's estimates were far too high, and his second visit showed the property nearly exhausted. With great humiliation, he confessed that his first report was "a very exaggerated and misleading one," based primarily on information from the manager, who had deceived him as to the extent and richness of the ore "and on almost

107. For cautions to the engineer, see *Mining World, 2* (February 20, 1881), 163; *M&SP, 66* (January 28, 1893), 52.

108. Raymond testimony, July 25, 1877, The Richmond Mining Company of Nevada et al., Appellants, *v.* The Eureka Consolidated Mining Company, Appeals from the Circuit Court, District of Nevada, to the U.S. Supreme Court p. 204.

109. R. H. Stretch to unidentified, Leadville, February 24, 1880, copy, Hague MSS, Box 15; James D. Hague to Samuel Emmons, July 25, 1880, copy, Letterbook 4, ibid.; *Helena Herald,* August 3, 1880; *EMJ, 30* (August 7, 1880), 85; *Mining World, 1* (May 1881), 10; *M&SP, 42* (April 9, 1881), 232, *119* (October 11, 1919), 504.

every other point."[110] Ashburner was at the twilight of his career, and the episode, while it did not reflect on his integrity, did nothing for his professional reputation.

Thomas A. Rickard was another whose life work was affected by his too-trusting reliance on others. In 1899, Rickard examined the Independence mine near Cripple Creek, and on the strength of his report the property passed into British hands. Five months later, it was disclosed that he had "seriously overestimated" the ore reserves, using unchecked figures from subordinates at the mine. This, and another experience, in which Rickard overvalued the Camp Bird mine, only to be contradicted by John Hays Hammond, helped convince Rickard that he should forsake a career as an active engineer in favor of editorial work.[111] "So long as I examined relatively small mines for individuals, my work as a consulting engineer was delightful," he wrote later. "But when I was called to report upon mines the purchase of which involved millions of dollars, I found that I had to be mixed up in a kind of business for which I had a decided distaste."[112]

Not only should the examining engineer be wary of relying on others, he must always be on guard against deliberate fraud and misrepresentation. "Dressing" a mine for examination was a relatively simple matter, but often difficult to detect. By pushing a stope so long as waste or low-grade ore prevailed, but halting when good ore was reached, a mine manager could set the stage for a favorable examination, although there was no way of telling how far pay rock might actually extend. One mine in the Wood River country of Idaho was so well

110. William N. Symington, *Report on the Quantity and Value of the Ore in Sight in the Robinson Mine, Summit County, Colo.* (December 26, 1881), pp. 1–4; James D. Hague to William Ashburner, New York, November 10 & December 13, 1881, January 4, 1882, copies, Letterbook 20, Hague MSS.

111. For different points of view of these episodes, see Stratton's Independence, Ltd., *Prospectus,* May 11, 1899; *Directors' Report,* May 1, 1899–June 30, 1900; Henry R. Wagner, *Bullion to Books,* p. 33; Wiltsee, "Reminiscences," pp. 124–24a; J. H. Hammond, *Autobiography, 2,* 483–94; Rickard, *Retrospect,* pp. 75–76; *EMJ, 70* (December 1 & 15, 1900), 632, 693; London *Financial Times,* December 8, 1900.

112. Rickard, *Retrospect,* pp. 81–82.

"dressed" that a reputable engineer estimated $450,000 in ore developed, whereas subsequent working proved there was only $125,000 worth.[113]

Especially in a small mine, it was essential that the expert see everything, for occasionally the owner had taken the heart out of the ore body through a secret working, leaving telling cross-cuts blocked off with timber or drifts caved in to obscure the evidence. Both John B. Farish and Alfred Wartenweiler encountered such situations in separate Colorado mines and exposed them.[114] To avoid embarrassment, it was recommended that the mine manager be asked to sign a statement that he had informed the examining engineer of all existing workings, appending to this a map or short description of all inaccessible workings.[115]

Usually, efforts to exclude an expert from any part of a mine were to keep him from negative evidence, but occasionally the object was to hush up indications of a rich discovery in order that his client, a prospective buyer, might not take up his option. Examining a mine in the San Juans, Albert Ledoux found that the owner had quietly restruck the previously lost vein and no longer wished to sell; hence Ledoux was not to be shown everything. When Harold Titcomb reported in 1904 on property of the New York and Nevada Company near Ely, Nevada, he quite correctly took into account the negative comments of Mark Requa as an interested party: "Mr. Requa's pessimistic statements should, in my opinion, be taken *cum grano salis*. It might be his plan to secure some option on the Copper Flat property." [116]

But the most common effort to misrepresent a mine's value was positive and was either by salting the property beforehand or by tampering with the expert's samples. Mine salting "was

113. Rickard, *Sampling,* pp. 147–48.

114. John B. Farish, "Ways That Are Dark," *M&SP, 110* (April 10, 1915), 578–79.

115. Rickard, *Sampling,* p. 31.

116. Albert R. Ledoux, "Technical Reminiscences—III," *M&SP, 110* (June 12, 1915), 906; Harold Titcomb to A. Chester Beatty, Salt Lake City, January 29, 1904, copy Harold Titcomb MSS, Box 3.

among the first of the occult sciences" introduced into the West, and even in the twentieth century it refused to become one of the "lost arts," in part because of the difficulties of bringing conviction. Many engineers insisted that they could not be salted, but history was against them, and it is safe to say that most came up against some of it during their careers. It was said that there were "probably 100 different ways of salting and 99 ways of finding it out." [117] Thus, the capable expert could never relax; although he might not prevent salting, if careful, he could discover it before too late.

Vigilance and a suspicious mind were essential, for the ways of those who sought to graft precious metal where nature had not intended it to be were wondrous and diverse indeed. "The best safeguards are an alert eye and a strong right arm," said Herbert Hoover, who advocated as much direct supervision of sampling work as possible. "Write 'Caveat emptor' (in quotes and italics) on the front of your note-book, and on the back inscribe the proverb, 'it is better to be sure than sorry,' " advised another.[118]

One protection commonly used was the "dead-head" sample. The expert might fill a few bags secretly with a known quality—with ore the exact assay of which was recorded or with worthless rock or even cinders—so that any salting would be evident. In one case, perhaps apocryphal, the expert used ashes from the company furnace, and when they assayed rich in gold, he sardonically advised his client to buy instead the mine that had produced the coal for the furnace, because "it appeared to be a surprisingly good gold-silver property." [119]

In addition to "dead-head" samples, most engineers used a system of duplicate or even triplicate assays by independent assayers unknown to one another, and they sought to guard their samples carefully. Canvas and burlap sample bags were sometimes used, but unless guarded day and night by trusted em-

117. C. S. Herzig, *Mine Sampling and Valuing,* p. 125. See also unidentified clipping, November 7, 1863, Bancroft Scrapbooks, *94,* no. 1, 167.

118. Hoover, *Principles,* p. 7; "Salting Samples," *M&SP, 94* (April 6, 1907), 416.

119. *EMJ, 94* (October 5 & 12, 1912), 627, 677.

ployees, these were not safe. Samples of the Mudsill mine at Fairplay, Colorado, were salted, probably by the injection of a fine silver powder forced into the sealed burlap bags by a syringe. Sacks of samples waiting on a Juneau wharf under the eye of a watchman were salted by three "tourists," who asked the guard a multitude of questions about them, all the while poking them with canes loaded with chloride of gold. In another case, samples were taken from a carpetbag and replaced by better ones, the culprit removing the bottom and sewing it up again while the engineer was asleep.[120]

Many examining experts preferred leather bags that would show punctures readily and that could be locked or sealed with a special wax. Others used metal cases or tubes equipped with Yale locks.[121] Early in the 1880s, John B. Farish examined a mine near Candelaria, Nevada, carrying his samples in a strongbox sealed with green tape and an odd-colored wax. Even though he carried the box with him to a small town some distance from Candelaria and expressed it to the assayer, his samples were doctored—by the express messenger at the railroad station. Fortunately, Farish had carried a few additional samples in his suitcase as an added precaution.[122]

For safety's sake, the examining engineer trusted no one at the mine outside of his own crew, and even there he could not always be certain. On record is one instance of salting by an assayer who used a pestle that had a small core of gold; in another, a helper smoked cigarettes into which gold dust had been inserted, deliberately dropping his ashes into the samples. A Mexican foreman, desiring to retain his position, salted the samples of a young American engineer and ruined a budding career.[123]

120. Mudsill Mining Company, Ltd. et al. *v.* Watrous et al., *Federal Reporter, 61* (1894), 164–90; G. L. Sheldon, "Salted Mines," *EMJ, 96* (December 13, 1913), 1113; Rickard, *Sampling,* pp. 27–28. For another version of the Mudsill affair, see Siringo, *Cowboy Detective,* pp. 74–84. A general treatment of the problem is Charles M. Dobson, "Mine Salting," *Cosmopolitan, 24* (April 1898), 575–83.

121. Hoover, *Principles,* p. 7; Rickard, *Sampling,* pp. 27, 130; O. H. Packer, "How Mines Should Be Examined," p. 457.

122. Farish, "Ways That Are Dark," *M&SP,* CX (April 10, 1915), p. 576.

123. J. H. Hammond, *Autobiography, 1,* 168–69; interview with Ira B.

Engineers in charge of drilling test holes in dredge property had special problems. In Oregon, even while the experts watched, a drill operator managed to salt the samples by letting gold run down into the drillings from a small vial up his sleeve; at Oroville, an employee was able to smear gold mixed with pipe clay on the inside of a piece of casing, so that in the drilling process it was washed into the samples.[124] To prevent salting at the end of a shift, it was customary either to set a watchman at night or to dump barren tailings into the drill hole and lower the heavy bit into it, but his was no deterrent to another means of falsifying samples in dredge ground—namely the deliberate recording of shallower test holes than were actually drilled to make test averages seem better than they actually were.[125]

More than one engineer invested his own capital in a mine on the basis of salted samples, and more than one reputation was besmirched in the process. Perhaps the most famous instance was the infamous diamond scandal of 1872, in which the victim was Henry Janin, an engineer of ability and integrity, who, ironically, had made his reputation "By condemning almost every new scheme he was called to report on."[126] Hired to examine "fabulous" diamond and ruby deposits in an unidentified part of the West (subsequently identified as northwestern Clorado), Janin brought forth a scintillating report on this "wonderfully rich discovery . . . one that will prove extremely profitable," and exercised his option, as partial pay for his services, to subscribe a substantial number of shares in the new company at favorable rates.[127] Soon, however, the entire profession was shocked when geologist Clarence King visited the "fields" and exposed the whole as a

---

Joralemon, San Francisco, April 6, 1964. See also S. M. McClintock to George D. Louderback, Buston, Washington, October 16, 1904, Louderback MSS.

124. *M&SP, 121* (October 16, 1920), 546; Weatherbe, *Dredging for Gold,* pp. 31–32.

125. Ibid., pp. 30–31, 32.

126. Thurman Wilkins, *Clarence King,* p. 159; London *Times,* December 2, 1872.

127. For the report, see Henry Janin to S. L. M. Barlow, New York, June 26, 1872, copy, Barlow MSS, Box 81.

clever salting swindle, with Janin the prime dupe.[128] Undoubtedly, Janin's reputation suffered, but he did not die "of a broken heart" because of it, as one writer insists.[129] Rather he weathered the storm of criticism, paid back his profits, and continued his engineering career.[130] As for Clarence King, his reputation was assured. "Who's the King of Diamonds, & isn't he trumps?" chortled one of his friends.[131] King was besieged with opportunities to examine property, and, according to another friend, "he never charges less than $5000 to look at a mine." [132]

Another celebrated case was the Bear's Nest (Some called it the "Mare's Nest") on Douglas Island in Alaska. In 1888, the property was sold in London at an inflated price, after three engineers, including Robert M. Brereton, had made independent reports on it. Brereton then became manager, and it was he who subsequently exposed the fiasco, but not until a new 120-stamp mill was nearly completed. Only then was it revealed that quartz outcrops and diamond-drill cores had been dosed with sulphate of iron and then chloride of gold and that ore from another mine had been carefully bedded along the vein—a process that reputedly took two years. At the time, Brereton insisted privately that he had not invested any of his own money in the Bear's Nest and that he had kept his personal friends out of it; but later, in retrospect, he complained that he had lost $60,000 in the swindle and had to build anew his professional reputation.[133]

128. For details of the affair, see Rickard, *History of American Mining,* pp. 380–96; Wilkins, *Clarence King,* 159–68; AIME *Bulletin,* no. 53 (May 1911), xxviii–xxxvi; *M&SP, 25* (November 30 & December 14, 1872), 344, 380; Goetzmann, *Exploration and Empire,* pp. 452–57.

129. Charles H. Leckenby, *The Tread of Pioneers,* pp. 140–41.

130. Henry Janin to S. L. M. Barlow, January 24, 1873, Barlow MSS, Box 86; Charles Hoffmann to Josiah D. Whitney, Oakland, December 15, 1872, Charles Hoffmann MSS; for Janin's defense, see Henry Janin, *A Brief Statement of My Part in the Unfortunate Diamond Affair* (San Francisco, 1873), pp. 1–32.

131. Josiah D. Whitney to William Brewer, Cambridge, January 5, 1873, Brewer-Whitney MSS, Box 4.

132. Charles Hoffmann to Josiah D. Whitney, San Francisco, December 27, 1872, Hoffmann MSS.

133. *EMJ, 48* (September 7 & December 14, 1889), 200, 519 *49* (April 26,

Even if the examining expert successfully completed his report without encountering trickery, his troubles were not necessarily over. He might still have to contend with an owner, vendor, or promoter who refused to accept a negative verdict and might be willing to argue, cast aspersions on the expert's honesty or ability, or even withhold part of his inspecting fee. One attitude was reflected in J. C. Murphy's "Sad Tale of the Turned-Down Prospector":

> I gets a letter yesterday.
> It says: "Dear Mr. Brown,
> I very much regret to say
> I've turned your prospect down."
> Now don't that skin you? say, I know
> Just why that geezer laughed—
> I never offered him no dough,
> And he was out for graft! [134]

An owner of California property wrote Fred L. Morris, after an adverse report: "Also, I will be glad to tell you *personally* my opinion of your company and its judgment in turning down this proposition up here, but, of course can't do the subject justice in a letter without violating the U.S. postal regulations." [135] Ellsworth Daggett once wrote a negative report on a Cripple Creek mine owned by or bonded to Eben Smith, who considered Daggett's statement "simply absurd," and proceeded to form a company anyway and to issue a pros-

1890), 465; *M&SP, 69* (July 28, 1894), 50; Robert M. Brereton, *Reminiscences of an Old English Civil Engineer,* pp. 37–38; Robert M. Brereton to James D. Hague, Victoria, British Columbia, October 9, 1889, Hague MSS, Box 11.

Another famous episode was the salting of the Mulatos mine in Sonora before its sale to several California mine operators, who passed it on to a British concern. The sale was ultimately rescinded in the San Francisco courts, but not before several American engineers had their names dragged through the mire. For the Mulatos and other instances of salting, see *EMJ, 51* (March 7, 1891), 279, 109 (May 8, 1920), 1074; *M&SP, 28* (January 25, 1879), 50; San Francisco *Stock Report,* December 22, 1879; H. J. Jory to Samuel B. Christy, San Francisco, February 12, 1892, Christy MSS; I. B. Hammond, *Reminiscences,* p. 72; Emory Fiske Skinner, *Reminiscences,* p. 300.

134. Quoted in *EMJ, 114* (October 21, 1922), 717.

135. W. H. Christie to Fred L. Morris, Trinity Center, California, May 10, 1913, Morris MSS, Box 6.

pectus that claimed 70,000 tons of ore in sight. If this were true, wrote Daggett, who still retained a sense of humor even after wrangling with Smith, it would indicate "an unprecedented state of affairs." This amount of ore, he said,

> would have to project as a slab of ore one foot thick and 181 feet high over 4000 feet horizontally into the air over and across Elkton gulch and probably into Squaw mountain on the other side—why, out in Elkton Gulch they would have to trestle up into the air three or four hundred feet to mine it.[136]

Sometimes, when their reports were not pleasing to their clients, examining experts had difficulty in collecting their fees, although many prudently required an advance or, in dubious cases, the deposit of funds in a designated bank as a guarantee.[137] J. S. Phillips had to go to court to get his pay in 1868, Joshua Clayton encountered the same trouble in 1874, and twenty years later, Daniel Barringer and Richard Penrose ran afoul of E. H. Harriman, after an adverse report on mines near Tucson left the railroad magnate in a bad humor. When Harriman complained about the verdict and its cost and demanded additional information before he would pay, Barringer urged his partner to "treat him roughly"—"Handle him without gloves." Barringer threatened to put the bill in the hands of a collection agency or to sue "that scoundrel Harriman." "A hog like this should be told in plain language what decent men think of him." But Harriman eventually paid in full,[138] and, in subsequent years, Barringer would

136. Eben Smith to J. Parker Whitney, Denver, December 3, 1895, copy, J. Parker Whitney to James D. Hague, Boston, December 4, 1895, and Ellsworth Daggett to James D. Hague, Salt Lake City, December 14, 1895, Hague MSS, Box 19. For other examples of owners' or even managers' complaints, see Fred L. Morris to T. H. McDerby, San Francisco, June 17, 1912, Morris MSS, Box 1; Arthur D. Foote to George B. Agnew, March 5, 1926, copy, A. D. Foote MSS.

137. Janin & Smith to John D. Elwell & Company, San Francisco, October 29, 1908, copy, Charles Janin MSS, Box 5.

138. Unidentified clipping, Bancroft Scrapbooks, *94*, no. 2, 303; Joshua Clayton to A. H. Phillpotts, Salt Lake City, April 12, 1874, copy, Joshua E. Clayton MSS, (Calif.) 2; Barringer to Richard Penrose, January 22, & 30, February 7, 11, & 12, & March 5, 1895, copies, and Barringer to Harriman,

regularly do examination work for him without further difficulty.

Another problem for the expert had to do with the use made of his report. The employer who hired Robert Stanton to examine the Atlanta mine simply "stole the first copy & hid it," so its adverse nature might not come to light.[139] Sometimes a vendor or promoter misused or perverted the report, in the process distorting the expert's professional opinion. Joshua Clayton complained in 1871 that a promoter had misrepresented a report of his and had attached an outrageously high price to the property in question. "If he had succeeded in selling them at *his* figures it would have *ruined my status* as a mining engineer," said Clayton.[140] Others protested that promoters sometimes took liberties with the status of their consulting engineers. F. W. Werlitz, who had examined the Major Budd mine in 1887, was listed in the prospectus got out by St. Louis promoters as manager of the Anaconda Copper Company, which he was not. George Maynard became consulting engineer for a tunnel company in Mexico in 1905, but resigned in disgust when he discovered the concern using his name and official title on its prospectus even before he had an opportunity to visit the property.[141]

Occasionally, a promoter deliberately changed the wording of the engineer's report. Thus, in a report that originally had stated, "The Bull Whale property has considerable young timber for lagging purposes," the word "lagging"—i.e. planking for flooring or roof support in a mine—was changed to "logging." In another sentence, "The Pipe-Dream is not extensively faulted as previously supposed," the promoter omitted the word "extensively" from his printed circular.[142]

January 30, 1895, copy, Letterbook A, Barringer MSS. See also Seeley W. Mudd to A. Chester Beatty, Los Angeles, June 24, 1910, copy, Titcomb MSS, Box 5.

139. Entry for April 25, 1904, field notes 23, Stanton MSS.

140. Joshua Clayton to Henry Hughs, Treasure City, Nevada, May 19, 1871, copy, Clayton MSS, (Calif.) 2.

141. *Helena Herald,* August 1, 1887; George W. Maynard to editor, Altar, Mexico, November 6, 1905, *EMJ, 80* (November 18, 1905), 934.

142. Clyde M. Becker to editor, Deming, September 27, 1915, *M&SP, 111* (October 23, 1915), 620.

More than one vendor or promoter lost track of the time sequence in his use of an engineer's report. Benjamin Silliman complained in 1877 that a recent prospectus of the Arizona Chief Gold and Silver Mining Lode used his description as he had given it in 1864, although subsequent exploration had been negative. In 1906, Rossiter Raymond protested that a circular soliciting stock subscriptions in the Arlington Mining Company, with property near Yosemite, included quotations from reports he and James D. Hague had made independently in 1874—thirty-two years before.[143] Citing another such case, Thomas Rickard believed that, without express sanction of the examining expert, no report should be quoted more than one year after it was written; for small mines, the period should be six months. But Rickard realized that legal recourse was impractical, that apologies were never seen by the proper people, and that education of the investing public against such misrepresentation was the only salvation.[144]

Rarely, either in real life or in fiction, has the mining engineer been cast in the role of villain, and probably most were honest. Yet, constantly, leading engineers preached the need for higher ethical standards. The "professional man of the highest type has the inheritance of the old aristocracy, expressed in the motto '*Noblesse oblige*'—the compulsion to be honorable," insisted Rickard in 1907.[145] But the profession was made up of human beings, with all the faults and foibles of the race, and a few tended to create a stereotype that an unthinking public might apply to all:

With leather leggins and doe-skin suit he rides at a furious pace,
And the anxious miner weighs every word and tries to read his face.
He has few scruples but takes many drams, and he'll take a double fee

143. *EMJ, 24* (July 21, 1877), 44; Rossiter W. Raymond, "Ancient History and Modern Investments," *EMJ, 71* (March 10, 1906). 458.

144. Thomas A. Rickard, "The Life of a Report," *EMJ, 79* (March 16, 1905), 523.

145. *M&SP, 95* (October 19, 1907), 474–75.

From the man who buys and the man who sells, and serves
them both loyally.[146]

Occasionally, an engineer was exposed in his rascality, for as one of the profession put it, "The moral code of some men is the penal code." [147] When George W. Howe was convicted in 1905 of promoting investment in mines to which he had no title, his promotional literature included a glowing report by Lucien M. Turner, "C.E., M.E., F.G.S. and Sc.B." It soon became clear, however, that Turner and Howe were friends and that Turner had never visited the property, but had written his report from clippings supplied by Howe.[148]

In a more celebrated case—that of the Lost Bullion Spanish Mines—settled in the U.S. Circuit Court at Denver in 1907, eleven men, including two mining engineers, were convicted of fraudulent use of the mails. In this example of *opéra bouffe* mining, one engineer had declined to make a report after viewing the property; another made an examination, but found his samples doctored and his report garbled; whereas a third, who said he was "tired of making examinations for other people," took 200,000 shares of stock in exchange for a glowing report that spoke of "mountains of untold wealth" and "millions of tons of ore in sight, all ready for shipment." Court evidence indicated that the "mine" was little more than an old cave, and the only undisputed metal actually found in it was a bundle of baling wire! [149]

Most inspecting engineers, however, took seriously their responsibility as a bulwark between the promoter and the investor and attempted to produce an objective report; a man with a reputation to preserve or a reputation to make could hardly do otherwise. But the uncertainties of mining and the infallibility of humans often combined to produce differences of opinion among technical men. "Among mining experts (and

146. *Silver Standard* (Silver Plume, Colorado), September 10, 1887.
147. *M&SP, 95* (September 7 & October 27, 1907), 286, 522.
148. *EMJ, 80* (November 25, 1905), 988.
149. *EMJ, 84* (September 14, 1907), 508; *85* (January 11, 1908), 131; *M&SP, 95* (August 31, 1907), 253; San Francisco *Chronicle,* December 27, 1907.

you can take those of highest standing at that) there is usually a very wide variance in their judgment on the same thing," noted Victor Clement in 1887.[150] Thus, even if the examining engineer completed a fair, impartial report, free from pressures and misrepresentation, rare indeed was he who had not at least some miscalculations chalked up against his record.

Pointing out that engineers were not inclined to publicize their mistakes, one member of the fraternity insisted in 1916 that most had a "string of failures to their credit" much longer than their successes.[151] Of errors of judgment, none were so common as miscalculating ore reserves, for the old adage that no miner could see beyond the end of his pick continued to have real meaning. Thomas A. Rickard was one of the experts who lauded the Seven Star Group of mines in Arizona in 1892, predicting the "promise of a future which would place it in the rank of the very greatest mines in the United States." Within the year, the Seven Stars enterprise had "gone up the spout," its title misrepresented, its workable ore nonexistent.[152] George Maynard made an excessively high valuation of a property in Mexico for the "Vanderbilt crowd" about the same time, but a second examination by another expert showed Maynard to be in error.[153] In the late 1890s, Fred Bradley, one of the finest engineers in the country, gave the Esperanza, also in Mexico, an excellent report, but a follow-up inspection by Ross E. Browne gave far lower estimates, which resulted in the purchase of the property at $3 million, rather than the $6 million originally asked. Even John Hays Hammond was not infallible. On the strength of his enthusiastic endorsement of the Nipissing in Canada, the Guggenheims bought in, only to lose substantially when the property failed to live up to expectations.[154]

150. Victor Clement to Simeon Reed, Wardner, July 17, 1887, Reed MSS.

151. Frederick F. Sharpless to editor, New York, December 5, 1916, *EMJ, 102* (December 23, 1916), 1106.

152. Thomas A. Rickard, *Report upon the Seven Stars Group of Mines, Yavapai County, Arizona* (August 10, 1892); Eben Olcott to James D. Hague, New York, August 17, 1893, Hague MSS, Box 20.

153. Eben Olcott to Charles M. Rolker, November 11, 1891, copy, Letterbook 18, Olcott MSS (N.Y.).

154. J. H. Hammond, *Autobiography, 2,* 511, 515.

If the engineer risked his reputation when he took up examination work, he also risked his constitution, for mine inspection, properly done, was hard and sometimes dangerous labor, often carried out under adverse weather conditions. In the spring of 1916 William H. Lanagan arrived in Arizona to do some sampling and was impressed neither with the environment nor the work:

> Tombstone makes no hit with me. It would take a salamander to get through a summer at Las Pas without going any farther south. I slept under the stars, with nothing on but a sheet. The warm weather hadn't started at that time. When it gets warm you have to get up every hour and soak the sheet in cold water and wrap it around you in order to get some sleep. Fine country. I drank four and a half gallons of water in seven hours, and then was dry as a bone most of the time. Taking samples in the bottom of a dry shaft, with the dust so thick you can cut it, is a fine occupation.[155]

But a month later, Lanagan was acclimated: "I am a regular salamander now; hot weather doesn't have any effect on me at all." [156] In November of the same year, Lanagan was in Montana and wrote back his negative impressions of a dredge property inspected under slightly different circumstances.

> I hate to hand you anything like this on Thanksgiving Day, but it can't be helped. I know that it would take some placer proposition to make a hit with a man panning in ice-cold water, with his feet wet and freezing, his arms numb up to the elbows, his windward ear full of sleet and his nose running like a leaky faucet, but even now, after a swim at the Hot Springs, a shave, clean clothes, and three dollars worth of alleged food in the diner, the outlook is no rosier.[157]

155. William H. Lanagan to Fred L. Morris, Kingman, June 15, 1916, Morris MSS, Box 1.

156. Lanagan to Morris, Los Angeles, July 14, 1916, ibid.

157. Lanagan to E. B. Corbett, Billings, November 30, 1916, copy, Morris MSS, Box 7. In the winter of 1879, Nathaniel Shaler examined a mine above Leadville in deep snow and twenty-below temperatures. "My wits seem frozen

Even when the weather was more temperate, the job could be arduous. Eben Olcott described part of his experience in Nevada in 1892, when he had "to descend 1400 feet by ladders, see what I can see down there & pull myself up again." Another engineer recalled sampling in a Leadville shaft 310 feet down by straightaway ladder. With his pockets full of ore, his main haversack crammed with samples, and several smaller bags hung round his neck, he was fortunate, after a terrible struggle, to make it back to the surface.[158]

Falls were not unusual. John B. Farish, George W. Maynard, and Sewell Thomas all survived such accidents, but others were killed outright or injured fatally by plunging down shafts or winzes.[159] Once, while descending in an iron bucket to inspect a mine in Mexico, Raphael Pumpelly caught the seat of his trousers on a spike, was lifted out bodily, but fell back in the bucket unharmed.[160] Some were caught in cave-ins, and the fortunate, such as Charles Janin, escaped with only fright or minor injury.[161]

Abandoned mines were especially risky. Loose timbers and rotten ladders were constant hazards, and engineers were sometimes forced to wiggle along on their stomachs through debris-filled areas, candle in hand and fearful of bad air. Two examining experts at a Montana mine narrowly escaped disaster in 1888 after being exposed to "a nauseous dose" of gas that left them feeling "as though they had been dining of sulphur matches." John Farish had to abandon efforts to inspect a Colorado mine in 1889 because of foul air.[162] Another po-

---

in me," he wrote his wife, "I find it hard to fix my mind on work or hold impressions." Quoted in Shaler, *Autobiography,* p. 297.

158. Eben Olcott to Euphemia Olcott, Pioche, May 15, 1892, Olcott MSS, Box 2, (Wyo.); Henry E. Wood, "Some Assaying Experiences at Leadville in 1878," p. 14, Henry E. Wood MSS, Box 3.

159. *EMJ, 64* (July 3, 1897), 12, *70* (October 20, 1900), 465, *82* (September 1, 1906), 409, *119* (October 4 1919), 488; Thomas, *Silhouettes,* p. 150; Helen D. Croft, *The Downs, the Rockies—and Desert Gold,* p. 229.

160. Pumpelly, *My Reminiscences, 2* 650. See also Wood, "Some Assaying Experiences," p. 15; P. B. McDonald, "Queer Mine Accidents," *M&SP, 112* (February 12, 1916), 244.

161. Charles Janin to Chester Purington, September 14, 1912, copy, Charles Janin MSS, Box 5.

162. *Helena Herald,* July 25, 1888; John B. Farish to Isaac Ellwood, Denver, June 27, 1889, Ellwood MSS, Box 102.

tential danger in old workings came in the form of rodents, reptiles, or insects. One engineer examining an abandoned property near Tombstone found a host of horned toads, rattlesnakes, "about a thousand rats—and a half a dozen of the d—st Gila monsters you ever saw." "I wouldn't give a cuss—for the mine—except as a side-show—or a menagerie," he added. "It's worth a hundred thousand dollars for that—a regular bonanza." [163]

Sometimes the inspecting engineer had to contend with water. In 1892, John Hays Hammond and Ernest Wiltsee traveled nearly a hundred miles from any railroad to examine a mine in eastern Nevada, the two of them taking turns running alongside their runner-equipped buckboard to keep from freezing. Once at the property, they made a perilous descent to the water-filled twelfth level, which they inspected in a collapsible boat, with the water over the head of the diminutive Hammond, and stayed so long underground that word of disaster had gone out by the time they finally reached the surface.[164]

From time to time there were other unpredictable hazards. James D. Hague was once arrested on charges of breaking open a mine while trying to examine it. Hammond had to fight a local assayer before being allowed to proceed with his examination, while patrons of a nearby saloon poured out to cheer his opponent. Another engineer arrived in Arizona to inspect property in 1880 in the midst of labor troubles, just in time to be taken hostage for three days by irate miners protesting the late arrival of their payroll.[165]

Engineers all recognized that the tropics were unhealthy

163. Quoted in *M&SP, 43* (August 6, 1881), 86. Thomas H. Jenks spent several months recuperating from complications stemming from the bite of a poisonous reptile incurred while examining a mine in Mexico in 1921. *M&M, 3* (January 1922), 25.

164. J. H. Hammond, *Autobiography, I,* 173–75; Wiltsee, "Reminiscences," pp. 18–20. In the same year, Eben Olcott reported on a mine near Pioche: "I had to travel through a very long drift on which there was 5 ft. 3 in. of water then work 6 hrs. in our wet clothes & come out again in the same way." Olcott to Katharine Olcott, Pioche, July 1, 1892, Olcott MSS (Wyo.) Box 2.

165. Arnold Hague to James D. Hague, Washington D.C., June 28, 1897, Hague MSS, Box 2; J. H. Hammond, *Autobiography, I,* 57–58; Theodore Van Wagenen to editor, *EMJ, 119* (July 8, 1922), 50.

and that in addition to malaria and typhoid they might fall prey to "cork fever" if they tried to "carry the climate about with them in a bottle."[166] For reasons of health, Louis Janin refused to venture into the higher altitudes in winter or into Mexico during the rainy seasons,[167] but many were less cautious. It was typical of Arthur D. Foote, for example, to push himself on a strenuous horseback trip of 160 miles to examine a Colorado mine in 1879, or to inspect an Idaho property in a blizzard while half sick some fifteen years later—and in both instances to fall ill of overexertion.[168] Raphael Pumpelly was struck by typhoid while examining a mine above the timber line in the Rockies in 1898, but managed to ride back to Cripple Creek for medical attention. There, while recuperating, he dictated his report to his clients, but "It took all day, for I kept dropping off, and had to be revived with strychnine."[169]

Early engineers, especially in the 1860s and 1870s, sometimes faced hostile Indians. In Arizona to examine mines in late 1863 or early 1864, Louis Janin and his companions were attacked by Apaches, and in a running fight two of the party were killed and the rest deserted, leaving the young engineer and one associate to fend for themselves. Janin managed to escape unscathed, finish his work, and return to San Francisco, where in addition to writing up his reports, he could announce that he had been "refreshing myself by visiting the young ladies."[170] William Ashburner also had some thrilling experiences to relate as a result of a brush with Indians while on a Southwestern examination trip in 1865, and Rossiter Ray-

166. Frank Owen to editor, Akrapong, Ashanti, December 31, 1900, *EMJ, 71* (February 23, 1901), 240.

167. Louis Janin to James D. Hague, San Francisco, November 9, 1891, Hague MSS, Box 12.

168. Mary H. Foote to Helena Gilder, Leadville, June 13, 1879, M. H. Foote MSS; Mary H. Foote to James D. Hague, Boise, September 19, 1894, Hague MSS, Box 2.

169. Pumpelly, *My Reminiscences, 2,* 674.

170. Ross Browne thought "Young Janin behaved with greater coolness with a Henry rifle" on this occasion. At one point, Janin turned to his companion, who was braced for a last desperate stand, and said, "Colonel, I can't see them very well—lend me your specs!" J. Ross Browne, *Adventures in the Apache Country,* pp. 151, 216–23; Louis Janin to Juliet C. Janin, San Francisco, April 12, 1864, Janin family MSS, Box 10.

mond, whose skull cap concealed a head as hairless as a billiard ball, once "had a lesson in the tactics of Indian warfare, under the hostile tutorship of 'Sitting Bull.'"[171] J. M. Robinson had a narrow escape while doing inspection work in New Mexico's Black Range in 1881. "Robinson's bald head saved him," quipped one editor, but two men in the group had been killed, and New Mexicans failed to see any humor in the situation.[172]

In many instances, travel itself had to be reckoned as an additional peril. Ideally, it was agreed, the reporting engineer should travel first class and stay in the best hotels, for, as one expressed it, "it is always desirable for an engineer to travel as a gentleman, and not to appear cheap in any way."[173] Rossiter Raymond sometimes used a private railroad car on his inspection trips, but on one occasion a British engineer signed on as a member of a work crew to get free transportation from Winnipeg to the Canadian Rockies.[174] Unfortunately, however, many mine examinations were in remote, out-of-the-way places, far from rail lines and decent accommodations—circumstances that invariably caused the consultant to set a higher fee. As Louis Janin put it in 1891, the future might bring improved transportation facilities, "but the expert of today must suffer in the flesh—and his client in the pocket."[175]

A stagecoach was still a stagecoach, whether in 1860 or half a century later. Louis Janin described a 200-mile trip in Nevada in 1863 as "a very trying and disagreeable one," with the stage overcrowded and once upset.[176] Three years later, his brother Henry, inspecting mines in Idaho, traveled by steamer to Portland, then overland to the Owyhee district under condi-

171. Emelia Ashburner to William Brewer, Stockbridge, Massachusetts, August 7, 1865, William Ashburner MSS; *Rossiter Worthington Raymond* (1910), p. 47.

172. *Mining World, 1* (February 1881), 12.

173. Lynwood Garrison, "Professional Customs," *M&SP, 95* (November 2, 1907), 552.

174. Walter R. Ingalls, "Rossiter W. Raymond," *EMJ, 107* (January 18, 1919), 142; McCarthy, *Incidents,* pp. 131–32.

175. Louis Janin to James D. Hague, March 21, 1891, Hague MSS, Box 12.

176. Louis Janin to Juliet C. Janin, Empire City, September 11, 1863, Janin family MSS, Box 10.

tions far from pleasant. "I accepted the 'job' as there are indeed 1000 reasons (i.e. $.) why I should do so," he wrote afterward. "I am pained to say that I feel I did not charge enough."[177] A third brother, Alexis, made a "rather rough" examination trip to Arizona in 1879. En route, he spent thirty-six hours on the Southern Pacific, then seventy-five hours of combined stagecoach and open wagon travel, all the while subsisting on fat pork, beans ("Arizona strawberries"), and "canned truck."[178] John Hays Hammond, being a short man, was able to curl up and sleep on mail pouches in the boot of the coach on some of his inspection trips, and he once made a two-week buckboard journey, carrying his own water, to examine a worthless prospect in Death Valley.[179] Even as late as 1916, a California engineer going to look at dredge ground at Dixie, Idaho, had to swing down by stagecoach "over a pretty rough road" a good hundred miles from any railhead.[180]

Often, even a stagecoach or wagon would not take the expert where he wanted to go. J. S. Phillips once made a thirty-day horseback trip in Nevada, and when he examined quartz veins on Vancouver Island, he traveled by Indian canoe in a driving rain, with nothing to eat but salt clams and stewed herring spawn.[181] Much later, to inspect a steamshovel manganese prospect in Washington, Robert B. Stanton traveled eighty miles by boat from Seattle, then nine miles by stage to Cushman Lake, three-quarters of a mile by rowboat, and walked the final five miles to the property.[182]

177. Henry continued to make such excursions, for as he explained in 1870, "Frequent trips into the sage brush & inhaling of a fair amount of alkaline dust, are necessary to me to preserve any self respect & to pay my hotel bills." Henry Janin to Louis Janin, Salt Lake City, June 20, 1866, and Henry Janin to Juliet C. Janin, Hamilton, Nevada, August 12, 1870, Janin family MSS, Box 6.

178. Alexis Janin to Juliet C. Janin, San Francisco, March 28, 1878, ibid., Box 5.

179. J. H. Hammond, *Autobiography, 1,* 90–91.

180. Ray Humphrey to Fred L. Morris, Elk City, Idaho, July 23, 1916, Morris MSS, Box 1.

181. *M&SP, 37* (August 24, 1878), 114.

182. On the day of his return, Stanton recorded in his notes: "Awful day. Hard climb up mt. 1000 ft to see Rowe's fake manganese mine. Raining all day. Left camp 1:30 P.M. Walked 5 miles in rain to Hotel Antlers. . . . Horrible fuss

In Alaska or the Yukon, the expert expected to have to travel by snowshoe or dog sled, as William H. Lanagan did in 1907,[183] but even in Colorado, Idaho, or Oregon he might have to employ near-Klondike transportation techniques in the winter. Thomas A. Rickard made a number of professional trips in the San Juan mountains on skis or snowshoes and was once swept off the trail by a snowslide. "It was a healthy and vigorous life," he commented later.[184] Another engineer, Walter H. Hill, once skied twenty-seven miles from Florence, Idaho, to inspect the Jumbo mine, then broke a toboggan trail another thirteen miles.[185] Ray Humphrey wrote back from Burnt River, Oregon, on New Year's Eve, 1916, that it had taken him four days to get there by team and sleigh from Baker: "Was some trip but got here with everything and did not break anything. Tipped over a few times and went through the ice a couple of times." [186]

Undoubtedly, the coming of the automobile would be of great benefit to the examining expert, just as Thomas Rickard predicted in 1905, after Herbert Hoover and J. H. Curle had used the gas buggy on an extended inspection trip in Western Australia.[187] Curle would be the first to take a motor car into Death Valley to examine mines, and the engineer in general was among the first in the West to take advantage of the new means of locomotion. The automobile greatly increased his mobility and his efficiency and enabled the expert to carry more samples. One ingenious engineer even inspected an abandoned shaft by tying 200 feet of rope to the rear of his car, looping it over the frame of an old hand hoist, and, with a friend backing his vehicle up, had himself lowered down the shaft and back up again. Still, while the automobile was a real

on boarding boat. Dangerous job, and insulted by officers & crew." Entries for September 23, 24, & 30 & October 1 & 3, 1910, field notes 22, Stanton MSS.

183. Fred L. Morris to William H. Lanagan, March 20, 1907, copy, Morris MSS, Box 1.

184. Rickard, *Retrospect,* pp. 54–55, 61.

185. Robert C. Bailey, *Hell's Canyon,* pp. 170–71.

186. Ray Humphrey to Fred L. Morris, Burnt River, December 31, 1916, Morris MSS, Box 1.

187. *EMJ, 79* (May 25, 1905), 1002.

asset, it was no substitute for the horse or pack mule in penetrating the wilderness.[188]

At the mine, on the inspection job, the expert often faced primitive or at best, third-rate living accommodations, spiced sometimes with a bit of excitement, but as often as not with dullness and boredom. True, in booming mining camps, the engineer might come into contact with a boisterous and at times lawless crowd. From Deadwood, Dakota, on the Fourth of July, 1887, Arthur D. Foote noted that "The streets are full of waving, shouting, drunken men, staggering around, fighting & shooting as men will with no law to controll [*sic*] them." [189] About the same time, Alexis Janin was finding life in an Arizona mining town a decided contrast to the Paris he had known a few years earlier:

> Everybody carried a big revolver and was prepared to shoot on small provocation. During my stay a cold blooded murder was committed and there being no county officials in the neighbourhood nothing was done about it for a few days. Finally we organized and arrested the suspected parties and gave them a trial, my "bed room" serving as a court of justice. At this time a Methodist preacher came around and he was engaged to preach a sermon over the grave of the dead man, but a drunken Irishman swore that no Protestant sermon should be preached there and attempted to shoot the minister while he was performing the ceremony. So we took him too, tied him up with the other men whom we held for the murder, and sent the whole gang off to the county jail at Prescott.[190]

188. For comments on the use of the automobile and its value to the engineer, see Curle, *Shadow-Show,* p. 4; Claude T. Rice, "Automobiling in Nevada," *EMJ, 82* (October 6, 1906), 632, *109* (May 1, 1920), 1027; *Mining World, 25* (July 14, 1906), 37; *Black Hills Engineer, 13* (1925), 208–09; Thomas, *Silhouettes,* pp. 142–43; Gressley, ed., *Bostonians and Bullion,* p. 115 (see also illustration between pp. 92–93); Schmidt, "Early Days," 19.

189. Arthur D. Foote to Helena Gilder, Deadwood, July 4, 1878, M. H. Foote MSS.

190. Alexis Janin to Juliet C. Janin, San Francisco, March 28, 1878, Janin family MSS, Box 5.

"Jack" Hammond witnessed a shooting scrape on his very first day in the town of Bodie in 1880,[191] but probably life for most examining engineers on a "job" was more prosaic. In 1910, Dr. James Douglas, one of the brightest lights in the profession, confessed:

> I have made four trips a year to the Pacific Slope since 1880. I have never seen a man shot. I have never carried a pistol myself, or found the slightest need to do so. I have never seen a man hanged. I have never been "held up" in a train or a stage-coach or in anything else. In fact, the journey is commonplace, and there is no romance at all about it.[192]

Not many experts had the use of an army ambulance and the sparkling company of John C. Frémont and his charming wife, as did George Maynard, when he examined Arizona property in 1879.[193] More typical, perhaps, was the experience of Joshua Clayton, who made an inspection trip to Parrott City, Colorado, in 1877, camped out most of the way and was only too happy to reach his destination and to sleep on the dirt floor of a log cabin.[194] Certainly more representative was the experience of Eben Olcott, who, after roughing it for days on a Nevada trip, was delighted to spend a quiet Sunday reading in a plain little hotel in Pioche.[195] Or possibly Ellsworth Daggett spoke for a cross section of his profession in 1899, when he described an Oregon town where he stayed on examination work:

> Gold Hill is a typical old dead California mining camp where they make you thoroughly uncomfortable and wretched as long as you have to stay with them. A continual rain or raw fog caused me more suffering from cold

191. J. H. Hammond, *Autobiography, 1,* 86.

192. Quoted in *Rossiter Worthington Raymond* (1910), p. 47.

193. George W. Maynard, "The Verde Mine, Arizona," *EMJ, 87* (March 13, 1909), 560.

194. Entries for June 16 & 22, 1877, Clayton MSS (Calif.), *5*.

195. Eben Olcott to Euphemia Olcott, Pioche, May 15, 1892, Olcott MSS (Wyo.), Box 2.

than did the Tomboy trip. There was absolutely no place to be hired where one could work or write of an evening, and the only good thing to say of the hotel was that it was cheap.[196]

Yet, the examining engineer kept busy, and, despite complaints, most seemed to enjoy the rugged aspects of their work. The wife of one who had just completed a long inspection journey some 600 miles above Victoria, British Columbia, in 1866, found her fears for her husband's health and safety unjustified—"He had a fine time, grew as fat as possible and quite provoked me by coming back in such fine health and spirits when I had worried myself quite thin about him." [197] Mary Hallock Foote could also wonder about her "dear old boy," away in central Idaho in 1878.

> I get a letter every day, in the hideous envelope of the "Leading Hotel of the Wood River Country"—the last one was written just after a fifty mile ride, that day, on horseback, and a supper such as the hotel of the Wood River Country wd be likely to offer—and *yet* in the best of spirits! [198]

Eben Olcott termed his trip into Utah and Nevada in 1892 "simply glorious work," and Ernest Wiltsee looked back with some nostalgia to the days when he was a young "hard-riding, hard-working mining engineer" able to make a sixty-mile drive by spring wagon over the Harquahala Desert in 130 degree weather to inspect claims north of Yuma.[199] And Harvard's Nathaniel Shaler spoke glowingly of the escape experting work afforded him from his academic routine. It gave him a chance to exercise his imagination and come to grips with the

196. Ellsworth Daggett to James D. Hague, Salt Lake City, December 6, 1899, Hague MSS, Box 16.

197. Emelia Ashburner to William Brewer, San Francisco, May 14, 1866, Ashburner MSS.

198. Mary H. Foote to Helena Gilder, Milton, New York, June 7, 1878, M. H. Foote MSS.

199. Eben Olcott to Katharine Olcott, Salt Lake City, March 22, 1892, Olcott MSS (Wyo.), Box 2; Wiltsee, "Reminiscences," pp. 157, 168.

elements, and he was pleased with the strong, self-educated mining men with whom he was thrown in contact. He once recounted being met in Montana by a miner decked out in what was deemed an appropriate garb in which to welcome a "Professor"—black suit, patent-leather shoes, and tall silk hat, and jestingly, he liked to say that "whatever he might be in Cambridge he was a great man in the Rockies." [200]

A more tangible benefit, of course, came in hard cash—payment for services performed. The amount of the consulting fee varied considerably, depending on the particular job, the circumstances, and the reputation of the expert. If fees were high, as they often seemed to the public, it was for a number of reasons. In the first place, the consultant experienced unpredictable periods of feast and famine. Part of the time, he might have no work; part of the time, he might have to turn work away. If unseasonable weather did not prevent inspection work, it might at least interfere and delay. Louis Janin complained in mid-1862 that the heavy rains and floods that halted all mining enterprise in California left him unemployed for some time.[201] In December 1891, Janin again commented, "There is rain in the City and there is snow in the country. 'Blessed be the Lord,' say the farmers, and 'durn the luck,' say the traveling Experts." [202]

Moreover, at least in part, the size of an engineer's reputation determined the size of his fee. Knowledgeable mining men, both trained and untrained, came to realize that cut-rate experts were in the long run expensive, that higher fees paid to engineers of real standing was money well spent. As Daniel Barringer told a prospective client in 1905, when he recommended Pope Yeatman, "Unless you get the advice of some high class man such as this, and are willing to pay him a good

200. Shaler, *Autobiography,* pp. 336–38.

201. Louis Janin to Louis Janin, Sr., Virginia City, Nevada, July 5, 1862, Janin family MSS, Box 10.

202. Louis Janin to James D. Hague, San Francisco, December 8, 1891, Hague MSS, Box 12. For other comments on delay owing to weather, see J. S. Phillips to editor, San Francisco, November 30, 1868, *M&SP, 17* (December 5, 1868), 354; T. B. Ludlum to L. Bradford Prince, Oakland, February 26, 1895, Prince MSS.

big fee for his opinion, I am afraid you will remain in the dark as to the real value of your property." [203]

Normally, the examining engineer received a predetermined fee, unless the duration of the work was indefinite, in which case he might be paid on a per diem or salary basis. Expense funds were expected in advance, and it was customary to ask a retainer of from 10 to 50 percent of the total before the work commenced, although clients sometimes complained.[204] Legitimate expenses were to include all transportation and living costs, including tips, medicines, and servants. Those of one engineer included stamps, envelopes, field books, and even a "monkey-wrench." [205] Occasionally, a client agreed to provide entertainment and expenses for wining and dining, if this was necessary to the expert's obtaining information about a property.[206]

It is difficult to talk in terms of an average fee. Eben Olcott wrote a friend in New York in 1881:

> As to my fees for examinations & reports on mining properties, many circumstances control their amount *e.g.* location of mine, accessibility, time required to reach it & the facilities there present for making the necessary investigations, the party desiring the services, &c. More still depends on the nature of the report required. If it be one of information to owners as to value & method of working a property they own or even to advise investors in an established Co. I generally charge $500. to $600.

203. Daniel M. Barringer to W. M. Morgan, July 11, 1905, copy, Letterbook K, Barringer MSS. See also Vigouroux, ed., *Diary of a Mining Investor,* p. 15; Harry J. Newton, *Pitfalls of Mining Finance,* p. 152.

204. Eben Olcott to John H. Cathrae, New York, August 14, 1893, copy, Letterbook 20, Olcott MSS (N.Y.); Henry Villard to James D. Hague, New York, June 5, 1890, Hague MSS, Box 12; Samuel Insull to Charles Janin, Chicago, August 3 & 19, 1912, Charles Janin MSS, Box 5.

205. Ray Humphrey to Fred L. Morris, Whitney, Oregon, July 6, 1916, Morris MSS, Box 1.

206. E. C. Watson to F. E. Hoover, Goldfield, October 7, 1905, copy, Goldfield Companies MSS, Box 1; Lynwood Garrison, "Professional Customs," p. 552. For other letters summarizing what the expert should expect by way of expenses, see *M&SP, 95* (October 5 & 19 & November 2, 1907), 428, 490–91, 551–53.

and expenses provided the work does not take more than two weeks. If however a buyers or sellers report is sought on which depends a heavy investment and is likely to influence the floating of a new Co. I would charge considerably more.[207]

Over a dozen years later and now better established in the profession, Olcott said that "for years it has been my rule not to make reports west of the Mississipppi River for less than $1500.00 and expenses, limiting the time consumed to one month."[208]

The elite commanded very high fees, though few were so fortunate as Benjamin Silliman, who examined the Emma mine in Utah late in 1871 and received $5,000 as an advance, plus an additional sum of from $10,000 to $20,000 when his report was completed, "depending on the estimate I might form of the value of the service."[209] John Hays Hammond, whose watchword, according to his card, was "Accuracy, Dispatch and Reasonable Charges," earned $500 for his first examination; some twenty years later, as consulting engineer for a firm in Colorado, he made $50,000 a year, but this involved more than one inspection. Hammond admitted that he charged "all the traffic would bear, or almost the limit," on the grounds that his reputation was at stake.[210]

Rossiter Raymond once made a quick visit to a mine in Colorado and wrote a brief report for $200 and expenses, all the while gathering mineral statistics for his next report to the federal government; a few years later, he would receive $5,000 for his few hours' work in putting together the misleading and soon discredited report on the Chrysolite.[211] Winfield

207. Eben Olcott to John M. Knox, Jr., Lake City, Colorado, July 5, 1881, copy, Letterbook 3, Olcott MSS (N.Y.).

208. Eben Olcott to W. T. Sanger, New York, October 19, 1894, copy, Letterbook 22, ibid.

209. Silliman testimony, "Emma Mine Investigation," *House Report,* no. 579, 44th Cong., 1st Sess. (1875–76), 126.

210. *M&SP, 44* (January 7, 1882), 13; Stratton's Independence, Ltd., *Annual Report,* year ending June 30, 1901; J. H. Hammond, *Autobiography, 1,* 108–09, 148.

211. See fragment of document (n.d.), Adelberg & Raymond MSS; *M&SP, 42* (April 9, 1881), 232.

Scott Keyes was already familiar with the Richmond mine in Nevada when he examined it for the English market in 1871 and was able to complete the job in a very short period—"Perhaps an hour, I don't know. Not a very long time," he testified later. His fee was $2,000.[212]

But most examinations required more time and greater expense on the part of the expert. When James D. Hague agreed to examine the Anaconda and its attending mines and works at Butte and Anaconda for James B. A. Haggin in 1899, the fee he set was $12,500. But this was a major undertaking resulting in a detailed seventy-nine-page report, and out of his stipend, Hague had to pay his associate, Ellsworth Daggett, who directed the work, and a number of assistants.[213] Hague also did a report on a mine in the Coeur d'Alene for Henry Villard in 1890, charging $5,000 and expenses, and one on several other Idaho properties for E. H. Harriman in 1895, receiving a total of $6,000. Eben Olcott charged a client $2,461.01 in 1893, and defended it as a "perfectly reasonable" assessment.[214]

When the fee was on a per diem basis, well-known engineers in the 1890–1914 era charged $50 a day and expenses, even when, as Charles Janin admitted to a friend in 1913, "Mining is rotten dull out here." [215]

Sometimes, it was possible to get professional engineering advice from reputable banking houses, who paid for such services and passed reports on to their clients at reduced rates. Or costs could be lessened by employing more obscure local engi-

212. Keyes testimony, July 24, 1877, Richmond v. Eureka, p. 173.

213. J. B. Haggin to James D. Hague, New York, September 28, 1889, Hague MSS, Box 11, copy, Box 14.

214. Henry Villard to James D. Hague, New York, June 5 & July 2, 1890, ibid., Box 12; E. H. Harriman to James D. Hague, Chicago, June 22, 1895, ibid., Box 11; Eben Olcott to A. R. Meyer, February 9, 1893, copy, Letterbook 19, Olcott MSS (N.Y.).

215. Charles Janin to Chester W. Purington, February 3, 1913, copy, and H. A. Millard to Charles Janin, New York, February 18, 1913, Charles Janin MSS, Box 5; Louis Janin to James D. Hague, San Francisco, November 9, 1891, Hague MSS, Box 12; Daniel M. Barringer to F. E. Bond, May 26, 1908, copy, Letterbook N, Barringer MSS.

neers at lower rates and travel expenses.[216] Or a younger expert or one who did examination work as a sideline might prove more reasonable. For example, Henry E. Wood, whose livelihood came mainly from his assaying and sampling works, was willing to report on a mine at Ouray in 1890 for $560. After paying his assistant and other costs, he cleared $350 and enjoyed his "charming jaunt," especially the day of fishing en route back to Denver.[217]

Occasionally, an engineer might reduce his rates, as did John B. Farish in 1889, when he quoted Isaac Ellwood of DeKalb, Illinois, a special figure, in order that he "might become known to the mining investors of your section."[218] Charles Janin once wrote a friend that he had made an examination trip to Arizona, "where the 'mine' looked so bad that I discounted the fee agreed upon with my client to his great delight." "Perhaps I am a fool," Janin continued, "anyway I have a sign 'Please kick me' pinned to my coattail."[219]

The inspecting engineer did not always settle for a flat fee, per diem, or salary for his services. As an editor of 1871 noted, some clients were "silly enough or dishonest enough" to tempt them with promised payment in company stock or a fee contingent upon the sale of the property.[220] That such practices did persist throughout the era is clear, although the ethics of them bothered many of the profession. At the time of the Lost Bullion Spanish scandal in 1907, Thomas A. Rickard—the conscience of the profession—warned against accepting

216. *EMJ, 87* (January 16, 1909), 173.

217. Henry E. Wood to Belle Wood, Denver, September 25, 1890, Wood MSS, Box 1.

218. John B. Farish to Isaac L. Ellwood, Baker City, Oregon, May 16, 1899, Denver, May 29, 1889, Ellwood MSS, Box 102.

219. Charles Janin to H. Foster Bain, San Francisco, March 31, 1919, copy, Charles Janin MSS, Box 8.

220. *EMJ, 12* (November 14, 1871), 313. For examples of the contingent fee and the use of stock in lieu of examination fees, see San Francisco *Bulletin,* March 15, 1880; *Silver Standard,* March 5, 1887; Theodore B. Comstock, "Responsibility in Mining Enterprises," *EMJ, 27* (March 22, 1879), 200; Lindemann *v.* Beldon Consolidated Mining Company, *Colorado Appeals, 16* (April 1901), 342; William M. Read to Alfred H. Oxenford, Capitola, California, September 25, 1900, copy, Read MSS.

stock in lieu of inspection fees, "for that way lies the road to perdition." "Avoid contingent fees as you would a direct bribe," he cautioned, "for they undermine professional integrity and excite the cupidity of which none of us is wholly devoid."[221]

Probably most engineers shunned contingent fees from vendors, but were willing to accept them or an interest in shares from buyers. It was not uncommon for an expert to give his services for a fee or salary, plus an interest in any undertaking that developed on the basis of his advice. When Victor Clement examined Enos Wall's Utah copper property for Joseph R. DeLamar in 1899, for example, he was to receive a one-eighth interest in any holding acquired by DeLamar as a result of his recommendations.[222] Such agreements were especially common where the expert was employed as a "scouting" engineer on a regular basis for an exploration or development company. Daniel Barringer offered Henry M. Blackmer this type of arrangement with the Commonwealth Exploration Company in 1910.[223] Seeley W. Mudd took no salary from the Guggenheim Exploration Company, but received, as he put it, "rather large compensation for anything that might be found."[224] With the same firm, John Hays Hammond had been paid a handsome salary, plus an interest in property taken up on his advice—an arrangement so lucrative, according to one account of his life written for the youth of America,

221. *M&SP, 95* (September 7, 1907), 286. Rossiter Raymond told the class of 1906 at Lehigh that there was nothing wrong with making a vendor's report for a fee payable if the mine sold on the strength of the report, provided that the arrangement was made clear to all at the beginning. "Professional Ethics," *EMJ, 81* (June 23, 1906), 1200.

222. Parsons, *The Porphyry Coppers,* p. 52. For other examples, see entry for September 26, 1892, field notes 15, Stanton MSS; Eben Olcott to W. T. Sanger, New York, October 19, 1894, copy, Letterbook 22, Olcott MSS (N.Y); Charles Janin to Jonathan Bourne, Jr., San Francisco, September 20, 1907, copy, Charles Janin MSS, Box 5; Claud B. Andrews to R. A. Jones, Los Angeles, January 28, 1925, copy, and C. R. McIntosh to A. B. Renehan, Santa Fe, November 14, 1925, Renehan-Gilbert MSS.

223. Daniel M. Barringer to Henry M. Blackmer, April 14, 1910, Commonwealth Exploration Company, Letterbook 1, Barringer MSS.

224. Rickard, ed., *Interviews,* p. 397.

that his income was set at $1 million a year, or "about four cents a second." [225]

But Hammond and Mudd were exceptional engineers, able to name their own terms. Robert Brewster Stanton was neither so well known nor so well paid by the United Mining and Development Company of America, which he served as a "scout," seeking promising property in 1903 and 1904. Nor was the UM&DCA Guggenheim-backed: it existed on slim capital, operating from day to day, hoping to "strike something big." For his part, Stanton received $100,000 in stock outright, with his monthly salary—beginning at $300—based on the amount of stock sold to the public.[226] It was clear from the time he was sent West in spring 1903 to find property "that would not cost too high to start, and such as would produce some net income as quick as possible," that this was a shoestring operation, although Stanton was assured that adequate expense funds would be provided. Most of the summer found Stanton scouring the West, usually without money to pay for his assays or even his lodging. The following spring, he "went over the whole thing without gloves" with company officials and resigned.[227]

Stanton's experience notwithstanding, the development of the exploration company late in the nineteenth century was a significant innovation in the mineral industry, especially when solidly financed. One of the first, the Exploration Company, Ltd., a London enterprise, was founded in 1886 with Rothschild capital and managed by two American engineers, Hamilton Smith and Edmund DeCrano, through whom it maintained connections with leading western experts including James Hague, Ernest Wiltsee, Thomas Mein, and Fred

225. J. H. Hammond *Autobiography, 2,* 502–03; Wildman, *Famous Leaders of Industry,* p. 123.

226. When $500,000 worth of shares had been sold, Stanton's salary would rise to $500 a month; when $1 million had been marketed, it would go to $750. Entry for February 18, 1903, field notes 20, Stanton MSS.

227. See entries for April 6, 15, 22, 24, & 30, May 15, 16, & 27, June 3, 17, & 25, October 20 & 26, 1903, April 30, & May 12 & 13, 1904, ibid.

Bradley.[228] Of equal, if not greater importance were the General Development Company and the Guggenheim Exploration Company, the latter with the high-priced Hammond at the head of its far-flung scouting forces between 1903 and 1907. Over a two-year period—1910 and 1911—weary "Guggie" experts considered a total of 1,605 mine propositions, rejecting 1,263 outright. Preliminary examinations were made on 268, complete examinations on 74, and, in the end, 3 were deemed worthy of purchase.[229]

By the time of World War I, the exploration companies and the great mining corporations with their own corps of engineers had taken much of the examining work from the independents. "The big fellows have their cruisers everywhere all the time and the regularly organized staff is assigned to the work of examination," complained one westerner, who pointed out that the free-lance reporting that remained was of the "chicken-feed type"—"small, intermittent and unprofitable." [230] The inspecting engineer was becoming salaried; young men found it more difficult to break into the field; and many preferred the security offered by the large corporate organization.[231] This was in keeping with American industrial development in general.

228. See Exploration Company, Ltd., *Memorandum and Articles of Association,* November 1, 1886, Companies Registration Office, Board of Trade, Bush House, London; *The Statist* (London), *43* (March 25, 1899), 476; Exploration Company, Ltd., *Report for Year Ending December 31, 1901; San Francisco Call,* August 30, 1903; *EMJ, 70* (July 7, 1900), 16; *M&SP, 110* (April 3, 1915), 507; Hamilton Smith to James D. Hague, London, June 1, 1895, Hague MSS, Box 19; J. H. Linrach to Hague, London December 18, 1895, ibid., Box 11.

229. "Some Reasons Why Mining Languishes," *EMJ, 93* (May 25, 1912), 1017.

230. *EMJ, 92* (September 9, 1911), 840, *98* (July 4, 1914), 5.

231. *EMJ, 85* (February 29, 1908), 466, (November 20, 1915), 856.

CHAPTER 5

"I Am Standing the Responsibilities If Anything Goes Wrong"

If you can wear your dress suit, sack or jumper
And look at ease in each one just the same;
If you can take the job of "Supe" or pumper
Or any other man that quits the game;
If you can set a bone and tie up sinews,
Or later preach a sermon for the dead;
If you can talk like Webster, Clay or Depew
And turn a dinner table on its head

—Wayne Darlington[1]

The old-time mining engineer, jack-of-all-trades that he was, was bound by no clear-cut lines of demarcation within his profession. Statistics are inconclusive, but it is safe to assume that most engineers at some time in their careers took a fling at mine or mill management or perhaps both.[2] A not uncommon pattern was for a young engineer to work his way up to superintend a milling plant before going on to manage a larger operation, including both mining and milling, then to turn to consulting practice. Many, of course, did little or no consulting, but spent their professional lives as specialists in charge of plant and property.

A good manager was vital, especially for an undeveloped mine. "A prospect resembles a young child rich in possibilities, but hedged around with all the uncertainties of immaturity," said Thomas A. Rickard. "The promising prospect may succumb to the measles of bad management, or the whooping-

1. From "The Mining Engineer," *EMJ, 97* (September 26, 1914), 587.

2. A survey of graduates of the mining curriculum of Harvard's Lawrence Scientific School, for the years 1897 to 1905 inclusive, indicates that fully one third were engaged as managers or superintendents at the time of the poll. *M&SP, 95* (August 10, 1907), 173.

cough of inexperience."[3] Even at an established, producing mine, an able manager might spell the difference between success and failure, the more so when ores were complex or of only medium or low grade.

One of the common problems of the industry, at least in the nineteenth century, was the frequency with which absentee mine owners—themselves ignorant of mining techniques—sent out equally uninformed and unqualified managers. "Counting-room superintendents," "uneducated quacks, retired sea-captains or broken-down stock-speculators" took their place alongside camp cooks, cowboys, or book peddlers in charge of property.[4] Or what of the "jolly dogs"

> —usually nephew of the president, or the son of the head director—excellent masters of the billiard cue, with uncommon pride in high boots and spurs, whose champagne bills were charged to "candles," and whose costly incense to Venus appeared on the books as "cash paid for mercury"?[5]

Such managers too frequently displayed a talent for extravagance, ineptness, and occasionally dishonesty, and in their wake left undeveloped property and disgruntled investors faced with "Chinese dividends"—calls for more capital. Not that inexperience was necessarily tantamount to failure: some such managers, excellent, adaptable businessmen who selected capable subordinates who knew mining, were indeed successful, but the reverse was more often true.

Nor does this mean that the trained mining engineer was invariably superior. A young Freiberg or Columbia product was always a risk until he had married practical experience with his theoretical background (often at company expense), but he was less a gamble than the manager who had neither technical education nor experience. And plenty of trained, seasoned en-

3. "Mining Risks," *EMJ, 75* (April 4, 1903), 510.

4. *AJM, 1* (July 14 & September 15, 1866), 248, 386 *6* (August 8, 1868), 88; *M&SP, 41* (October 2, 1880), 209; *The Miner, 1* (March 1866), 5; *The Chicago Tribune,* April 10, 1869; Amasa McCoy, *Mines and Mining of Colorado,* p. 35.

5. Samuel Cushman & J. P. Waterman, *The Gold Mines of Gilpin County, Colorado,* p. 33.

gineers proved to their client's satisfaction—or dissatisfaction—their inability to operate a mine efficiently, and the fault did not always lie with the property. Some sent out exaggerated reports of ore values, mistook pyrites for pay ore, or overpaid for equipment. Being human, they sometimes made errors of judgment, for as one of them said, "It seems simply a question of time when one of us slips." [6]

But the educated and experienced manager could avoid many of the pitfalls that trapped the uninitiated. He was aware of the common causes of mine failure and realized that a mine was a wasting asset and must return costs of ground, development, and plant over a relatively short time. He knew the value of systematic sampling and the folly of "picking the eyes out" of a mine to show a quick profit without regard for future development.[7] He recognized, too, that one of the greatest common blunders was the premature erection of costly milling and reduction equipment before the richness of the ore had been sufficiently established. Most western areas went through periods of "process mania," especially in the 1860s, as companies began to encounter unfamiliar rebellious combinations of ore. Fantastic mills, "more fitted for boudoirs or saloons than for the purpose of mining," were constructed, many of them "conceived in sin and born in iniquity"; even more of them "conceived in ignorance and born in incompetency." [8]

To an investor or director, an impressive new mill was tan-

6. Carl A. Stetefeldt to Samuel Christy, New York, September 24, 1886, Christy MSS. See also *EMJ, 13* (March 26, 1872), 201; London *Times,* November 4, 1872; Montana Company, Ltd., *Final Report of the Committee of Inquiry To Be Presented at the Extraordinary General Meeting of the Shareholders, To Be Held on Thursday, the 19th Day of March, 1885,* p. 17.

7. For discussions of common reasons for mine failure, see *Report of J. Ross Browne on the Mineral Resources of the States and Territories West of the Rocky Mountains* (1868), pp. 12–13; "Causes of Mine Failures," *M&SP, 56* (May 9, 1903), 295; Richard M. Stretch, *Prospecting, Locating and Valuing Mines,* pp. 2–15.

8. The literature on this point is extensive. See *MI&R, 16* (September 26, 1895), 114; James P. Whitney, *Colorado, in the United States of America,* pp. 56–57; San Francisco *Post,* December 27, 1879, clipping in Bancroft Scrapbooks, *51,* no. 3, 1073; Cushman & Waterman, *Gold Mines of Gilpin County,* pp. 32–33; Bayard Taylor, *Colorado: A Summer Trip,* 56; James F. Rusling, *Across America: or, The Great West and the Pacific Coast,* p. 67.

gible evidence of a mine. One old California mining shark explained that whenever he landed an eastern buyer for a property, in order to keep his customer happy he always advised the immediate construction of a milling plant, on the basic principle of "what is a home without a mother, or a mine without a mill." [9] Perhaps another reason was that manufacturers of mining machinery never suggested further testing and development work and frequently gave a kickback to the engineer or manager recommending purchase of their equipment.[10] Whatever the cause, one expert estimated in 1908 that 30 percent of all capital wasted in mining was owing to precocious erection of milling plants.[11]

Yet, from the beginning, sound mining men gave constant warning: "Nobody cooks hares till the hares are caught. . . . First get your ores!" [12] Undoubtedly some trained engineers, such as young Louis D. Ricketts, fell into the trap from time to time, but with the increasing incidence of technical men in advisory and managerial posts the frequency of such disasters decreased sharply.[13]

But the transition was a slow one; the college-bred mining engineer at first made inroads more rapidly in the consulting field than in managing. In the 1880s, according to Waldemar Lindgren, the trained engineer "was a somewhat rare specimen as a superintendent of mines," but by the turn of the century he was a common fixture at important properties.[14] No longer could a manager of ordinary talents work his way up the ladder. "The fact remains," said the editor of *Mining and Scientific Press* in 1915, "that nearly every successful mining

9. Reginald W. Petre to editor, Baltimore, May 25, 1903, *EMJ, 75* (June 6, 1903), 849.

10. Henry B. Clifford, *Rocks in the Road to Fortune,* pp. 78–79.

11. Ibid., p. 77.

12. *AJM, 7* (April 24, 1869), 264. See also *M&SP, 4* (February 22, 1862), 2 *41* (October 23, 1880), 264; *MI&R, 16* (August 15, 1895), 38; Clarence King to J. J. Higginson, Washington, D.C., April 20, 1880, copy, Letterbook 4, King MSS; William M. Read to W. C. Baker, Capitola, California, March 10, 1907, copy, Letterbook (1900–07), Read MSS.

13. Bimson, *Ricketts,* p. 13.

14. *EMJ, 75* (May 9, 1903), 702.

operation of consequence, old or new, is today in the hands of experienced technically trained men."[15]

The manager, like the consultant, won his position on reputation. Such prominent engineers as James D. Hague and Louis Janin were constantly recommending promising young men for manager's posts. Janin and Hague were both asked to suggest a likely prospect for a position in Western Australia in 1897. Janin thought the job an excellent one, though he admitted he knew little of the region. But, "as for the creature comforts," he said, "these are generally adequate wherever there is plenty of money."[16] Apparently Herbert Hoover, whom he recommended, agreed, for he took the post and found it a golden opportunity.

As already noted, few fledgling engineers—not even a Herbert Hoover—stepped into a managerial position without first gaining some experience. It was relatively easy to attain the rank of assistant superintendent or assistant manager after a little "seasoning," but the jump to manager often required several years of on-the-job training under the watchful eye of a veteran.[17] Or sometimes the young engineer bridged the gap by leasing and operating property on his own for a time.[18]

Down through the years, certain major mines gained the reputation of training both mine and mill managers. Hague's North Star at Grass Valley was one. Another was the Standard Consolidated at Bodie, where a good deal of pioneering work in electrical transmission and in the use of the cyanide process was done, and where such men as R. Gilman Brown, Charles W. Merrill, Theodore Hoover, and R. Chester Turner gained at least a part of their professional

15. *M&SP, 110* (January 16, 1915), 96.

16. Louis Janin to James D. Hague, San Francisco, June 16, 1897, Hague MSS, Box 12.

17. Perhaps an "average" schedule was that of Ernest A. Wiltsee, a Columbia graduate: Coastal Survey measuring the flow of the Hudson, the steel mills of Pittsburgh, metallurgical jobs at Pueblo and Denver, then three years as assistant manager at the North Star in California, then a full-fledged managership. Wiltsee, "Reminiscences," pp. 13–15.

18. *EMJ, 52* (August 29, 1891), 239.

preparation.[19] In Idaho, the Bunker Hill & Sullivan played a similar role, with Victor Clement, Fred Bradley, Albert Burch, and Stanley Easton among the important managers there. The North Bloomfield in California was another, with such prominent engineers as Hamilton Smith, Hennen Jennings, and Henry C. Perkins passing through en route to greater eminence.[20]

The work of the mine manager was many-sided. Basically, his task was to take charge of a property, direct exploration and development, and make provision for the extracting, processing, and even sale of its ore.[21] In seeking out ore and following its course, he had to have some knowledge of geology, mineralogy, and surveying. In extracting it, he had to know something of tunneling, drifting, shafting, and other underground work. In addition, he had to be enough of a mechanical engineer to plan for the use of hoisting, ventilating, or draining equipment, and some means of underground transportation. He must be able to weigh the virtues of steam power versus electric or air drills versus electric drills. In order to supervise treatment of ore, he had to be something of a chemist and metallurgist, with an understanding of milling or perhaps reducing or smelting machinery. Frequently, he assumed the role of a civil engineer and built roads, tramways, or flumes. At the same time, he was a business manager, constantly concerned with costs, purchasing, accounting, ore sales contracts, or perhaps leases. He was the supervisor of a large labor force and concurrently responsible to his own corporate or individual employer, whose rights, both property and legal, he was expected to protect. On top of all this, he also served a public relations function and was expected to "live an exem-

19. *M&SP, 114* (March 17, 1917), 360.

20. Rickard, ed., *Interviews,* p. 226.

21. Exploration refers to the gaining of knowledge of the size, shape, position and value of an ore body; development is the driving of openings to and in an ore body for the purpose of extracting it. James F. McClelland, "Prospecting, Development and Exploitation of Mineral Deposits," in Robert Peele, ed., *Mining Engineers' Handbook, 1,* 405.

plary life so as to command the confidence and respect of the camp and the company." [22]

Obviously, it took a man of broad background and experience to fill such a position. When Simeon Reed, owner of the Bunker Hill & Sullivan, sought a manager in 1887, he specified an expert "that knows *all* about runing [*sic*] A large Mining property," including deep tunneling, construction of a concentrating works to be operated by water power conducted via an 8,000-foot flume, operation of a cable tramway, timbering, and underground techniques.[23] The man he hired, Victor Clement, then managing the Empire mine at Grass Valley, sent a brief professional biography emphasizing his "thorough scientific training" and nearly a dozen years of experience in the field:

> My experience embraces—bookkeeping, assaying, mine and land surveying, general engineering; I am likewise familiar with the details of the metallurgy of Gold and Silver ores. I have assisted in erecting and run chlorination, lixiviation, amalgamation and smelting works. I have had experience with different classes of concentrators, with a view of leaching or smelting the products. I have a complete knowledge of the economical development of mines. At present here I conduct all the underground and surface work myself. My experience has been in remote countries, where skilled labor was usually scarce, and thus necessity has made me very familiar with all the minor details of the profession.[24]

The ideal manager, though his duties were primarily those of planning and supervision, with details left to his subordinates, should be familiar with the particulars of all operations. He "should know how many cars a man will tram a shift

22. "The Mining Engineer, His Functions and Importance," *Northwest Mining Journal, 3* (June 1907), 87.

23. Simeon Reed to James D. Hague, Portland, May 25, 1887, Reed MSS.

24. Victor Clement to Simeon Reed, Grass Valley, May 24, 1887, ibid. For Clement's career, see Flora Cloman, *I'd Live It Over,* pp. 65–226.

under existing conditions; how many tons of rock a mucker should shovel in a shift; how many feet of ground should be drilled by hand or with machines in a stated time." [25] He should

> be able to run an engine, know how to run a mill in all its branches, know when each stamp is doing its duty, detect a loose mortar-hole, cut out any timber for a shaft or drive or elsewhere, sharpen a pick or drill; in fact, he must be a miniature encyclopaedia.[26]

The better managers took this catholicity to heart, as they did the injunction of one of their contemporaries in 1912, that "a mine cannot be managed from an office chair." [27] Some ignored such advice, but the successful managers made it a point to know personally what was going on day by day at the property under their supervision. Charles Lyman Strong, in charge of the Gould & Curry on the Comstock, visited the company's numerous works (at times, seven or eight mills) daily, but refused to delegate authority and overwork broke his health.[28] When Victor Clement first came to the Bunker Hill & Sullivan, he insisted on doing all underground surveying personally. Good surveyors were rare, he explained, and the work enabled him to learn more about the mine. Besides, "there is one point about the working which if known to outsiders might involve this Co into considerable trouble." [29] Later, at the same property, Albert Burch made it a point to go through the workings twice a day,[30] and Frank Sizer, while manager of the Empire on Montana, was underground every day or two, sampling, surveying, measuring stopes, or generally keeping himself up to date on progress.[31]

25. *M&SP, 86* (June 13, 1903), 378.
26. *M&SP, 75* (October 30, 1897), 406.
27. S. H. Brockunier to editor, Gaston, California, July 8, 1912, *EMJ, 94* (August 17, 1912), 294.
28. *DAB, 18,* 146; *EMJ, 37* (February 16, 1884), 118–19.
29. Victor Clement to Simeon Reed, Wardner, October 24, 1887, Reed MSS.
30. *M&SP, 121* (August 28, 1920), 297.
31. See entries for January 25 & 28, February 1, April 9, & July 11 & 14, 1887, Sizer diary.

But a large mine might employ from 300 to 1,000 men,[32] and though a manager might keep a hand in, he had to depend on carefully selected subordinates. Moreover, the administrative and managerial organization might vary considerably with the size and complexity of the operation. Thus, the manager of a small enterprise was often in direct charge of both mine and mill, supervising a mine foreman, a mill foreman, and an assayer, perhaps, with a second round of subordinates—shift bosses, mechanics, and so forth at a lower level.[33] The structure of a larger property might be more intricate: sometimes the manager was over an assistant manager or superintendent who was concerned with the entire operation; sometimes there was both a mill superintendent and a mine superintendent; or a substantial mine might be divided into several sections, three or four underground levels comprising a section, each under an assistant superintendent.[34]

But whatever the arrangement, any manager worth his salt understood the need for capable subordinates and was quick to replace those of doubtful worth with men of known ability. Often, when a manager assumed a new position, he brought with him key personnel—superintendents, foremen, shift bosses, or bookkeepers.[35]

In opening up and developing a mine, it was the manager (or sometimes a consulting engineer) who planned the strategy and determined the location, shape, and size of underground openings. A shaft might be vertical or inclined or perhaps a combination of the two. But in planning it, the manager had to take into account a number of factors: the depth, nature of the hoisting apparatus, type of subterranean transportation, the number of men to be employed, the presence of water, the nature of the ground, and the capital outlay. He

32. *DAB, 18,* 146; Hester Ann Harland, *Reminiscences,* mimeographed, pp. 30–31.

33. Kett, *Autobiography,* pp. 19–20.

34. Ralph M. Ingersoll, *In and Under Mexico,* pp. 43–44.

35. See Pierre Humbert, Jr., to L. Auerbach, San Francisco, June 1901, copy, Asbury Harpending MSS, Box 7; Victor Clement to Simeon Reed, Grass Valley, June 14, 1887, Reed MSS; Edgar Rickard to Samuel Christy, Cornucopia, Oregon, May 28, 1897, Christy MSS; Kett, *Autobiography,* pp. 44, 45–54, 56.

had to keep in mind the desirability of a balanced winding engine, with the weight of a descending skip or cage helping hoist a loaded one. The need for an escape route, room for ventilating and water pipes and for electrical wires meant at least three compartments per shaft.[36]

It was the manager who determined where tunnels, drifts, crosscuts, stopes, and winzes should be placed and what precautions needed to be taken as work proceeded. It was he who decided on the mode of timbering and on the method of caving and breaking ground most appropriate for a particular mine, all the while keeping in mind costs and long-range development plans.

Part of the manager's job frequently was to design, purchase, and install machinery. In the last part of the nineteenth century, the mechanical appliances of mining tended to improve more rapidly than mining methods themselves, and mine managers were sometimes accused of paying more attention to the equipping of a property and the treatment of ore than to the underground work. Occasionally, but not often, a mechanical engineer would be put in charge of an entire mine operation, but such specialists were in short supply in the West and were generally attached to large concerns, in addition to the normal experts on mining. Thus, the average mine manager of whatever background had to bear the burden of putting in the necessary machinery in most instances. And at least one mechanical engineer in 1915 believed that the smaller western mines particularly consistently displayed some "horrible examples" of mechanical engineering practice.[37]

The manager or consulting engineer did the best he could. He either purchased ready-made equipment, which was then modified for his purposes, or he set forth the nature of his problem, with specifications, leaving the manufacturer to come up with suitable machinery.[38] If he were not too scrupulous,

36. Hoover, *Principles*, pp. 70–82; McClelland, "Prospecting Development and Exploitation of Mineral Deposits," in Peele, ed., *Mining Engineers' Handbook*, *1*, 487–89.

37. *EMJ*, *99* (May 22, 1915), 919; *M&SP*, *48* (June 14, 1884), 397; *Mining World*, *33* (December 3, 1910), 1050.

38. *M&SP*, *112* (June 3, 1916), 811.

he was happy to accept a commission from the manufacturer on the transaction, although the bluenoses in the profession decried the all-too-common practice.[39]

Because the manufacturers of mining equipment were concentrated in a few cities, such as New York, Chicago, Milwaukee, Denver, and San Francisco, the manager might have to travel to study or order machinery. M. G. Gillette bought equipment in Chicago in 1881 for the mine he supervised in New Mexico; Robert Peele, manager of an Oregon concern a few years later, condemned the $40,000 worth of machinery installed by his predecessor and decided to go to California to look at some new apparatus.[40] Arthur D. Foote was sent to the Calumet & Hecla in Michigan in the early 1890s to study electrical pumping equipment to be adapted at Grass Valley.[41] While managing the Hoskaninni Company, a dredge enterprise at Glen Canyon in the Colorado River, Robert B. Stanton traveled to Montana and Idaho to view dredges in operation, and in addition made several trips to Milwaukee and Chicago to investigate their manufacture and that of engines, ice-making machines, and power plants. When the contract was let to the Bucyrus Company, he spent several months in Milwaukee supervising the construction of the dredge.[42] Once it was finished and shipped unassembled on two special trains to the railhead at Green River, Utah, Stanton directed its unloading and hauling, first by twelve-horse wagon teams, then by barges and scows, to Glen Canyon, where he was in charge of building the dredge hull and assembling the whole.[43]

Victor Clement was responsible for ordering machinery for the Bunker Hill & Sullivan and, in 1891, handled the letting of

39. *EMJ, 29* (January 31, 1880), 81, *53* (January 16, 1892), 104; J. C. Bayles, "Professional Ethics," AIME *Trans., 14* (1885–86), 611; *Mining Science, 70* (July 1914), 36–37.

40. B. F. Grafton to Frank Hess, Washington D.C., July 15, 1881, copy, Ivanhoe Mining Company, Letterbook July 23, 1881–December 1885, in Robert G. Ingersoll MSS, Box 94, Library of Congress (cited hereafter as Ivanhoe Letterbook 1); Robert Peele to Eben Olcott, Cornucopia, Oregon, December 17 & 27, 1889), Olcott MSS (N.Y.).

41. Mary H. Foote to Helena Gilder, 1893 or 1894, M. H. Foote MSS.

42. Entries for August 21 & 27, 1897, September 18 & November 2, 1899, & February 3 & March 12, 1900, field notes 19, Stanton MSS.

43. Entries for June 24, 28, & 30, & July 9, 13, & 16, 1900, ibid.

contracts for construction of a new mill.[44] When powder smoke slowed down work in one of the tunnels, he took the initiative in ordering a large ventilating fan, just as he bought a small compressed air hoist costing $550 without consulting his employer. Anticipating a price rise in blasting powder because of a combination of explosives manufacturers, he stockpiled an eight months' supply—again on his own responsibility.[45] When he found he had made an error in specifications for machinery ordered from the Risdon Iron Works in San Francisco, Clement philosophically accepted the blame:

> I make no pretension at being free from mistakes—mining is in many ways an experimental science, and mistakes are open to anyone. Usually, in judging for the effectiveness of any new scheme, I study it well from the sides of its points of failures—I try to be conservative—even then, a person will slip up sometimes.[46]

Like other managers, Clement sometimes found himself supervising the construction of ancillary equipment and projects. In order to obtain lower insurance rates, he sketched out a system of pipes, hydrants, and fire hose for the Bunker Hill plant; he directed the building of more than a mile and a half of flume to carry water for power; and he superintended the construction of an aerial tramway to haul ore from the mines.[47] But hardly had the latter been completed when a snowslide "shook up the whole Sullivan Mountain," causing miners several hundred feet under ground to "drop their tools & run." In the process, it carried away the Bunker Hill ore bins, crushed part of the tunnel compressor building, and left the upper half of the tramway in splinters. That repaired, the tramway subsequently broke, and several heavy buckets came crashing down, creating "a kind of stampede among the residents below," and leaving Clement with the task of renovating

44. Victor Clement to Simeon Reed, Wardner, October 8, 1891, Reed MSS.
45. Clement to Reed, Wardner, August 19, 1889, ibid.
46. Clement to Reed, Wardner, February 1, 1889, ibid.
47. Clement to Reed, Wardner, July 25, 1887 & November 3, 1889, ibid.

$5,000 worth of damage and of purchasing the right of way underneath to avoid future lawsuits.[48]

Other mine managers, including George Attwood, also built aerial tramways, and one, F. A. Benjamin, in charge of a property in Nevada in 1870, suddenly found himself with the added duties of managing the White Pine Water Works, just purchased by the major shareholder of the mining company.[49] In Canada, just before World War I, manager Robert Livermore had the task of supervising the draining of a lake in order to prevent the flooding of the mine.[50]

In remote areas, the manager had to improvise. When Raphael Pumpelly went out to the Santa Rita mines of Arizona in 1860, fire brick was not available, so he had to build primitive furnaces of sun-dried mud and fire them with charcoal made from mesquite, all the while keeping a wary eye open for hostile Apaches and Mexican bandits.[51] Another manager, many years later, had to use the utmost ingenuity to weld a broken main shaft of his Colorado mill—or pack it nearly 300 miles for repairs; others, in emergency situations, took over specialized tasks from their subordinates.[52]

Sometimes the manager surveyed and constructed roads, as did Robert Stanton for the Hoskaninni project in 1899.[53] Arthur D. Foote was sent to Baja California in 1893 to supervise an onyx mine and to "take general charge of sending supplies down and getting onyx up from our landing some 250 miles down the coast, and of the quarrying 50 miles inland from the landing, and seeing to the Contractor who does the hauling."[54] Foote, who had been languishing on an abortive irrigation project in Idaho, was overjoyed. Less so his wife:

48. Clement to Reed, Wardner, February 7, 1890; May 8, October 8, 1891, ibid. See also Eben Olcott to M. P. Boss, February 17, 1890, copy, Letterbook 16, Olcott MSS (N.Y.).

49. *Who's Who in Mining* p. 3; W. Turrentine Jackson, *Treasure Hill,* p. 147.

50. Gressley, ed. *Bostonians and Bullion,* p. 172.

51. Pumpelly, *My Reminiscences, 1,* pp. 198–200.

52. *MI&R, 17* (January 16, 1896), 324; Richards, *Robert Hallowell Richards,* pp. 114–115.

53. Entries for December 14 & 15, 1899, field notes 19, Stanton MSS.

54. Samuel F. Emmons to James D. Hague, San Diego, March 30, 1893, Hague MSS, Box 11.

"The work is mining, road-building and sea-coast, piers, &c. All is new and undeveloped. . . . But it doesn't sound very domestic, does it?"[55]

Nor did the job last very long. When the quarries shut down, James D. Hague employed Foote at the North Star mine in Grass Valley and gave him a free hand to design and install a steel pipeline and air-compressor plant, the first of its kind and one that attracted wide attention. Next Foote was put in charge of reopening the North Star, an intriguing and professionally rewarding process:

> It has been a weird thing, re-opening this old mine. After the main shaft was pumped out and they got down to the lower levels it was like a watery Inferno. Black stagnant lakes in each one of the stopes, and silent canals black as the Styx leading away through the rock where the old drifts had been driven.[56]

Foote's engineering work on the North Star, including the driving of a new shaft joined to an old incline by means of a 1,600-foot raise, would bring recognition and praise among mining men.[57]

Frequently, the manager negotiated contracts for the sale of ore or ore concentrates; often, if metal prices were low, he stockpiled ore until the price rose.[58] Victor Clement kept a watchful eye on New York lead and silver prices, and on both the railroad company and the reduction company that handled the concentrates; both, he insisted, used a sliding scale, raising rates as mining profits increased or lowering them if profits dropped. Moreover, Clement challenged the reduction company's deduction of weight for excess moisture content.[59] In

55. "Well, it does my heart good to think my dear old boy is afloat again. It will take ten years off his life." Mary H. Foote to Helena Gilder, 1893 M. H. Foote MSS.

56. Mary H. Foote to Richard Gilder, Grass Valley, July 9, 1896, ibid.

57. *EMJ, 76* (December 17, 1903), 917.

58. Rickard, *Retrospect,* p. 64; Fred Bulkley to J. B. Wheeler, December 19, 1892, copy, Mining and Smelting Company, general manager's reports, in D. R. C. Brown MSS.

59. Victor Clement to Simeon Reed, Wardner, September 6, 1887; February 7, 1888, Reed MSS.

1888, when the price of lead fell "at a fearful rate," Clement concluded that the only solution was to curtail production completely until the market improved or a new contract could be negotiated for reduction. The prevailing agreement was with Samuel Hauser of Montana and stipulated that the Bunker Hill & Sullivan deliver its entire output, with no stockpiling. It called for *"all our products,* said Clement. But, "if our mill does not run, we can have no products—we can therefore get out of it in this way without liability." [60]

Apparently, later contracts were more flexible, although transportation difficulties were partly responsible, for in the next year, Clement put in a bin at the mill capable of holding 10,000 tons of concentrate for shipment to reduction plants.[61] But Clement was convinced that unless a smelter were built in the immediate vicinity, transportation costs would drive the Bunker Hill & Sullivan out of business. Like other managers, he made the arrangements for shipping,[62] and from the beginning adopted a hard line with the railroads, for, as he said, "It will never do to let them ride on us, and not kick, as is vulgarly said." [63] Clement was one of the supporters of the new Mine Owners' Association being proposed in the Coeur d'Alene late in 1889. "We will act in a body if necessary when dealing with Smelters & Railroads for incoming & outgoing freights," wrote Clement. "Will also endeavor to regulate many abuses in the labor question, &c." [64]

Clement's successor, Fred Bradley, fought these same interests through the MOA and outside of it. On his own initiative, he increased or decreased the output of ore, depending on metal prices, and virtually halted production at one point in 1893 when silver had plunged to an unprecedented low.[65] Bradley assumed responsibility for negotiating ore contracts

60. Clement to Reed, Wardner, October 26, & December 6 & 8, 1888, ibid.

61. Clement to Reed, Wardner, January 2, 1889, ibid.

62. See, for example, Fred Bulkley to J. B. Wheeler, January 2, 1893, copy, Brown MSS.

63. Clement to Reed, Wardner, July 11 & 12, 1887, Reed MSS.

64. Clement to Reed, Wardner, November 23, 1889, ibid.

65. Fred Bradley to N. H. Harris, Wardner, July 3 & August 23, 1893. Bradley MSS.

let on the basis of competitive bids, a task that presupposed some detailed knowledge of the market situation on his part.[66] That Bradley felt it important to keep well posted is indicated by his working arrangement with a Denver ore shipper's agent to be kept informed by a confidential weekly letter on the state of the smelting industry.[67]

In addition to ore contracts, managers also concluded agreements with lessees or renewed old ones, if a company was willing to lease ground. As manager of the Aspen Mining and Smelting Company in the late 1890s, Frank Bulkley spent most of his time supervising lessees, of which his company had fifty-three in 1899, and making sure that they lived up to their contracts.[68]

Some managers handled certain legal matters connected with the property, although if more than routine were involved, most would resort to a lawyer. Usually, even where companies hedged in their managers with all sorts of restrictions, they gave them power of attorney to do whatever was necessary to obtain government patents.[69] On behalf of the Hoskaninni Company, Robert B. Stanton in 1897 surveyed and marked out possible dam and power sites, along with 145 placer claims, in Glen Canyon.[70] Stanton also took care of the legal requirement of providing proof of labor on these. In Salt Lake City early in 1901, in one day he "Hired a whole typewriting establishment; had written 126 pages of letters & 64 pages of proof of labor. Completed this by 6:30 P.M. . . . A very hard three days work." And when the company's dredge proved unable to save the find gold from the river gravel and the enterprise was wound up later that year, Stanton was appointed receiver.[71]

Sometimes, in cases where surveying or recording errors

66. Bradley to Victor Clement, Wardner, May 13, 1893, ibid.

67. Bradley to W. S. Wigginton, Wardner, August 12, 1893, copy, ibid.

68. Frank Bulkley to J. B. Wheeler, October 9, 1899, copy, Brown MSS.

69. Story B. Ladd to Samuel Chittenden, Washington, December 24, 1881, copy, Ivanhoe Mining Company Letterbook, December 24, 1881 to April 22, 1884, Ingersoll MSS, Box 95.

70. Entries for October 2–12, 1897, field notes 19, Stanton MSS.

71. Entries for January 13, 1900; September 4, 5, & 23, 1901, ibid.

had been made, a manager made locations in his own name for fractions at the end of property lines,[72] assigning them to his company to avoid trouble with neighbors, although occasionally the less scrupulous sought to turn such a situation to his own profit.[73] Victor Clement loyally made entries for the Bunker Hill & Sullivan, and when he realized that the firm's claim was not secure on part of its property, he suggested working through an outsider to buy up all the claims along the vein for several miles beyond the property line.[74] Clement seems to have had authority to negotiate for ground in dispute, and he worked hand in hand with company attorneys in a variety of cases that, with his normal duties, kept him perpetually busy. "If I find time enough to take my meals I am doing well," he grumbled.[75]

Fred Bradley also wrestled with numerous legal problems at the Bunker Hill & Sullivan. He acquired federal patents on a number of claims, negotiated with the Northern Pacific Railroad for additional land, and testified in cases ranging from conflicting property claims to indemnity for men accidentally killed on company premises.[76] All the while, he complained about the company's lawyer, Weldon Heyburn, whom he called "a financial genius of a peculiar order . . . a mighty persuasive worker in securing over-drafts." [77]

Sometimes a legal hassle erupted into physical conflict. In 1903, Reno Sales, geologist for the Amalgamated Copper Company, and George T. McGee, superintendent of the Butte and Boston Company, sneaked illegally into the Michael

72. *Mining Science, 61* (January 13, 1910), 36.

73. In 1869, Peter Brandow, manager of the Pocotillo mine in Nevada, discovered a clerical error in the recording of his company's property. He immediately filed claim on the property and had it recorded in the names of friends who conveyed it to him. Resigning his post, he was about to sell the claim in San Francisco when the company brought suit and through an injunction halted negotiations. *White Pine News,* March 20 & April 15, 1869, cited in Jackson, *Treasure Hill,* p. 114.

74. Victor Clement to Simeon Reed, Wardner, July 16, 1887, July 9, 1888, February 7, 1890, Reed MSS.

75. Clement to Reed, Wardner, September 25, 1889, & May 25, 1891, ibid.

76. Fred Bradley to N. H. Harris, Wardner, February 8, 12, August 9, 1893, Bradley MSS.

77. Bradley to Harris, Wardner, December 13, 1893, ibid.

Devitt workings to prove that Augustus Heinze was removing ore from disputed areas in spite of a court injunction. Later, Sales and several others were blasting in the disputed zone (also in violation of a court order) and were met with a counterblast from the opposition, who used burning rubber to produce a thick smoke and "grenades" made of dynamite equipped with cap and six inches of fuse set in a small tomato can.[78] A few years later, in Oregon, manager Pierre Humbert, Jr., of the Cornucopia Mines was involved in a shooting "scrimmage" as part of a dispute in which his company was engaged. Company flumes were cut, but an injunction appears to have settled matters.[79]

As the custodian of company property, the manager was responsible to see that supplies were not wasted or stolen and that ore and processed metal were protected against theft. Sloppy management could be expensive: supplies purchased by the manager of an English firm in Colorado should have lasted six months; with free access and no system of accounting, they were exhausted in thirty days.[80] In its early days at Marysville, the Montana Company, Ltd., lacked a storehouse and any real means of keeping an inventory. As a result, according to an 1884 investigating committee, "Large amounts of mercury, amalgam, and retorted metal had been stolen, and there was no means of ascertaining the amounts nor the times of the thefts." [81]

In areas known for especially rich ore deposits, "high grading," or the stealing of valuable ore, was a common phenomenon in all periods. Usually the miners were the culprits, but occasionally a manager was implicated. At Cripple Creek in the 1890s, in some of the twentieth-century Nevada gold camps, and in the tungsten mines of California during the World

78. Reno H. Sales, *Underground Warfare at Butte,* pp. 31–41, 56–57.

79. John E. Searles to Pierre Humboldt, Jr., New York, August 7, 1905 and report by "#16," apparently a detective, to Asbury Harpending, Cornucopia, Oregon, August 21, 1905, Harpending MSS, Box 8.

80. *The Colorado Miner* (Georgetown), June 25, 1887.

81. Montana Company, Ltd., *Report of Messrs. N. Story-Maskelyne and J. R. Armitage to the Board of Directors,* November 12, 1884, pp. 8–9.

War I boom, "high grading" became a problem of epic proportions and added an increased watchdog dimension to the manager's functions.[82]

The mine manager was also in charge of keeping accounts and records. Perhaps he employed a clerk or bookkeeper for the mechanics of this work; perhaps not. In any event, it was his responsibility to conduct business, purchase and pay for supplies, meet payrolls, handle correspondence, and render an accounting to his corporate superiors. In the early period, mine accounts lacked standardization and often were merely "a matter of ill assorted jottings in the manager's pocket memorandum book." [83] Managers frequently devised their own "systems": Victor Clement thought his "superior to anything I have seen on the coast or elsewhere." [84] On the other hand, many companies stipulated a prescribed accounting form that was both intricate and time-consuming. Generally this included entries for mining and milling expenditures, working costs, mineral output, with a breakdown of all expenses by department. Monthly and even weekly returns were expected, giving the physical progress of the enterprise, details of ore extraction, metallic content and recovery, labor force employed, construction and equipment, and the results of underground development.[85]

Some managers personally drew up the checks for payment of their miners, and most complained of the heavy demands on their time required by paperwork. One Comstock manager is known to have worked on accounts and correspondence nearly every evening. Frank Sizer normally devoted several full days a month to these tasks while in charge of a Montana mine in

82. *Georgetown Courier,* January 25, 1883; Colorado United Mining Company, Ltd., *Annual Report,* year ending July 16, 1879; *EMJ, 34* (July 22, 1882), 42, *76* (October 17, 1903), 572–73, *82* (November 24, 1906), 988, *93* (January 13, 1912), 137, *103* (January 20, 1917), 164; *M&SP, 111* (September 25, 1915), 464; Siringo, *A Cowboy Detective,* pp. 231–32.

83. *M&SP, 96* (June 7, 1908), 754; *EMJ, 107* (February 22, 1919), 367.

84. Victor Clement to Simeon Reed, Grass Valley, June 14, 1887, Reed MSS.

85. See *Seventh Annual Report of the Gould & Curry Silver Mining Co.,* pp. 13–20; Aspen Mining and Smelting Company, G. M. R., Brown MSS.; J. H. Jefferys to George T. Coffey, London, December 11, 1900, George T. Coffey MSS.

the 1880s.[86] But Sizer, like other foward-looking managers, realized well the value of applying simple business and accounting rules in the profession. How else could financial progress be charted and mine owners assess their own situation and their administrators'?

Often the task of running a boardinghouse for the mine laborers was part of the manager's duties. In 1879, Joshua Clayton responded to company criticism of the way he handled purchases for the boardinghouse, insisting that his job was to build a mill and develop a mine, not to keep track of the price of beef, butter, or coffee.[87] Victor Clement also clashed with an official of his company in 1887 over questions of ordering boardinghouse provisions,[88] but especially in remote areas, it was recognized that an adequate boarding arrangement was necessary in order to attract and keep good men. But "a mine boarding house should never make a profit," according to experienced mining men.[89]

In the course of his work, the mine manager dealt with people on a number of different levels and in a variety of relationships. He was directly concerned when a consulting engineer was brought in to advise his company; he had a responsibility and often a personal association with corporate officers and shareholders; and he was cast in the role of a major employer and supervisor of labor. Moreover, he might also have to deal with the wives of his subordinates in the isolated communities that developed around one or a few mines. Thus, the manager had to be something more than a technical expert, an experienced miner, and a businessman. He had to have what Samuel Christy called "a certain freedom and openness of mind," [90] to be something of a practicing psychologist, skilled in human contacts. Ralph Ingersoll, a young New York engineer with

86. Louis Janin to Adolph Sutro, Virginia City, March 11, 1867, Sutro MSS, Box 6; *DAB, 18,* 146; entries for January 15 & 22, 1887, Sizer diary.

87. Joshua Clayton to unidentified, September 5, 1879, and Clayton to John Hossack, Eureka, Nevada, September 26, 1879, copies, Clayton MSS (Calif.), 2.

88. Victor Clement to Simeon Reed, Wardner, October 31, 1887, Reed MSS.

89. W. L. Fleming, "Handling Miners under the Wage System," *EMJ, 88* (August 14, 1909), 319.

90. Samuel Christy to Franklin Booth, Berkeley, June 25, 1899, Christy MSS.

experience in Mexico and the Southwest, wondered "why more mine superintendents are not chosen for foreign diplomatic posts."

> No class of men in the world can be more silent; for the value of the mine, the statistics of its operation, and its geology are locked in the superintendent's head, and no man on earth can pump them out of him. With that as a fundamental characteristic, the head of a big mine has a training in diplomacy and etiquette equal to that of the greatest masters in Europe. . . . You do not need kid gloves with which to handle people, in such [mining] circumstances; you need silk ones of the very finest quality, and equipped with these, you must work with the most delicate touch. . . . By all laws of logic, superintendents should end either at the Court of St. James or in the insane-asylum![91]

In dealing with a consulting engineer brought in to look over his shoulder, the manager indeed had to be a diplomat if there was to be no abrasion. The traditional relationship between these two was "like the typical bishop and the dean, each jealous of the other's prerogatives and keen to detect any trespass on each other's authority," although presumably in the twentieth century more tolerance prevailed.[92] Enoch Kenyon, manager of property in Boulder County, Colorado, in the early 1890s, clashed openly with George Maynard, his firm's consulting engineer, and decided to leave, adding a postscript, "Maynard is a 'Jackass.'"[93] And about the same time, although Eben Olcott thought that Fred Bradley and the consulting engineer of the Bunker Hill & Sullivan, Christopher Corning, "pull well together," it was clear that Bradley did not want Corning in his backyard. John Hays Hammond, one of the kingpins in the company, was fully aware of this senti-

91. Ingersoll, *In and Under Mexico,* p. 226.
92. *M&SP, 114* (March 17, 1917), 359.
93. Enoch Kenyon to Eben Olcott, Ward, Colorado, October 20, 1891, Boulder, November 12, 1892, Olcott MSS (N.Y.).

ment, and, believing Bradley "almost indispensible at this juncture," he threw his weight against Corning:

> Mr. Corning is a geologist and not a practical mining engineer, but has an unfortunate way of meddling, so I know, having advised you of this, you will discourage any overtures Corning might make to renew his position with the Co. as Consulting engineer. There are many Corning's but few Bradley's.[94]

One common task of the manager was to serve as host and to escort through the plant and mine, if policy permitted, public figures, corporate directors, and shareholders. In the show mines of the Comstock, this was a significant and undoubtedly a time-consuming function, but even at lesser properties it was an unavoidable one.[95] The president of the unsuccessful Ivanhoe Mining Company informed the manager in New Mexico late in 1881 that the A.T. & S.F. was considering running a branch line into the region and that he should "do all in your power" to make railroad officials comfortable "at the expense of this Co." [96]

Visitation by directors or investors often meant additional trouble and headache, as the mine put on its "holiday attire" for the occasion.[97] As one novelist wrote, the entire town "drew a deep breath and stood at attention"; engineers shined their boots and donned a clean shirt; and the mine itself "was cleaned and oiled up like a watch." [98] When the inspection was over, the manager heaved a sigh of relief. Of a grumbling, fault-finding group of New York officers at the Colorado River operation of the Hoskaninni Company, Rob-

94. Eben Olcott to James Houghteling, May 16, 1893, copy, Letterbook 20, ibid.; John Hays Hammond to D. O. Mills, London, January 11, 1897, copy, Letterbook 3, and N. H. Harris to Hammond, San Francisco, June 26, 1896, Hammond MSS, Box 3.

95. Entries for July 15, October 29, & November 23, 1887, Sizer diary.

96. E. H. Paine to Samuel H. Chittenden, Washington, D.C., November 30, 1881, copy, Ivanhoe Letterbook 1.

97. Often this same "holiday attire" was donned when the manager announced that he would visit the workings. Claude T. Rice, "Prevention of Accidents in Metal Mines," *EMJ, 87* (February 6, 1909), 298.

98. Frank Waters, *The Dust within the Rock,* p. 396.

ert B. Stanton could write in April, 1901: "The whole New York Party started home—thank God—this A.M. Their visit has been the greatest trial & the greatest *mystery* of my whole years work." [99]

Stanton was merely documenting a generalization made by a successful manager, a character in a novel written in 1903 by Frank Nason, himself an engineer: "You've got a hard row of corn. When you tackle a mine you've got to make up your mind to have everyone against you, from the cook-house flunky to the president of the company, and the company is the hardest crowd to buck against." [100] Nason knew whereof he wrote. The literature of mining is full of clashes between mine managers and corporate officials. Louis Janin wrote from the Comstock in 1867 that the mine under his charge was looking poor, with no new ore in sight. "The Trustees are on the rampage," he reported. "I am 'blowed up' by nearly every mail." [101] Nearly a decade later, James Hague wrote a friend of the disadvantages of mine management—"Superintending mines may do—but as a rule you have to belong to the trustees who are ringmasters on Cala street—& that is not pleasant." [102]

All managers, of course, were responsible to the owners of their property, whether individuals or corporations, but their degree of autonomy varied. Some were closely bound by detailed instructions, both in terms of the work at the mine and their expenditure of funds. From the property of the Ivanhoe Mining Company, for example, M. G. Gillette was required to submit to the directors a regular monthly report, as well as a special statement "every Tuesday" on a form provided by the Washington office. Operating funds had to be requisitioned, also on a special form, subject to approval by company officers. Moreover, he was instructed to ship at least ten tons of ore per

99. Entry for April 25, 1901, field notes 19, Stanton MSS.

100. Frank Lewis Nason, *The Blue Goose,* p. 134.

101. Louis Janin to Adolph Sutro, Virginia City, March 11, 1867, Sutro MSS, Box 6.

102. James D. Hague to Samuel Emmons, San Francisco, November 4, 1875, copy, Letterbook 6, Hague MSS.

week for reduction, and his successor, Samuel Chittenden, was given even more detailed orders, with little leeway, for the development of the property.[103]

Most managers, however, were given considerable latitude as to development of the mine and even expenditure of funds. The superintendent of one California dredge company was sent money "in quantities of five to ten thousand dollars at a clip, giving him practically 'carte blanch' [*sic*] on the whole game," according to one observer in 1913. Unfortunately, these funds were expended recklessly, payrolls were padded, machinery was greatly overpriced, and the manager neglected his duties, "chasing everything that wore skirts in the Trinity Center country." [104]

Such abuses represented one extreme, just as the rigorous restrictions imposed on Gillette and Chittenden represented another. Such limitations were often resented by managers as undue meddling by corporate officials. Safeguards to the company, they accepted, but they demanded freedom to direct actual mining operations as their technical skills dictated. Joshua Clayton resigned a position in Nevada in 1871 because the company refused to leave mining matters to his discretion, and he excoriated his employers a few years later on the same count, insisting that "In this part of the U.S. we can't always have things done just as they are in New York." [105] Victor Clement early clashed with the secretary of the Bunker Hill & Sullivan and took this occasion bluntly and candidly to lay down the ground rules under which he would work. Simeon Reed, the major owner, was to inform him of all new programs directly: "it will save many annoyances—I can carry out those instructions without anyone interfering—that is what I am here for." If Reed disliked his approach, he should

103. H. E. Paine to M. G. Gillette, Washington, D.C., November 2, 7, & 8, 1881, Paine to Samuel Chittenden, Washington, D.C., December 1, 1881, January 1, 1882, and Story B. Ladd to Chittenden, Washington, D.C., January 2 & 25, 1882, copies, Ivanhoe Letterbook 1.

104. O. C. Perry to O. B. Perry, January 12, 1913, copy, Morris MSS, Box 2.

105. Joshua Clayton to D. A. Jennings, Treasure City, Nevada, January 7, 1871, Clayton to unidentified, September 5, 1879, and Clayton to John D. Hossack, Eureka, Nevada, September 26, 1879, copies, Clayton MSS (Calif.), 2.

say so plainly, Clement said, for "I am standing the responsibilities if anything goes wrong."

> Under no circumstances do I wish to be placed in a situation where my actions are fettered and thus become placed in any equivocal position. . . . I have enough annoyances here without having a new source. I did not make out that order in a haphazard-sort-of-a-way—I consider myself competent, and if not you should let me know.[106]

But it took an astute and self-assured engineer to manage company officials and investors—particularly in the face of nonpaying property.

> If you can hold a board of cross directors
> In happiness against their gauzy schemes;
> If you can dodge the wrath of the electors
> Till dividends will flow as in their dreams;
> If you can make a mine pay from the grass roots
> No matter what the time or place or year;
> Then on my soul until the final blast shoots
> We'll add the title "MINING" when we
> call you "ENGINEER."[107]

At the same time, especially where managers were also shareholders, relations with corporate officials might be cordial and communications excellent. In 1893, Fred Bradley, then in charge of the Bunker Hill & Sullivan, apologized to the president of the company for the fullness of his letters:

> My letters may contain too many details—it is not my intention to force your attention to such matters, but to keep you fully informed as to what we are doing.

106. Victor Clement to Simeon Reed, Wardner, October 31, 1887, Reed MSS. For other examples of managerial difficulties with directors or major stockholders, see Gressley, ed., *Bostonians and Bullion,* pp. 167–68, 171; Wiltsee, "Reminiscences," p. 16; Robert M. Brereton to James D. Hague, London, January 9, 1886, and W. D. Bourn to Hague, San Francisco, May 27, 1890, Grass Valley, November 11, 1896, & n.p., November 14, 1896 January 17, 1897, Hague MSS, Box 11.

107. From Wayne Darlington, "The Mining Engineer," *EMJ, 98* (September 26, 1914), 587.

> I note your willingness to leave to me all matters pertaining to local management, and appreciate the opportunity that I have. I am giving the Company's affairs my entire time, and am ambitious to help make the property pay dividends.[108]

The ideal manager manifested loyalty not only to his corporate superiors, but also to the stockholders, real and potential. He would not, as some were accused of doing, either through ignorance or deliberately in order to retain his job, lead directors to believe that a mine was doing better than it actually was.[109] John Hays Hammond cites the manager of an unprofitable property in Mexico who kept the mails filled with wordy statements of great future expectations. Said Hammond, "He usually guessed wrong the first half of the month, and apologized for his mistake the second half." [110] Some believed that the law should require an accurate managerial report,[111] but respectable managers assumed that responsibility on their own and were willing to lay the cards on the table, face up.

> I'll never throw dust in a stockholder's eyes,
>   Said I to myself, said I;
> Nor hoodwink an expert who's not overwise,
>   Said I to myself, said I.
> If I'm working a mine and the ore "peters out,"
> Or its future is somewhat a matter of doubt,
> I'll tell everybody they'd better keep out,
>   Said I to myself, said I.[112]

Ideally, too, the manager kept aloof from stock manipulation involving his own mine and divulged no information to

108. Fred Bradley to N. H. Harris, Wardner, August 18, 1893, Bradley MSS.
109. Poorman Gold Mines, Ltd., *Circular to Shareholders,* June 28, 1901.
110. Hammond, *Autobiography, 2,* p. 503.
111. In 1880, a Colorado newspaper suggested that the state legislature make it a penitentiary offense for a manager to report falsely to directors or shareholders on matters of receipts, expenditures or general conditions of a mine. *Georgetown Courier,* cited in *Mining Record* (New York), *7* (April 24, 1880), 386.
112. From J. C. Bayles, "Said I to Myself," *EMJ, 35* (March 3, 1883), 113.

jobbers or speculators. One young man complained in 1890 of his inability to get inside information from the superintendent of the Bodie Consolidated, who was, he said, "such an old 'smothie' [*sic*] that even his best friend could not get a reliable opinion or 'Pointer' from him." [113]

But not all managers were such paragons of virtue. Especially in the balmy days of the Comstock, some played an active role in speculation. Well known are the stories of Comstock superintendents who, when a rich strike was anticipated, closed off that part of the mine, except for a picked crew, until the proper buying and selling of shares could take place, with substantial profit for sharp operators.[114] One early Virginia City manager is supposed to have telegraphed his directors one morning—"We have struck a ledge!—Particulars by Pony." The stock rose fantastically, and those "on the inside sold out." At night came the promised "Particulars by Pony": "As to the ledge we struck this morning, *there's nothing in it,*" at which point the stock plunged downward, and the knowing ones would buy in again. If the story is apocryphal, so must be the insistence that this dodge was played so often that it became a standing joke among the San Francisco operators.[115]

At Leadville, George Roberts, "a small-sized individual, with a big head, a quick eye, and a pleasant smile," and—it should be added—"a mining operator in the large sense of the word," was supposed to have teamed with managers Winfield Scott Keyes and George Daly to manipulate shares of their companies in 1880. Keyes was charged with running the Chrysolite into debt in order to pay unearned dividends and sustain high share prices. Both he and Daly were accused of fomenting labor trouble and precipitating a strike. The resulting fall of Chrysolite and Little Chief stock, it was believed, enabled them to sell short and make a profit. The extent of these charges was never proved, but feeling in Leadville ran high,

113. Charles Veasey to Henry M. Yerington, Quebec, May 15, 1890, Yerington MSS, Box 2.

114. *Annual Mining Review and Stock Ledger* (San Francisco, 1876), p. 2.

115. *American Mining Gazette, 2* (December 1865), 714.

and Keyes soon returned to San Francisco and Daly to New Mexico, where he was killed by Apaches.[116]

Although the mine manager had to deal with his own employer, with consulting engineers, with boards of direction, visitors, and the public in general, he was also an intermediary standing between capital and labor. As such, he had to keep in mind a dual responsibility, not only to management's interest, but in addition to the small army of muckers, trammers, smelters, and others who toiled to extract ore from the bowels of the earth and convert it into shining metal.

With respect to his labor force, the mine manager's responsibility was also a dual one. On the one hand, his object was to produce minerals as cheaply and efficiently as possible; on the other, he was charged with working the mine under his control with the least possible likelihood of loss of life or injury to personnel. This combination of safety and efficiency, essential to continued operation, could only be a cooperative matter, but the initiative for working out an acceptable balance lay in the hands of the manager.

That mining was at best a dangerous occupation is indisputable. Insurance statistics in 1911, at a time when the industry was conscious of the safety factor, indicated that mining was more hazardous than soldiering.[117] Thus, the successful mine manager learned not to skimp, but to take detailed precautions against underground accidents. Timbering should follow closely behind tunneling and should err on the side of safety. "Be sure that you get your timbers strong enough," admonished Rossiter Raymond in 1868. "If your supports happen to be *too large,* as a rule, no one will ever be able to prove such to be the case. If, on the contrary, they are *too small,* they will very soon be their own witness to the fact." [118] Regular and systematic inspection of the tunnel roof should be made, and it

116. See Carlyle Channing Davis, *Olden Times in Colorado,* pp. 249, 261; *Mining Record* (New York), *7* (May 22, 1880), 483, *8* (August 14 & September 4, 1880), 145, 219; *EMJ, 30* (August 7, 1880), 85; *Pacific Coast Annual Mining Review* (San Francisco, 1878), p. 48.

117. *EMJ, 90* (April 29, 1911), 843.

118. Quoted in *AJM, 6* (October 3, 1868), 216.

was the duty of the manager to see that minimum explosive charges were used and that a suitable magazine and thaw house be maintained to ensure maximum safety. He was to caution the utmost discretion in dealing with missed shots and to authorize periodic inspections of air compressor valves, for the explosion of gas or burning grease in the receiver or pipeline could produce noxious gases. The manager should also see that his men were equipped with the proper tools, that only safety lamps were used, and that underground workers were warned against riding on cars on loaded trips and against electrical wiring and apparatus. He was to make sure that ventilating equipment was in order, that wooden buildings and rubbish were back from the mouth of the tunnel to minimize the fire hazard, and that prudence was exercised when driving toward possible flow of water. If enlightened, he saw to it that at least one of his subordinates was instructed in the proper methods of resuscitation.[119] He was expected to enforce the ban against liquor that prevailed in most mines, for more than one accident occurred when a drunken hoist operator ran the bucket or platform up into the sheaves.[120]

Sometimes mine safety depended on cooperation by managers of connecting properties. In 1880, Charles Rolker, the new manager of the Chrysolite at Leadville, had to cope with a major underground fire. He believed he had it isolated, but it flared out anew, killing two men, after the manager of the adjoining Little Chief, in defiance of protests and a court order, kept his property open and started up a large ventilating blower.[121]

It was well known that a rich mine could make a manager's reputation and that a rich mine also made for an extravagant manager. Serious men agreed with the profane Californian credited with the classic statement "D--n a mine that won't pay with *bad* management!"[122] The Comstock Lode was the

119. David W. Brunton & John A. Davis, *Modern Tunneling,* pp. 319–21.
120. Claude T. Rice, "Prevention of Accidents," p. 298.
121. *EMJ, 30* (November 20, 1880), 330, 334.
122. Quoted in James D. Hague to Ellsworth Daggett, San Francisco, October 1, 1873, copy, Letterbook 6, Hague MSS. A later description of a good

prime example, according to some engineers, who insisted that it "did more harm than good to legitimate mining" by encouraging "the idea of the sudden acquisition of wealth without work, of finding ore without systematic search, of forming share-mongering companies on mere expectations." [123]

Thus, a Comstock manager, though successful, was not necessarily regarded as a good one. When Simeon Reed was seeking a manager for the Bunker Hill & Sullivan in 1887, he was warned against taking an old-timer from the ranks: "You don't want an old Comstock mining Supt. They would bust an iron globe wide open & make a mine of Solid gold squeal for more mud." [124] On the other hand, an inferior mine could drag down the reputation of even the best manager. "Where a poor, low-grade mine is carefully handled at the lowest possible expense, by a careful and experienced man, and still fails to pay, the unfortunate 'super' gets few thanks and small credit for his effort." [125]

The measure of a manager's success was, of course, in the efficiency of his production. He was, as one of the fraternity expressed it, "interested in the handling of men so as to make money out of them"—for the company.[126] His task was to hire good men and keep them working at maximum productiveness. For such professionals as Herbert Hoover, who estimated that labor costs made up from 60 to 70 percent of underground expenses, "the whole question of handling labor" could be reduced to one word—"efficiency." [127]

It was his ability to translate this word into practice and to handle low grade property, where every cent counted, that made the American manager so much in demand on a global scale at the end of the nineteenth century. Sometimes, he

---

copper mine was "one that will not only pay dividends, but will also withstand mismanagement and rascality and still pay dividends." *EMJ, 95* (February 1, 1913), 293.

123. Rickard, *History of American Mining,* p. 110.

124. Del Linderman to Simeon Reed, San Francisco, May 25, 1887, Reed MSS.

125. "Causes of Mine Failures," *M&SP, 56* (May 9, 1903), 295.

126. D. B. Huntley to Samuel B. Christy, New York, May 15, 1887, Christy MSS.

127. Hoover, *Principles,* pp. 161–62.

might hire cheaper labor, but the experienced manager realized the risk inherent in undercutting existing wage scales. Victor Clement wrote Simeon Reed at Portland in 1887: "If you *can* get good reliable men for the concentrator at low terms as Mr. Winch spoke of, it would be well to send them on—that is providing they will not rebel after they get here and find out the rate of the wages of the camp." [128] Fred Bradley, who left college for financial reasons to take charge of the Spanish gold mine in the 1880s, there established a record in production costs that was held up in the profession as a model of low cost operation.[129] Throughout his career, Bradley constantly sought means of reducing expenses, but he was also a pragmatist who knew when to give and when to stand firm. When manager of the Alaska Treadwell in 1921, he wrote one of the large shareholders that

> Our present wage scale, one and a half time for over time is intolerable; but it is not policy, in the face of no material change as yet in the war habit of living, to make any changes or reductions until after the local jury trials in some pending slide damage cases have been completed; besides we are in the unfortunate condition of having no working capital that would be necessary to complacently face a labor or any other disturbance.[130]

The able manager might be able to maintain wages, but reduce manpower and still increase production. Hoover cites the case of a California mine at which the new manager cut the number of miners by one third, yet, through greater coordination of effort and a more effective application, achieved the same level of output.[131]

The manager had the ultimate voice in hiring and firing.

128. Victor Clement to Simeon Reed, Wardner, October 24, 1887, Reed MSS.

129. Cost for both milling and mining was a total of 52.2 cents a ton for November 1887 and under 60.6 cents a ton for the period from September 1887 to March 1888. *EMJ, 45* (May 5, 1888), 324.

130. Fred Bradley to Bernard Baruch, Treadwell, Alaska, October 13, 1921, copy, Bradley MSS.

131. Hoover, *Principles,* p. 165.

Even if he surrounded himself with trusted assistants, he frequently kept an eye on the performance of his entire crew, down to the lowest laborer. Victor Clement discharged a number of teamsters in 1887 because they abused their mules; Frank Sizer saw to the firing of a man caught asleep on the job.[132] As manager of a silver mine in the Cobalt region when World War I broke, Robert Livermore was ordered to shut down the property and discharge many faithful employees, then classified as "enemy aliens." At the risk of his own position, he disregarded these instructions, kept the nucleus of his work force intact, and in a few weeks, when the initial confusion and panic subsided, the mine proceeded as usual.[133]

Although foremen and shift bosses had the most intimate contact with the day laborer, the manager, if he hoped to be successful, had to have a "feel" for his men and an understanding of the "art of bossing," as a Columbia professor put it in 1893, attempting to explain why so many managerial positions had gone to practical men in previous decades.[134] His job required good sense, an excellent memory, infinite tact, and the innate ability to deal with people. In later years, some thought he also needed training in sociology, psychological conditions and attitudes, and the history of trade unionism.[135]

The good manager recognized, as did one Scotsman in the Southwest, that miners "were not all Methodists," but that, handled properly, "they are as a rule good fellows and splendid workmen."[136] By the 1880s and 1890s they were also highly diverse in terms of ethnic or national background: among 329 employees of the Bunker Hill & Sullivan in 1894 were 84 Americans, 76 Irishmen, 27 Germans, 24 Italians, 23

132. Victor Clement to Simeon Reed, Wardner, October 24, 1887, Reed MSS; entry for December 27, 1887, Sizer diary.

133. Gressley, ed., *Bostonians and Bullion,* p. 175.

134. Henry Silliman(?) to Samuel Christy, New York, June 26, 1893, Christy MSS.

135. *Mining and Engineering World, 37* (September 21, 1912), 519; *EMJ, 86* (July 18, 1908), 131; *M&SP, 121* (November 6, 1920), 651–52; Sam A. Lewisohn, "Industrial Leadership and the Manager," *Atlantic Monthly, 126* (September 1920), 416.

136. Quoted in Arizona Copper Company, Ltd., *Annual Report,* year ending September 30, 1900.

Swedes, 19 Englishmen, 14 Scots, 12 Finns, 11 Austrians, 8 Norwegians, 7 Frenchmen, 5 Danes, 2 Swiss, and one each of Spaniards, Portuguese, and Icelanders.[137] Such an admixture made the manager's task more complex and meant that he had to delegate more to foremen and shift bosses who were bilingual.

The manager recognized, too, that, except perhaps for the sailor, the miner was one of the greatest wanderers in the country. In the late nineteenth century, the "ten-day miner" was a common phenomenon throughout the West.

> He worked with reasonable diligence at one job until he had accumulated a stake large enough to take him to another camp. Then he would either voluntarily "call the old hole deep enough" and quit; or he would soldier on the job until the foreman "gave him his time."

Some ten-day miners were simply gaining experience, but many were less ambitious and were motivated by restlessness and the desire for a change of scenery as they made a regular circuit —"Butte, the Coeur d'Alene, Utah, Nevada, Arizona, Colorado, and back to Butte."[138]

Efficiency demanded a stable work force, and mining firms sought to minimize turnover with improved working and living conditions, as well as recreational facilities in remote areas. Even so, the manager of the Alaska Juneau mine in the far north could expect to lose nearly half his miners when spring brought good weather.[139]

The experienced manager knew what to expect from his men. He knew that some absenteeism was inevitable; he might be unhappy, but not surprised to find his drilling crew off rabbit hunting while the drill was being repaired.[140] He accepted the fact that the average miner was no follower of Carry Na-

137. *EMJ, 57* (March 3, 1894), 193.
138. Parsons, *The Porphyry Coppers,* p. 561.
139. Fred Bradley to Bernard Baruch, San Francisco, October 9, 1923, copy, Bradley MSS.
140. Entry for March 14, 1904, field notes 20, Stanton MSS.

tion and that this ought to be kept in mind. D. B. Huntley advised in 1894 that

> a few days should be allowed each month, after pay-day, for the average mine laborer to get drunk and gamble away his money. It is his "God given right, as a free-born American citizen, to do what he pleases with his own money." He should not be hampered in this respect. I believe he is a far better workman when "broke" than before.[141]

The good manager had his own techniques for getting the best from his men. He knew their capabilities and held them to them, sometimes attempting to build an esprit de corps by stimulating the establishment of a brass band or setting up a friendly competition between various jobs or groups of employees.[142] Some also sought to educate their men to get more out of their machinery in a working day, but the perceptive manager was careful not to overdo this.[143] One "working stiff" recalled the manager of a Colorado radium mine who "had not thrown his books away nor had he learned how to swing a pick or handle a muck stick, and he didn't know what it meant to do a day's work." He was continually "looking down the necks" of his workmen, driving them beyond their capacity, and the crowning blow came when he sat on the rimrock above the mill excavation and counted every shovelful per minute. From then on, his crew would do only the minimum.[144]

Perhaps few were as crude in his methods as J. W. Pender, who had worked his way up and spoke well the miner's tongue. He "always kept a pick handle alongside of his desk and boasted that if any workman started an argument, accompanied by a threat of physical violence, he would settle the dis-

141. D. B. Huntley to editor, Mullan, Idaho, December 22, 1894, *EMJ, 59* (January 12, 1895), 28. See also the tongue-in-cheek letter urging companies to own saloons and dance halls, pay their men weekly, and supply good liquor at high rates. *EMJ, 74* (July 5, 1902), 19.

142. Kett, *Autobiography*, p. 25; Hoover, *Principles*, p. 165.

143. *M&SP, 87* (September 19, 1903), 180; Letson Balliet, "Efficiency Engineering in Mining," *Salt Lake Mining Review, 14* (January 30, 1913), 19.

144. Frank Crampton, *Deep Enough* (Denver, 1956), pp. 210–11.

pute with a crack on the head."[145] Pender must have agreed with W. L. Fleming, a New York engineer, who believed that muckers and trammers, especially immigrants, required close and vigorous supervision. "The average 'Dago' trammer will push his car over a dirty track all day rather than clean it off," insisted Fleming. "Low classes of men require to be kept in fear of the boss. Even if they do good work they must be constantly told to do more. They need to be driven. It is the only system they understand and the only one that will be successful." At the same time, Fleming warned, it would be a mistake to try to bulldoze an intelligent miner.[146]

Undoubtedly it was one of the latter who was the subject of Berton Braley's "Bill and the Supe," in which the manager strolling through the mine finds a driller taking time out, sitting on a timber car, puffing away at a cigar:

> "Do you know," says the stranger, "who I am?"
> "I don't," says William, "nor care a damn!"
>
> "Well, I am the Superintendent here!"
> Bill's grin extended from ear to ear.
>
> "The Supe," he says, "of the hull big mine?
> That's bully," he says, "that's grand, that's fine;
>
> A mighty good job fer a man to git,
> If I was you I would tend to it!"
>
> Then Bill leans back on the empty car
> An' goes on smokin' his bum cigar.[147]

Bill's morale obviously needed no boost, but that of the average miner could be strengthened by providing for his material comforts—clean, comfortable bunks and decent board-

145. Kett, *Autobiography,* p. 21.
146. W. L. Fleming, "Handling Miners under the Wage System," *EMJ, 88* (August 14, 1909), 319–20.
147. Quoted in *EMJ, 93* (January 13, 1912), 105.

inghouses in particular. As early as the 1880s, imaginative engineers had urged the establishment of reading rooms and clubhouses as a means of helping to attract and retain reliable miners in remote regions.[148] By the twentieth century, isolation, increasing labor organization, and more enlightened management had combined to make these suggestions reality, as mining companies began to place more emphasis on "human engineering." By 1901, the Blue Bird mine at Butte had established a library for its workmen, and the Gold Coin at Cripple Creek had equipped a building for its miners with gymnasium, bowling alleys, billiard tables, reading material, and a bar that dispensed beer and soft drinks at nominal fees.[149] In subsequent years, commentators could pay tribute to the work of J. Parke Channing in Arizona, Albert Burch in California, E. P. Mathewson in Montana, and Stanley Easton in Idaho as examples of enlightened managerial efforts to establish workers' homes, parks, clubhouses and roads, oftimes in cooperation with local communities.[150] But such corporate welfare programs had to be legitimate. There was no room for bad boardinghouses, exploitive company stores ("company steals"), or cheap doctors—recent medical students, alcoholics or "horse doctors"—grudgingly provided by less altruistic companies.[151]

As the appointed spokesman for capital, it was the mine manager who incurred the wrath of his workmen if wages were reduced or if the payroll failed to arrive on schedule. When Charles Bonner, in charge of the Gould & Curry, was instructed to cut wages to three dollars by his company in 1864, irate miners chased him out of his office into hiding, an event commemorated at a local theater in rousing doggerel:

148. See report of J .H. Collins, in Alturas Gold, Ltd., *Directors' Report,* February 18, 1888.

149. Louis Janin, Jr., "A Miners' Club House," *EMJ, 72* (July 20, 1901), 67.

150. *M&SP, 116* (April 6 & June 1 & 15, 1918), 471–72, 761, 813, 825–26, 827. For other examples of corporate welfare programs, see Harland, *Reminiscences,* pp. 151–52; Daniel M. Barringer to Charles M. Williams, October 31, 1901, copy, Letterbook E, Barringer MSS.

151. Mark R. Lamb, "The Gentle Art of Appreciation," *EMJ, 91* (February 11, 1911), 326.

Mr. Bonner, the son of a gun
From Virginia City he had to run.
If we'd 'a got him, before he got away
He'd never 'a seen three dollars a day![152]

The manager of a Placerville, California, mine was killed when five disgruntled miners tried to rob the mill and office, after the owners had carelessly neglected to send adequate payroll funds.[153]

In the depression of 1893, when plunging silver prices had halted operation of the New Colorado Silver Mining Company, Ltd., and had closed the doors of its Georgetown bank, the harassed manager reported that his position "had been both unpleasant and dangerous," but that "after being in a stage of siege for ten days," he had been able to pay off the most pressing creditors, including the most obstreperous miners.[154] Thomas A. Rickard lost one manager's job when that panic swept across the West, but soon took charge of two Colorado properties, the Yankee Girl and the Enterprise. But the Yankee Girl fell two months behind with its payroll, and the miners understandably made trouble. After much delay, Rickard persuaded the major owners, Pittsburgh oil men, to send funds and then resigned, but he never forgot the episode.

> In those days it was a common trick to start work at a mine, and, if no rich ore was found, to let the payroll fall into arrears, followed by the closing-down of the mine, leaving all the employees high and dry. Where the men were not unionized, they could not act as a unit in self-defence. A lien could be placed on the plant and equipment, and this sometimes enabled the men to get what was due to them; but most of them were migratory, and

152. Richard H. Stretch, "The Comstock Lode in the 'Sixties," *M&SP, 123* (November 26, 1921), 741, *8* (April 2, 1864), 216; Miriam Michelson, *The Wonderlode of Silver and Gold,* pp. 178–79.

153. Harland, *Reminiscences,* pp. 46–48.

154. New Colorado Silver Mining Company, Ltd., *Directors' Report,* November 15, 1892, to May 31, 1894. See also F. W. Risque to R. C. Kerens, St. Louis, April 21, 1892, Elkins MSS.

often went away without taking any action to recover their wages. In some States a Labor Commissioner intervened successfully. To a mining engineer, it was no joke to have a group of unpaid men on his hands, even though he also was unpaid, because they, properly, looked to him, as the manager, to safeguard them.[155]

As the middleman between capital and labor, the engineer-manager was in an anomalous position, and his views on organized labor ran the gamut from moderate to extreme. Yet most, as one miner said, "were too far up in the clouds to have truck with ordinary hard-rock stiffs," and by both deed and pronouncement seem clearly to have leaned in the direction of capital.[156] There were exceptions, of course, but many mine managers fought unionization publicly and without restraint. They were, after all, corporate representatives, and some were investors and capitalists in their own right. But at the same time, because they could not function without a laboring force, they sometimes had to temper their outlook in the interest of practicality.

Rossiter W. Raymond, for example, was throughout his long career an indomitable and outspoken foe of unionization, who between 1892 and 1894 wrote at least fifteen articles in *The Engineering and Mining Journal* that made his position clear to all. Though he denied being opposed to a "fair" union, he was particularly bitter in his denunciation of the Butte Miners' Union and thought its methods "an un-American outrage as well as a blunder."[157] When labor unrest threatened the Pacific Northwest, Raymond accused labor of being willing to "extort the last atom of plunder which they can get by violence or threats" and damned the coalition of organized labor, politicians, and reformers as "a strange alli-

155. Rickard, *Retrospect,* pp. 53–54.
156. Crampton, *Deep Enough,* p. 51.
157. *EMJ, 44* (August 13 & September 3, 1887), 110, 165. See also *EMJ, 10* (November 8, 1870), 299; *29* (February 28, 1880), 147; McLaughlin, *The Tenderfoot Comes West,* p. 79.

ance of the corrupt, the cowardly, the Christian and the crank." [158]

Eben Olcott could be critical of management's handling of the Homestead lockout in 1892, but was much more sympathetic toward the labor policies of the Bunker Hill & Sullivan Company, in which he was a stockholder. Trouble there stemmed from "your abominable Butte Union," he insisted.[159] Olcott's friend, O. B. Amsden, manager of the Gold Bug Mining Company in Arizona, thought the Pullman difficulties of 1894 an "outrage" and was convinced that "the great portion of the laboring men of the Middle and Western States need lessons in 'loyalty.' " [160] Thomas A. Rickard exhibited little patience with the Western Federation of Miners and its aims and compared the Butte Miners' Union—"a gang of Irish-Austrian-Italian anarchists"—to the dread Mafia. Later he condemned "those pariahs of the industrial world," the tramp workers from whom the IWW recruited heavily, and charged that the union itself was an instrument for German propaganda.[161] Seeley Mudd agreed with many of his contemporaries in 1922 when he criticized the necessity of capital to show more concern for organized labor, just at a time when labor "has grown so arrogant and unreasonable that the sympathy of the mass of the people is swinging away from the unions." [162]

John Hays Hammond prided himself on possession of a photograph of labor leader Samuel Gompers inscribed "To the most conservative, practical, radically democratic millionaire I ever met, John Hays Hammond," and his speeches and writings indicated some real concern for the plight of the

158. *EMJ, 52* (August 29, 1891), 236; *54* (August p. 13, 1892), 146.

159. Eben Olcott to Euphemia Olcott, Pioche Nevada, July 10, & 17, 1892, Olcott MSS (Wyo.), Box 2; Olcott to Robert G. Brown, New York, September 13, 1894, copy Letterbook 22, Olcott MSS (N.Y.).

160. O. B. Amsden to Eben Olcott, Kingman, Arizona, July 18, 1894, ibid.

161. Rickard, *Retrospect,* pp. 56, 57, 300; Rickard editorials, *EMJ, 74* (December 13, 1902), 776, *76* (November 14, 1903), 726, *77* (June 9, 1904), 913; *M&SP, 116* (June 29, 1918), 877.

162. Rickard, ed., *Interviews,* p. 395.

miner and some concrete suggestions for improving labor relations. But Hammond's autobiographical account of the Coeur d'Alene troubles are biased and distorted, and his comments at the time were hardly sympathetic to the Bunker Hill & Sullivan strikers. "If I could see my friends out of either hole I would gladly sacrifice my own holdings before letting these miserable scoundrels dictate terms to the Company," he said in 1894.[163]

But Hammond was a moderate and saw the need for concerted action with labor. So was Louis Janin, who maintained an even temper, even when striking Comstock miners were closing down property and running managers out of town in 1864,[164] and Arthur D. Foote, who was supporting the eight-hour day at Grass Valley even before the Western Federation had demanded it.[165] Albert Burch was opposed to the walking delegate, but considered himself a "strong believer in the local union" as a means of collective bargaining, handling grievances, and fostering a closer relationship between manager and employees.[166] Herbert Hoover insisted that efficiency, rather than supply and demand, controlled wages, and that by 1909 unions had come to realize that their employer's rights also benefited them.

> Given a union with leaders who can control the members, and who are disposed to approach differences in a business spirit, there are few sounder positions for the employer, for agreements honorably carried out dismiss the constant harassments of possible strikes. . . . The time when the employer could ride roughshod over his labor is disappearing with the doctrine of "*laissez faire*" on which it

163. John Hays Hammond & Jeremiah W. Jenks, *Great American Issues*, pp. 75–83, 97–115; Hammond, *Autobiography*, *1*, 188–91; *2*, 699–700; Hammond to W. H. Crocker, November 10, 1894, copy, Letterbook 1, Hammond MSS.

164. Louis Janin to Juliet C. Janin, Ormsby County, Nevada, August 2, 1864, Janin family MSS, Box 10.

165. Mary H. Foote to Helena Gilder, Grass Valley, December 21, 1906, M. H. Foote MSS.

166. Rickard, ed., *Interviews*, p. 108.

was founded. The sooner the fact is recognized, the better for the employer.[167]

Like Hoover, J. Parke Channing saw the role of the engineer as an arbiter working to compromise the positions of capital and labor—to convince management of the long-range benefits of reduced hours, higher wages, and improved working and living conditions, and to convince labor to step up its output and efficiency. Channing recalled his introduction of light one-man drills to replace heavier two-man equipment in Arizona and the immediate opposition from drillers that developed. But once he explained that the ore body was not worth exploiting unless production costs could be cut, the crews came to accept the change.[168]

Perhaps by the 1920s, young salaried managers were more receptive to unionization than the old-timers, but any acceptance was cautious and qualified and born out of necessity. Their basic outlook was reflected in a resolution adopted by the Colorado Section of the AIME in 1920 stating that "everyone must have the right to join a union or not, and the use of force, violence, blacklist, boycott and lockout, as a means of influencing such individual action, is to be condemned as vicious and against public welfare." [169]

Most mine managers were called on to deal directly with strikes and union organization at some time in their careers. Some did so gracefully, others awkwardly; some met the issues on their own, but most combined with their fellow managers and operators to present a united front. Winfield S. Keyes, George Daly, and Arthur D. Foote were among a dozen or so managers who met at the Clarenden Hotel in Leadville in 1880 to unify their opposition to a general miners' strike.[170]

167. Hoover, *Principles,* pp. 167–68.

168. J. Parke Channing, "Man-Power," *M&SP, 116* (June 1, 1918), 760–61.

169. Quoted in *M&M, 1* (July 1920), 24.

170. Vernon Jensen, *Heritage of Conflict,* pp. 22, 24; Don L. & Jean H. Griswold, *The Carbonate Camp Called Leadville,* pp. 183–84; Davis, *Olden Times in Colorado,* p. 249; Frank L. Sizer, "Reminiscences of Leadville," *M&M, 5* (January 1924), 8.

Victor Clement and Fred Bradley took the lead in organizing the Mine Owners' Protective Association in the Coeur d'Alene, partly to combat union workers, and in nearly every period and region beset by labor troubles concerted action groups were formed by owners and managers.[171]

In the north Idaho upheavals of the early 1890s, the Mine Owners' Association sought to import scab labor and at times to starve workers into submission by closing down completely. In February 1892, Clement reported that all mines were closed and that "peace reigns serenely," but in May, John Hays Hammond noted that the strike was still in effect and growing more serious. "It will be necessary for us to 'down the union,' in order to prevent future troubles from that direction," he said.[172] A month later, Hammond could report that "we have 'done up' the miners' union," and work was being resumed with nonunion men. By mid-September, Hammond thought the matter was settled: "All mines are working non-Union Miners—A Law & Order League composed of 400 of the best men (not Miners)—merchants etc. is having a good moral influence in our district." [173]

But in the early part of 1893, Fred Bradley, now managing the Bunker Hill & Sullivan, urged the Mine Owners' Association to close down all properties. Wage demands were high and silver-lead prices low, with stockholders urging a curtailment of operations until the market rose—"Better the hope of a living dog than of a dead lion," wrote one.[174] The Bunker Hill & Sullivan did close, and Bradley believed the move effective: "It may be that if we remain closed on until next spring, that the community will be glad enough to see us start up

171. See Gressley, ed., *Bostonians and Bullion,* p. 101; Fred L. Morris to C. W. Evans, San Francisco, September 22, 1916, copy, Morris MSS., Box 1; *EMJ, 85* (January 11, 1908), 125; *M&M, 4* (August 1923), 404.

172. Victor Clement to Eben Olcott, Wardner, February 5, 1892, and John Hays Hammond to Eben Olcott, San Francisco, May 13, 1892, Olcott MSS (N.Y.).

173. Hammond to Olcott, San Francisco, September 12, 1892, ibid.

174. Fred Bradley to John Hays Hammond, Wardner, January 22 & 29, 1893, Bradley MSS; Eben Olcott to James Houghteling, New York, January 30, 1893, copy, Letterbook 19, Olcott MSS (N.Y.).

Boarding house and all. They might be willing to have us manage our own business and make all we could out of it." But Bradley lamented that two major mines, the Gem and the Frisco were still operating and would probably pay the $3.50 per day wages asked by the union, if pressures were increased.[175]

In August, Bradley, as President of the Mine Owners' Association, informed the Union Pacific and Northern Pacific railroads that the MOA had agreed to stand on a $2.50 wage scale and that, if freight rates were reduced, shipment of ore would be resumed.[176] The Bunker Hill & Sullivan began to hire men and to produce, but Bradley soon found that the lower wage scale could not be maintained.

> I am sorry to say, that I was forced by the dangerous aspect of affairs here, to the hasty conclusion that we can do no better for the present than to pay these wages [$3.50]. . . . I carried the matter almost to the point of dynamite, and saw that we would be unable to procure miners without importing them.[177]

Bradley believed, however, that this was a temporary solution. "If lead does not go up, and the men do not take a reduction, I will discharge half our crew." [178]

In mid-October, Bradley resigned as president of the MOA and served formal notice that the Bunker Hill & Sullivan was withdrawing to "go it alone." Soon violence erupted and the company's power house was dynamited, but Bradley insisted that press accounts describing the coming of federal soldiers greatly exaggerated the clash. "Nevertheless a reign of terror exists here, and I hope the U.S. troops now here will stay." [179]

Matters came to a head again in 1899, when the Bunker Hill & Sullivan was the only nonunion concern in the Coeur

175. Bradley to Clement, Wardner, May 6 & 14, 1893, Bradley MSS.
176. Bradley to N. H. Harris, Wardner, August 2, 1893, ibid.
177. Bradley to Harris, Wardner, September 27, 1893, ibid.
178. Bradley to Harris, Wardner, October 11, 1893, ibid.
179. Bradley to John A. Finch, October 17, 1893, copy, ibid.; Bradley to Eben Olcott, Kellogg, August 1, 1894, Olcott MSS (N.Y.).

d'Alene. Superintendent Albert Burch agreed to pay essentially the union scale, but refused to recognize union control, and discharged union miners when he announced the new scale. The mill was blown up, Idaho Governor Frank Steunenburg sent in troops, and in the end, amidst great bitterness, the Coeur d'Alene unions were for all practical purposes destroyed for many years.[180]

In some instances, individual mine managers or combinations of them were successful in meeting labor demands without violence, either through compromise or persuasion, although an occasional manager found himself hauled into court on charges of maintaining a blacklist.[181] Frank Bulkley managed to compromise a threatened strike at the Aspen Mining and Smelting Company in 1893, and Daniel Jackling, an inveterate foe of organized labor who at times even refused to talk with union representatives, went halfway a few years later by meeting union wage demands, but declining to recognize the local as a bargaining agency.[182] In the turbulent 1890s, Thomas A. Rickard, manager of the Enterprise at Rico, introduced the contract system, a time-honored practice borrowed from Cornwall. Under it, a miner's pay was based on the number of feet of advance in a level and the number of square fathoms of excavation in a stope.[183] When the local miners' union protested, Rickard was able to convince its members to give the system a fair trial and so averted the industrial turmoil then so prevalent in other Colorado mining centers. But to avoid trouble, Rickard joined the anti-Jewish, antiforeign American Protective Association, where the first person he

180. Jensen, *Heritage of Conflict*, pp. 74–75, 81–82; Rickard, ed., *Interviews*, pp. 99–100; "Coeur d'Alene Mining Troubles," *Senate Executive Document*, no. 142, 56th Cong., 1st Sess. (1899–1900), 42–53.

181. *M&SP, 70* (April 13, 1895), 227.

182. Frank Bulkley to J. B. Wheeler, October 9, 1893, copy, Brown MSS; Jensen, *Heritage of Conflict*, pp. 112, 269; *EMJ, 96* (September 13, 1913), 512.

183. Development work, stoping, and trucking were sometimes done on a contract basis. Pay could be based on the footage of holes drilled or heading advanced or on tonnage, cubic space, or square area broken. When coupled with a bonus system, it was deemed efficient. Hoover, *Principles*, pp. 165, 167.

saw when the initiation blindfold was removed from his eyes was the president of the miners' union![184]

Managers and owners were sometimes able to block union organization, as Daniel M. Barringer was successful in doing at an Arizona property in which he was interested in 1901. "Everything was moving along very smoothly," said Barringer, "when, like a bolt out of a clear sky, we found out that 81 or 82 out of 84 of our miners had joined themselves in a union with the ultimate purpose of dictating terms to us." Barringer and Brockman, his manager, made it clear that a union would not be tolerated, that it was "a menace to our security and the economical operation of the plant," and gave the miners a week in which to comply, then closed the mine, soon to reopen with nonunion labor. "I would rather see the mine closed down for a year than submit to the dictation of any miners' union as to hours, wages, etc.," said Barringer.[185]

Often the solution was not this simple, and countless managers experienced threatened or actual violence during the course of labor disturbances. E. B. Gage, of the Grand Central mine in Arizona, was saved from strikers who vowed to hang him when Nellie Cushman snatched him from danger and spirited him away in a buggy.[186] At the Cumberland Mining Company in Montana in 1891, Michael Cooney held off at gunpoint marchers striking for higher pay, although his foreman was roughly handled before fifty of the demonstrators were arrested.[187] At Ely, Nevada, John A. Traylor shot and killed three of a dozen union men who were determined to "run him out of town," and was acquitted on grounds of self-defense. "Mr. Trayler is evidently a mine manager who is not to be intimidated," commented one editor.[188] A few years later, according to his account, Hubert Eaton, then manager

184. Rickard, *Retrospect,* pp. 56–57.

185. Daniel M. Barringer to Richard Penrose, November 4, 1901, copy, Letterbook E. Barringer MSS.

186. John Clum, "Nellie Cushman," in Mary G. Boyer, ed., *Arizona in Literature* (Glendale, 1934), p. 375.

187. *EMJ, 51* (March 13, 1891), 334.

188. Schmidt, "Early Days," p. 13; *EMJ, 75* (January 17, 1903), 109.

of a mining company near Rawhide and later in charge of the swank Forest Lawn Cemetery in Los Angeles, was saved by Divine intervention from IWW dynamiters. "God took a hand" in the form of a red light girl singing "There's no Place Like Home" to a drunk—a song that so moved the Wobblies that they silently withdrew without accomplishing their mission.[189]

By their uncompromising stand against organized labor, some mine managers became marked men. A miners' strike protesting the contract wage system introduced by Arthur Collins at the Smuggler Union near Telluride erupted in 1901 into a bloody riot, in which several were killed or wounded, including George Nicholson, a Colorado School of Mines graduate who was shot through the head, and Charles Becker, superintendent at the Smuggler, whose arm was permanently shattered by a slug from a Krag-Jorgensen. Collins stood firm: he hired scab labor and openly defied the Western Federation. As president of the Colorado Mine Owners and Managers Association he was regarded by some as a champion of "the rights of property and good citizenship," and by others as "a man whose hate of unionism seemed innate" and who believed in the right of the corporation to run roughshod over its employees. On the night of November 19, 1902, while sitting with friends in the library of his home, Collins was fatally shot by an assassin outside the window, but no killer was ever convicted.[190]

Collins' successor, Bulkeley Wells, was even less friendly to

189. Adela Rogers St. Johns, *First Step Up toward Heaven*, pp. 73–74. For other episodes, see San Francisco *Bulletin*, June 5, 1878 & August 27, 1879; San Francisco *Chronicle*, November 4, 1879; *EMJ*, *100* (October 9, 1915), 607.

190. See *EMJ*, *71* (July 13, 1901), 47, *74* (November 29, 1902), 705, 721; Benjamin B. Lawrence, "Biographical Notice of Arthur L. Collins," AIME *Trans.*, *34* (1904), 835–38; Thomas A. Rickard, *Across the San Juan Mountains*, p. 42; Guy E. Miller, "The Telluride Strike," in Emma F. Langdon, *The Cripple Creek Strike, 1903–1904*, pp. 207–08; Mine Owners and Operators Association, *Criminal Record of the Western Federation of Miners from Coeur d'Alene to Cripple Creek, 1894–1904*, p. 12. At one point, Steve Adams confessed the shooting, but later repudiated his confession. James H. Hawley, "Steve Adams' Confession and the States' Case against Bill Haywood," *Idaho Yesterdays*, *7* (Winter 1963–64), 23.

the cause of labor. Characterized by a novelist as a "rival of Raisuli,"[191] Wells was a man of great personal courage and charm and was the center of mine operators' resistance to the Western Federation in the 1903–04 disturbances. When Telluride miners and mill men struck in September 1903 for an eight-hour day and abolition of the contract labor system, Wells closed down the Smuggler Union, and when the governor of Colorado sent in the militia, he was one of those in the forefront in deporting "troublemakers" and in importing scab labor. When, after several months of martial law, troops were withdrawn temporarily, Wells took the lead in forming a "Citizens' Alliance" to keep order. Young Robert Livermore, a fledgling engineer, served as a member of this and other "vigilante" groups (and later managed to shoot himself in the foot).[192] During the controversy, another manager, John Herron, literally sick with worry, left for California for his health.[193] Wells was primarily responsible for introducing the card system at the Smuggler Union as a device to blacklist known WFM men, and he gained additional renown for his role in helping arrest and whisk away to Idaho William Haywood and two other labor leaders in the Frank Steunenberg assassination case. Wells himself narrowly escaped assassination in March 1908, when a bomb exploded under his bed and blew him up to the ceiling; but amazingly he suffered only minor injuries.[194]

Fred Bradley may also have been earmarked for destruction by the Western Federation. On a November morning in 1904, he started out the door of his San Francisco apartment, lighting his cigar, and was blown out onto the sidewalk by a tremendous explosion. Bradley recovered and believed that gas leakage had caused the blast: indeed, the owner of the flat sued

191. Walter Hurt, *The Scarlet Shadow: A Story of the Great Colorado Conspiracy,* p. 362.

192. Gressley, ed., *Bostonians and Bullion,* pp. 101, 108, 152.

193. *EMJ, 76* (December 31, 1903), 1020.

194. Langdon, *Cripple Creek Strike,* pp. 211, 213; *EMJ, 76* (September 26, 1903), 455, *77* (March 17, 1904), 451, *84* (September 14, 1907), 506, *85* (April 4, 1908), 728; Gressley, ed., *Bostonians and Bullion,* p. 149.

and obtained damages from the gas company. Subsequently, however, Harry Orchard, the confessed assassin of Governor Steunenberg, asserted that he had been sent by the Western Federation to kill Bradley and that, failing to do so by poisoning his milk, he had attached a bomb to his front door.[195]

But organized labor was not the only hazard faced by the mine manager. By its very nature, mining was a dangerous business, and the manager, like the miner and the engineering consultant, was constantly exposed to a variety of perils. Violence, of course, was part of the early western tradition, especially when new, remote districts were being opened up. When Raphael Pumpelly came to Arizona in 1860 to reopen the Santa Rita mines, he first had to contend with a murderous blacksmith—an ex-convict from Australia; next he was ambushed by Apaches, who stole his horse; then he was one of the party who found the remains of a German engineer, Frederick Brunckow, his body pierced with a rock drill and flung down a mine shaft by Mexicans who fled with the run of silver bullion. After eighteen months in the heart of Apache territory, Pumpelly left the Southwest, marveling that he was "the only one of at least five successive managers of the Santa Rita who was not killed by Mexicans or Indians." [196]

But Indian danger in established mining communities was minimal after the very early years. Because mine managers had charge of the payroll and of bullion shipments, however, they ran a very real risk of being robbed or murdered in the process. Cyrus Gribble, manager of the Vulture mine and a man described as having "an immense stock of egotism and bullheadedness," was robbed and killed while taking $7,000 in

195. Describing the explosion, then attributed to gas, a local newspaper commented, "Had a bombshell exploded in his rooms the wreck could not have been more complete." *The San Francisco Call,* November 18, 1904. See also Fred Bradley to James H. Hawley, San Francisco, February 1, 1906, James H. Hawley MSS, Idaho Historical Society, Boise, and Harry Orchard *The Confessions and Autobiography of Harry Orchard* (New York, 1907), pp. 158–59, 160–65.

196. Pumpelly, *My Reminiscences, 1,* 200, 202–05, 218, 226. See also San Francisco *Post,* May 26, 1877, clipping in Bancroft Scrapbooks, Arizona Misc., Set W, *82,* no. 2, 387; Richard J. Hinton, *The Hand-Book to Arizona: Its Resources, History, Towns, Mines, Ruins and Scenery,* pp. 187, 190.

bullion into Phoenix in 1888, and a number of others met the same unfortunate fate.[197]

Most occupational hazards were less romantic, however, and were those normally associated with mining. Falls into shafts, cave-ins of tunnels, toxic gas, underground fires, and blasting or hoisting accidents all took their toll of managers, as of others connected with the industry.[198] In addition, lead poisoning or inhalation of mercury or other fumes common in mill work could be fatal, and, occasionally, a man died from drinking from a glass that had previously contained potassium cyanide.[199] More than the consulting engineer, the manager was exposed to the perils of working machinery, a not uncommon cause of accidental death or injury.[200]

Because mines were often located in rugged country, the resident manager's work exposed him to rigorous conditions for extended periods of time, whereas the consultant flitted in and out. Some mine managers remained at one location for as long as twenty or thirty years,[201] but this was exceptional, for by

197. Joseph Miller, ed., *Arizona Cavalcade,* p. 79, 81–83; *M&SP, 55* (March 24, 31, 1888), 185, 201; *EMJ, 47* (May 18, 1889), 461, *56* (September 30, 1893), 350.

198. For examples of blasting accidents, involving managers, see F. A. Benjamin to Adolph Sutro, Sutro, Nevada, May 25, 1874, Sutro MSS, Box 14; *Helena Herald,* March 14, 1888 & March 11, 1889; *EMJ, 61* (May 30, 1896), 523, *99* (February 12 & 19, 1910), 384, 432; *M&SP, 111* (July 10, 1915), 67. For falls and cave-ins, see *Helena Herald,* March 13, 1874; *The Amador Record* (special mining edition), April 1897, p. 14; *EMJ, 67* (May 27, 1899), 626, *79* (April 20, 1905), 787, *95* (June 7, 1913), 1167. For hoisting accidents and mine fires, see Charles Hoffmann to Josiah Whitney, Virginia City, February 5, 1877, Hoffmann MSS; *M&SP, 77* (November 26, 1898), 531, *95* (August 24, 1907), 224, *123* (November 26, 1921), 741–42; *EMJ, 64* (November 27, 1897), 642, *76* (November 14, 1903), 727, *97* (February 6, 1909), 298, *91* (June 10, 1911), 1173; *M&M, 1* (February 120), 28, *4* (April & October 1923), 174–75, 532.

199. *Helena Herald,* May 13, 1889; *EMJ, 49* (June 7, 1890), 645, *91* (May 13, 1911), 976; *AJM, 7* (June 12, 1869), 374; *M&SP, 70* (March 9, 1895), 151; H. J. Jory to Samuel Christy, Weaverville, California, April 20, 1891, Christy MSS.

200. *M&SP, 25* (October 19, 1892), 248, *114* (April 28, 1917), 598; *EMJ, 98* (July 25, 1914), 182; *M&M, 3* (April 1922), 40.

201. Arthur D. Foote spent thirty years at the North Star; Stanley Easton spent an even longer time at the Bunker Hill & Sullivan. Arthur B. Foote, "Arthur DeWint Foote," *Transactions of the American Society of Civil Engineers, 99* (1934), 1449; *Who's Who in America, 1962–63, 32,* 884.

and large, even the managerial side of the profession was characterized by its noticeable mobility. Mining itself was often uncertain, and supervisors turned over at a rapid rate; moreover, they came to recognize and accept the transitory nature of their work and their own indefinite tenure. Often, as when Victor Clement went to the Bunker Hill & Sullivan, no specific time limit was put on the engagement; his successor, Fred Bradley, was first appointed on a one-year basis, a not unusual arrangement.[202]

Sometimes, however, a manager was hired for a specific short-term purpose, perhaps to put up reduction works, open a mine and put it into producing shape, and then turn it over to a permanent supervisor.[203] In 1880, Arthur D. Foote was hired as an interim manager at Leadville to direct what was essentially a crash program of one company to beat another to pay dirt. Under the prevailing law, a valid claim was based on actual discovery. Thus, where claims conflicted, the party first reaching mineral gained possession. According to his wife, Foote "thoroughly enjoyed the excitement of the race, increased by all kinds of difficulties—water in the shaft, pumps freezing and bursting, heavy storms & c."[204]

Self-interest was basic, and better jobs with better pay helped make the profession a fluid one. Or sometimes personal disputes prompted managers to leave. Arthur Foote resigned from the New Almaden in 1877 because of differences with his superior—"a mean, corrupt, petty tyrant," in the eyes of Mary Foote.[205] Eben Olcott left a post in Colorado in 1881 after "falling under the bane" of the mine owner "through a most ridiculous & trivial cause without any grounds."[206] Albert Burch was discharged from the Bullion Beck & Champion

202. Victor Clement to Simeon Reed, Wardner, October 23, 1887, Reed MSS.

203. Joshua Clayton to Henry Hughs, Treasure City, Nevada, May 13, 1871, copy, Clayton MSS (Calif.), 2; Charles J. Moore to Eben Olcott, Leadville, March 11, 1894, Olcott MSS (N.Y.); Edgar Rickard to Samuel Christy, Cornucopia, Ore., May 28, 1897, Christy MSS.

204. Mary H. Foote to Helena Gilder, January 20, 1880, M. H. Foote MSS.

205. Mary H. Foote to Richard Gilder, Almaden, August 12, 1877, ibid.

206. Eben Olcott to Lewis Crooke, Lake City, July 25, 1881, copy, Letterbook 3, Olcott MSS (N.Y.).

mine in Utah in 1893, when the property changed hands and a bitter fight for control developed.[207]

Reasons of health were also common grounds for leaving, especially from high altitudes. More than one manager felt compelled to surrender a post on the Nevada desert or in the San Juan Mountains of Colorado to recuperate in California or some other moderate climate.[208] M. G. Gillette left the managership of a New Mexico property in 1881 because of his health, which was undoubtedly worsened by an Indian uprising in the Black Range.[209] Others moved not because of their own health, but because of that of some member of their immediate family.[210]

With the uncertainty of mining, not infrequently a manager found himself unemployed when the health of the property under his care failed. "I never open a letter of Arthur's without expecting to read that the Company have closed down; and he is out of a job," wrote Mary Foote in 1893.[211] J. S. Phillips came to Humboldt, Nevada, in 1867, to manage property for a New York firm, promptly condemned the mine and put himself out of work.[212] Both Eben Olcott and Thomas A. Rickard lost managerial positions in the 1880s when their respective production dwindled and the properties closed.[213]

Others were left jobless in times of financial distress, when sources of capital dried up,[214] or when some special circumstance brought mining operation to a halt. Failure to get adequate winter supplies into the Ouray country before the passes closed forced David W. Brunton to abandon a property in

207. Rickard, ed., *Interviews,* pp. 97–98.

208. Joshua Clayton to James Clough, Treasure City, Nevada, January 21, 1878, copy, Clayton MSS (Calif.), 2; *EMJ, 33* (April 22, 1882), 206, *51* (May 9, 1891), 561, *52* (August 29, 1891), 239.

209. H. E. Paine to Samuel H. Chittenden, Washington, D.C., November 30, 1881, copy, Ivanhoe Letterbook 1.

210. *DAB, 3,* 511; Rickard, ed., *Interviews,* pp. 97–98, 397.

211. Mary H. Foote to Helena Gilder, 1893, M. H. Foote MSS.

212. *M&SP, 14* (April 3, 1867), 235.

213. Eben Olcott to Euphemia Olcott, Las Delicias, Mexico, June 7, 1885, Olcott MSS (N.Y.); Rickard, *Retrospect,* pp. 36–37, 39.

214. Eben Olcott to Charles Rolker, May 17, 1893, copy, Letterbook 20, and Edward B. Durham to Olcott, Sawyers Bar, California, December 26, 1895, Olcott MSS (N.Y.).

1875.[215] But whatever the cause, the closing down of a mine often worked real hardship on the ousted manager. Mary Foote mentions one young engineer who was forced to leave New Almaden, "selling out at a sacrifice all their pretty new home gear—including china." [216] Another was especially bitter because he had left a decent position in Minnesota at considerable personal expense to accept a manager's post in Montana, only to be out of work two months later, when the company exhausted its funds.[217]

Sometimes, a man was specifically groomed to take over as manager. William Kett went to California in 1905 as an understudy to Lewis T. Wright at the Mountain Copper Company, but it was six years before he superseded him.[218] Once in a while a new manager found himself in an awkward position vis à vis his predecessor. He might have to clean up the other's mistakes, and more than one vowed to "investigate with a microscope" before making future commitments.[219] At Kerr Lake in 1912, Robert Livermore arrived to take charge, but found that Steward, the old manager, had been fired but refused to leave. "As a consequence," said Livermore, "I found myself a pseudo-guest in the house along with the Stewards who had no apparent intention of leaving, and their four children." [220]

Often a man undertook to manage several properties at once. Eben Olcott, for example, was in charge of two mines in Colorado when he was offered the managership of the Potosi Mining Company in Venezuela in 1881. At the time he was accused of manipulating stock, Winfield S. Keyes was the man-

215. Rickard, ed., *Interviews*, pp. 69–70.

216. Mary H. Foote to Helena Gilder, January 16, 1877, M. H. Foote MSS.

217. "B.B." to editor, Helena, February 24, 1911, *EMJ, 91* (March 25, 1911), 602.

218. Wright was able, but Kett found him "egotistical and dictatorial" and uninterested in teaching his subordinate. In turn, Kett's son was groomed for and eventually took over his father's post. Kett, *Autobiography*, pp. 115, 117, 160, 175.

219. Charles Munro to Fred L. Morris, Nikolaievst, July 11, 1911, Morris MSS, Box 2.

220. Apparently Steward had hoped to be reinstated, but in this he was unsuccessful. Gressley, ed., *Bostonians and Bullion*, p. 166.

ager of three important Leadville properties; and in 1899, if his letterhead is any indication, Seeley W. Mudd was managing the property of four companies, all in the same district.[221]

But not all companies or their shareholders were enchanted with the idea of sharing the services of a manager. When George Daly assumed charge of a second concern in 1878, he had to defend his action to the stockholders of the first, the Real Del Monte Mining Company. He needed the money, he explained. Not only did he have an ailing wife and four young mouths to feed, he had been forced to borrow $10,500 "to make up my losses on that unfortunate Alta deal of last winter." But Daly assured his critics that he was not neglecting the well-being of the Real Del Monte Company.

> I work early and late for the interests of the Company, and manage its affairs much closer than I would my own: in all attempts to injure or assail the Co's property, I have, without any regard for my personal popularity, or even safety, fearlessly stood up for our rights, and maintained them at all hazards. . . . That some stockholders will growl is *natural,* for some would growl if I worked for them 20 hours a day, for my board.[222]

A typical arrangement left the manager free to do additional consulting or supervising work so long as it did not interfere with his normal duties. An ambitious young engineer, such as Thomas A. Rickard, therefore, might in 1894 be in charge of three separate mines in as many Colorado towns and still maintain a consulting office in the McPhee Building in Denver.[223] Some mine owners were more restrictive. In 1905, when George T. Coffey was hired to manage the Yukon property of the Anglo-Klondyke Mining Company, Ltd., his con-

221. Eben Olcott to Somes & Company of London, Lake City, March 1, 1881, copy, Letterbook 3, Olcott MSS (N.Y.); *EMJ, 29* (March 20, 1880), 205; *Mining Record* (New York), *8* (September 4, 1880), 219; Seeley W. Mudd to S. B. Elkins, Leadville, May 26, 1899, Elkins MSS.

222. George Daly to Henry M. Yerington, Aurora, Nevada, September 13, 1878, Yerington MSS, Box 1.

223. *EMJ, 57* (May 12, 1894), 444.

tract stipulated that this was a full-time occupation and that he would do no consulting work for others unless given written or cabled consent by his directors—in which case the company was to receive half his fees.[224]

The basic salary of the mine manager varied greatly, of course, depending on the period, the size and type of property involved, and the experience and reputation of the manager. Quartz mining generally paid better than placer, but other generalizations are difficult to make, except that the range was wide.[225]

Bonanza managers on the Comstock in 1864 were rumored to have received as high as $40,000 a year, but this was an extreme.[226] More typical was the $8,000 paid to Louis Janin by the Gould & Curry in 1865.[227] In 1873, one of the Comstock superintendents received $416 a month, and, in 1881, Eben Olcott solicited a mill manager to go to Colorado at $3,000 a year.[228] When Thomas A. Rickard went to manage the Union mine in California six years later, he received $300 a month, plus a house, horse, and a Chinese servant; at the same time, Robert B. Stanton was in charge of an Idaho property at an annual salary of $4,200.[229]

The job Arthur D. Foote took managing an onyx mine in Baja California in 1893 paid a salary of from $250 to $300 a month, and the owner apologized for its being so low.[230] But Foote seems to have had no touch for money and spent most of his life working for less income than he and his wife believed he should be making. At one point, he told his employer, James D. Hague, that he thought he had been "on half-pay long enough even for a brother-in-law." In the late 1890s, while in charge of three mines and a payroll of more

224. Memorandum of agreement, Anglo-Klondyke Mining Company, Ltd. and George Thomas Coffey, January 16, 1905, Coffey MSS.

225. *EMJ, 65* (May 14, 1898), 586.

226. Entry for November 4, 1864, Brewer, *Up and Down California* p. 559.

227. Louis Janin to Juliet C. Janin, Virginia City, June 27, 1865, Janin family MSS, Box 10.

228. Abstract of payroll, March 1873, Sutro MSS, Box 21; Eben Olcott to J. B. Austin, May 9, 1881, copy, Letterbook 3, Olcott MSS (N.Y.).

229. Rickard, *Retrospect,* p. 37; undated field book 30, p. 102, Stanton MSS.

230. Samuel F. Emmons to James D. Hague, San Diego, March 30, 1893, Hague MSS, Box 11.

than a hundred men, he told a San Francisco mining man what his salary was and received the shocked response, "Jesus Christ!" [231]

A survey of eleven graduates of the Lawrence Scientific School who were mine managers in 1907 showed their average salary as $2,387, but these were young, relatively inexperienced men whose earning power had not yet advanced very far.[232] Name managers, such as Fred Bradley, Hamilton Smith, and Thomas Leggett, commanded much higher figures. In 1896, when John Hays Hammond was concerned about keeping Bradley at the Bunker Hill & Sullivan, he was willing to guarantee him $10,000 a year, with his attention to the mine to be half time.[233]

Managers who served more than one company were sometimes handsomely rewarded. When William Patton was in charge of five mines during the sunshine days of the Comstock, he received $1,000 a month from each, for a total of $60,000 a year—reportedly the highest managerial salary in the western mineral industry prior to 1898.[234] But this, too, was atypical. Nicholas Maxwell received $10,000 in 1873, split among three Utah companies; and in 1890, when Comstock salaries were no longer what they once were, R. P. Keating received $950 per month, contributed by three companies, and A. C. Hamilton was paid $800 a month, split among four concerns.[235]

Occasionally, a popular manager received an additional bonus, either in cash or in stock; sometimes he might receive a percentage of the mine's net profits over and above his salary. In at least one instance, a manager volunteered to take half salary until his company began to pay dividends.[236] If a firm

231. Mary H. Foote to Helena Gilder, Grass Valley, March 6, 1898, M. H. Foote MSS.

232. *EMJ, 83* (June 22, 1907), 1206.

233. John Hays Hammond to James L. Houghteling, December 17, 1896, copy, Letterbook 3, Hammond MSS.

234. *EMJ, 65* (May 14, 1898), 586.

235. London *Mining World, 3* (February 8, 1873), 283; *M&SP, 60* (January 18, 1890), 44.

236. *Helena Herald,* December 27, 1879; *EMJ, 53* (April 16, 1892), 432; London *Mining World, 3* (August 2, 1873), 278; London *Mining Journal, 41* (January 13, 1872), 27.

went defunct, he might have difficulty in collecting, for legally a manager was not entitled to a lien for his labor.[237] Except for a $300 advance, Raphael Pumpelly was unable to collect his $1,500 a year pay at the Santa Rita mines, so when he left Arizona in 1861, he took with him a twenty-eight-ounce bar of the last silver poured there.[238] Robert B. Stanton was unable to salvage even that much when the Colorado River placers he was managing went under in 1901.[239]

All in all, mine management paid reasonably well, but better for engineers of experience and reputation who handled large, complex operations, and better in boom times than, for example, during the 1864–65 mining bust, the aftermath of the general depression of 1873, the lean years of the early 1890s, when world silver markets were glutted, or the 1911–17 era, when turmoil in Mexico and World War I in Europe sent countless engineers to swell the supply in the West.

237. Boyle *v.* Mountain Key Mining Company, *Pacific Reporter, 50* (1897), 347.

238. Pumpelly, *My Reminiscences, 1,* 266.

239. Undated field book 30, p. 142, Stanton MSS.

CHAPTER 6

"We Must Overwhelm Them with Testimony"

The lawyers they explained the law,
The experts made the facts,
The judges they said "hum!" and "haw!"
And quoted various acts.

—Rossiter W. Raymond [1]

Heretofore, neither the trained geologist and engineer nor the application of mineral law has been given close scrutiny by historians of the West, and their combination—what one writer has called "legal geology" [2]—has been ignored completely. Yet "courtroom mining" has always been an important western industry, and increasingly, throughout the late nineteenth and early twentieth centuries, the technically educated man played a focal role. Just as the mineral industry brought forth an important school of highly influential lawyers tested in litigation before the bar, so it produced a hard core of technical experts experienced and well versed in legal battle.

The majority of mining engineers connected with the West sooner or later found some knowledge of mining law essential. Indeed, by the turn of the century, mining schools were beginning to include courses in mineral law in their technical curricula,[3] and veterans were insisting that no engineer could be a real mine expert without "a thorough knowledge of chemistry, assaying, geology, mathematics, surveying, and connaisance of several languages and of law." [4] James Hague put it even

1. From "The Law of the Apex," *EMJ, 38* (September 6, 1884), 155.
2. Charles H. Shamel, *Mining, Mineral and Geological Law,* p. 12.
3. "Mining Law for Mining Students," *Mining Reporter, 52* (September 21, 1905), 283.
4. "Who Is a Mining Engineer?" *Mining Reporter, 57* (February 7, 1907), 129. See also *Mining Science, 61* (January 13, 1910), 34–35.

more forcefully in 1904, when he insisted that "the fully qualified mining engineer and mining geologist . . . must be more than a practical or mechanical engineer, and more than a theoretical or practical geologist; he must be learned in the *law*." [5]

The average engineer, at one time or another in his career, found himself handling routine matters of acquiring or perfecting title under local and U.S. mineral statutes. He might locate fractional claims.[6] or, as Emmet Boyle, a graduate of the Mackay School of Mines and subsequently governor of Nevada, do the detailed work connected with applying for a federal patent. Boyle, who by 1913 considered himself "a sort of veteran now in the patent business," not only took care of making locations and applications for survey for his clients; he also spent time in the state library and the recorder's office in Carson City, checking for possible conflicting claims and plow ing through the numerous decisions of the commissioner of public lands.[7]

But it was in the courtroom, as an expert witness on technical matters, that the engineer and geologist were most closely associated with questions of mining law. The ambiguities of the statutes themselves no doubt encouraged litigation, and westerners seemed to thrive on it. Young Louis Janin, not long out of the great mining academy at Freiberg, chided his attorney-brother in 1862:

> I never knew of such a quarrelsome, law-loving people as the Nevadians. There seem to be a half-a-dozen claimants to each piece of property in the Territory, and each must go to law about it. We peaceful citizens think the lawyers keep up the dissensions, especially in mining

5. James D. Hague, "Mining Engineering and Mining Law," *EMJ, 78* (October 20, 1904), 627.

6. See pages 154–155, this book. Under federal law, after 1872 claims could be located by the locator posting a notice and description of the claim, with due registration to establish the right of possession. But he could not obtain final title until he performed a prescribed amount of work in improving the claim and received a government patent.

7. Emmet D. Boyle to George S. Sturges, Virginia City, January 14 & 21, February 28, & March 2, 1913, George S. Sturges MSS.

> claims, and often sigh out the wish that they were all hung.[8]

Poor mines—mines "as barren as a mule," as the saying went—posed no problem, but a successful property could be counted on to produce a crop of lawsuits directly proportional to the richness of the ore in its stopes, for litigation was often a form of blackmail, designed to force prosperous or potentially prosperous mine owners to buy out their less fortunate neighbors rather than risk prolonged and expensive legal action in defense of title. Rossiter Raymond expressed the dilemma one way:

> "Sure, many such things may combine
> To make your mine not yours, but mine.
> If you don't buy me, fear the worst!"
> That miner eloquently cursed,
> And said, "I'll see you—elsewhere first." [9]

Victor Clement, manager of the Bunker Hill & Sullivan in Idaho, explained it another way in 1890, when he wrote his employer:

> There are two characters to the value of mining properties—one mine may have a value, owing to its real intrinsic worth; another (having no intrinsic value) may have a value by being so situated as to harass the working of the really valuable mine—in mining camps one is looked upon as much of legitimate enterprise as the other.[10]

The types of cases involving the expert testimony of mining engineers or geologists varied. Occasionally, an engineer would be called on to examine and evaluate mines being held by a public administrator as part of an estate settlement.[11] Sometimes, he would be asked to make an expert valuation of

8. Louis Janin to Edward Janin, Empire City, Nevada, October 12, 1862, Janin family MSS, Box 9.

9. From Rossiter W. Raymond, "Lawyers and Experts," in Rickard, ed., *Rossiter Worthington Raymond,* 94–95.

10. Victor Clement to Simeon Reed, Wardner, January 11, 1890, Reed MSS.

11. Hammond, *Autobiography,* I, p. 44.

mines and to testify in tax cases.[12] Joshua Clayton gave lengthy testimony and prepared sketches for hearings involving the Emma mine controversy in 1877.[13] Sewell Thomas, an engineer and son of an outstanding mining lawyer, once spent several months examining Nevada mines and testifying on behalf of a U.S. attorney who was prosecuting "a slick, smart promoter," George Graham Rice, on charges of misrepresenting the property and using the mails to defraud.[14] Thomas also inspected underground workings and later gave testimony in damage suits in which lessees were charged with failing to put in the proper mine supports. "It was a wild and harrowing experience," said Thomas, "timbers broken and squeezed, slabs of roof rock peeling and falling, and the constant sound of timber everywhere giving way." [15]

A number of engineers appeared before the courts and before the U.S. land commissioner in the late 1870s to give evidence on behalf of hydraulic mining during the great California debris controversy. Among a stream of witnesses on both sides of the debris question in Keyes *v.* Little York et al., in 1878, was Louis Janin, who insisted that tailings from hydraulicking would not harm vegetation and indeed were actually beneficial to farm land.[16] At later hearings, William Ashburner testified that to bar hydraulic mining would undercut an investment of at least $100 million, and "would render 100,000 people homeless by destroying their means of support, as well as ruin the principal industry of eight or nine counties." [17]

Sometimes, the lawsuits were over the rights to mining machinery or patented mineral separation or smelting processes.[18] In the main, however, the largest and by far the

12. R. G. Hazard to James D. Hague, Peace Dale, Rhode Island, February 12, 1891, Hague MSS, Box 11.

13. "Went on the stand in the Emma case and kept on all day—only a starter—did good days work." Entry for March 16, 1877, Clayton diary, 1877, Clayton MSS (Calif.), 5. See also entries for March 17 & 21 & April 14, 1877.

14. Thomas, *Silhouettes,* pp. 107, 109.

15. "We completed the job in a week of work—and I never want another one like it," wrote Thomas, years later. Ibid., pp. 169–70.

16. Robert L. Kelley, *Gold vs. Grain,* pp. 108–09.

17. San Francisco *Chronicle,* October 14, 1879.

18. J. B. Randol to Samuel B. Christy, New Almaden, March 30, 1880,

most important market for the testimony of the mining expert was the great body of cases in which mining titles were in dispute under federal law, particularly the so-called apex law.

As evolved at the local level, confirmed by the federal statute of 1866, and made more explicit by that of 1872, mineral law gave the locator of a vein or lode on the public domain the right to follow it, "with its dips, angles and variations," downward to any depth beyond the sidelines, provided that its top, or apex, was enclosed within the end lines of the location.[19] From this proviso and its interpretation stemmed trouble for many an unsuspecting mine owner:

> The coast of his experience showed
> Full many a gallant ship
> Wrecked on those dangerous rocks, the Lode,
> The Apex, and the Dip.[20]

Designed to protect the rights of bona fide locators, the apex law presupposed a simple mineral vein between two walls, "like the slice of ham in a sandwich,"[21] a condition that did not always exist. The copper district around Bisbee in Arizona, for example, was largely a formation of replacement ore bodies in limestone with little in the way of lode or outcrop. Thus, because the apex law was unworkable here, owners by common consent eliminated the extralateral right feature and proceeded with a minimum of litigation. On the other hand, in many important districts, such as Leadville, ore deposits tended to be flat-lying "ore-shoots," with many faults and lying close to the surface, or in such areas as Butte, Coeur d'Alene, Bingham, Tintic, and Grass Valley, even though fissure veins predominated, a combination of geological conditions at the surface opened the way for additional types of en-

---

Christy MSS; Louis Janin to James Hague, San Francisco, December 8, 1891, Hague MSS, Box 12; *Who's Who in Mining,* p. 93.

19. See U.S., *Statutes at Large, 14* (July 26, 1866), 251; U.S., *Statutes at Large, 17* (May 10, 1872), 91.

20. From "The Law of the Apex," *EMJ, 38* (September 6, 1884), 155.

21. Courtenay DeKalb, "Some Defects of the United States Mining Law," AIME *Trans., 51* (1916), 284.

try—tunnel, mill, placer, and other—only enhancing the chance of numerous lawsuits.[22]

Much depended on a legal definition of the term "vein," "lode," or "ledge." If a vein leading down from the surface was lost near the vertical side wall of the claim and a similar vein of identical ore found below it or on one side of an adjoining claim, was this a geological continuation, or offshoot, of the first, or was it a completely separate vein? Was the great Comstock lode a single large irregular vein of practically unlimited boundaries ("The foot-wall is the diorite of Mount Davidson, and the hanging wall is Salt Lake City," said one old-timer),[23] or was it a number of narrow ore bodies separated by barren rock? An expert witness in a Colorado case once tried to give his arguments graphic form by standing up, touching the top of his head and saying, "If I stand upright, *here* is my apex; but," touching his rotund waistband, "if I lie flat on my back, *where* is my apex?" Actually, as James D. Hague pointed out, this analogy was much too simple:

> that witness's confidence in the top of his head as his apex would prove to be wholly misplaced, in the opinions of some lawyers and experts, according to whose hair-splitting definitions an apex might be found in the extreme end of each and every outcropping hair, however long, on the top of his head, all of which might, if so inclined, form a halo of outcrops, not all within the exterior limits of his side-lines or end-lines, dipping in all directions, at every conceivable angle, towards a common center beneath the surface. And if he happened to be bald-headed, would he forfeit his "top or apex," or lose his self possession, with the loss of his hair? [24]

Thus, the right of ownership under the apex law depended on how complex and controversial geological issues were resolved,

22. See Reno Sales, "Geophysical Mining Claims," in *Rocky Mountain Mineral Law Institute* (Albany, 1957), pp. 397–98.

23. Quoted in King, *History of the San Francisco Stock and Exchange Board,* p. 95.

24. Hague, "Mining Engineering and Mining Law," *EMJ,* 78 (October 20, 1904), 629.

and it was to provide technical "answers" to such questions of vein structure and formation that scientists and engineers were brought to the stand as paid expert witnesses.

In apex litigation, specialization was sharp, both in the legal and the technical realms. Outstanding lawyers, such as William A. Stewart, Harry Thornton, Charles S. Thomas, and Charles J. Hughes, Jr., had their counterparts in such geologists and engineers as Louis Janin, Rossiter Raymond, Clarence King, Joshua Clayton, and Horace V. Winchell, to mention but a few. Although there were many exceptions, with many technical men at one time or another trying their hand at it, a relatively small, elite group of highly skilled "professionals" tended to dominate, especially in the major litigations in which millions of dollars might hang in the balance. Generally, it was the "usual bunch," as an editor commented of the experts in one of the last of the major apex cases at Butte in 1915.[25]

The background of these men was that of mining engineers and geologists in general. A few, such as Joshua Clayton and George Robinson, were self-made and self-taught,[26] but most were graduates of leading colleges or schools of mines. Because so many before 1875 were trained at Freiberg and spoke at least passable German, a Nevada attorney in the 1860s could define a mining expert as "a man who looks wise, wears glasses and talks Dutch." [27] A few, among them Daniel M. Barringer, had formal law degrees in addition to their scientific background. Barringer, with an Ll.B. from the University of Pennsylvania, was co-author of a book on mining law [28] and regarded himself as an expert on "this abominable law of

25. *M&SP, 111* (November 13, 1915), 755.

26. Goodwin, *As I Remember Them,* pp. 236–37; *Northwest Mining Journal, 1* (August 1906), 36–37.

27. Quoted in William A. Chalfant, *Outposts of Civilization,* p. 80.

28. Daniel M. Barringer & John S. Adams, *The Law of Mines and Mining in the United States* (2 vols. Boston, 1897–1911). "It gives me much pleasure to fulfill my promise made to you sometime ago to send you a copy of my book on Mining Law," Barringer wrote a friend. "If you are ever troubled with insomnia I can highly recommend it as an almost certain cure." Barringer to John Hays Hammond, November 22, 1901, Letterbook F, Barringer MSS.

the apex." "I hope you will not consider me egotistical if I say that I feel that I thoroughly understand both its legal and geological aspects," he wrote a fellow engineer in 1895.[29] Such men as Barringer were the exception, however. Most courtroom experts studied law on their own, as did geologist Clarence King, whose services were very much in demand after his exposure of the diamond scandal in 1872. "Clarence King is in town," noted a San Francisco engineer in 1874. "He says that he is studying Law now, so as to perfect himself as a Mining Law yer, that is, to examine Experts in Mining Cases." [30] The ultimate dean of the mining experts, Rossiter W. Raymond, a man who had studied at Brooklyn Polytechnic Institute, Freiberg, and the Universities of Heidelberg and Munich, capped off his education at the age of fifty by studying law and was admitted to the bar in 1898. Raymond was already one of the most formidable technical witnesses in the country: his long exposition on the law of the apex, published serially in 1883–84, was standard; his series of lectures on the subject at Columbia Mines and his experiences in many of the classic western cases made him a man much sought after when litigation arose. "R. W. Raymond's advice in this matter ought to be worth something," became a common byword where apex questions were in dispute,[31] and opposing experts, such as Clarence King, admitted that his presence "does not add to my comfort." [32]

As was the case with many other functions of the engineer or geologist, many scientists were called from the groves of academe to testify. During the 1860s, when engineering and geology were yet in their infancy, evidence was given in one important Comstock case by at least three engineers, two geologists, a metallurgist, a botanist, and three chemists, including Professor Benjamin Silliman, Jr., of Yale.[33] In later years,

29. Barringer to A. S. Dwight, February 8, 1895, Letterbook A, ibid.

30. Charles Hoffmann to Josiah D. Whitney, San Francisco, May 13, 1874, Hoffmann MSS.

31. Fred W. Bradley to Victor Clement, Wardner, May 15, 1893, Bradley MSS.

32. Clarence King to James D. Hague, New York, February 3, 1899, Hague MSS, Box 12.

33. *Alta California,* August 29, 1864; *Territorial Enterprise,* February 6, 1868; Opinion of Referee, August 22, 1864, Gould & Curry Silver Mining Co.,

just any scientist would not do: a geological background was a must, and the courts drew on the staffs of the technical schools from all parts of the country, ranging from T. Sterry Hunt, veteran of the Canadian Geological Survey and professor at MIT,[34] to Samuel B. ("Cyanide Sam") Christy, of the University of California, who, despite reservations by the president of his institution, believed that for economic reasons "experting" was essential as a sideline activity.[35]

Although Josiah D. Whitney of the California Survey did take the stand in some of the Comstock litigation in 1864 (apparently without pay),[36] ordinarily personnel actively attached to state or federal geological surveys were not free to testify in mining cases in the United States.[37] At the turn of the century, however, a junior geologist of the U.S. Geological Survey, who in his official capacity had had access to some of William A. Clark's mines in Butte, resigned from the survey and subsequently used information thus gained as a witness against Clark in a local lawsuit. This breach of ethics not only shocked the director of the survey, but served to strengthen Clark's already firm antipathy toward geologists in general.[38] When dealing with the Butte area especially, the survey thereafter was careful to deal in generalities in its reports with properties still tied up by litigation, although it came under fire from time to time by partisans of one side or the other as the maze of lawsuits unfolded.[39]

Occasionally, an engineer failed to avoid a conflict of inter-

---

Plaintiff, *v.* North Potosi Gold & Silver Mining Co., Defendent (Virgina City, 1864), pp. 6–7.

34. Richmond *v.* Eureka, p. 275.

35. "If I can secure during my summer's vacation of three months enough expert work to help out a little towards putting my family beyond want in case of accident to myself I certainly would prefer to stay here." Samuel Christy to Henry Perkins, Berkeley, December 24, 1890, Christy MSS. See also Christy to D. B. Huntley, Berkeley, May 29, 1900, ibid.

36. Josiah D. Whitney to William Brewer, Virginia City, April 18, 1864, Brewer-Whitney MSS, Box 1.

37. Rossiter W. Raymond, "Biographical Notice," in *Clarence King Memoirs,* p. 359; Arnold Hague to James D. Hague, Washington D.C., May 7, 1900, Hague MSS, Box 2.

38. *M&SP, 116* (January 12, 1918), 41.

39. See *EMJ, 76* (September 19 & December 5, 1903), 436, 841, 845.

est. C. S. Batterman, who for several years had been employed by the Montana Ore Purchasing Company, was sued by that concern in 1897, on the grounds that he had brought to his new employer, the Boston and Montana Company, information to use in the legal struggles between the two firms.[40] Emmet D. Boyle in 1913 unwittingly found himself involved in a case in which two of his clients were adversely engaged and had to withdraw his services from both.[41]

When an apex suit was brought (or any other suit for possession of title), the mine manager usually played an important role in advising and in helping prepare the company's case. Sometimes he simply warned his employer against getting involved, suggesting instead an out-of-court settlement. "I tell you *law* is a bad thing about a mine, and if you can steer clear of it, you will be better off," one manager cautioned in 1878. An amicable understanding, in this particular case, he continued, would have many advantages, "whereas if you go to law, very heavy expense must be incurred, and the *result* still be uncertain." [42] Often the manager at the mine chose the experts for the impending suit, as did Fred Bulkley of the Aspen Mining and Smelting Company in 1893, and pledged himself "to keep expenses down as low as possible until we find that efforts in that direction will be availing, in which case, I propose to make a strong case without a flaw in it, which will make an addition to our expense." [43]

But often the manager's part was not merely a matter of advising for or against entering the courts or of selecting outside experts, but was a far broader and more time-consuming task, as indicated by a case study of the activities of a typical managing engineer in 1889–90, when property of his employer, the Bunker Hill & Sullivan Mining Company in the

40. *MI&R, 20* (September 9, 1897), 103.

41. Emmet D. Boyle to George S. Sturges, Virginia City, January 7, 1913, Sturges MSS.

42. George Daly to Henry Yerington, Aurora, Nevada, September 13, 1878, Yerington MSS, Box 1.

43. Fred Bulkley to J. B. Wheeler, January 30, 1893, copy, Aspen Mining and Smelting Company G.M.R., Brown MSS.

Coeur d'Alene was contested. The engineer, Victor Clement, a man of considerable experience and talent, took care of the preparation of the maps and models to be used in the courtroom.[44] He made recommendations for the hiring of expert witnesses from elsewhere, insisting that "if we rely on local experts our goose is cooked for more or less everyone about here is interested directly or indirectly in the 30 or 40 claims back of us" and would be anxious to pull down the Bunker Hill & Sullivan. So, said Clement, "we must prepare to give them a 'squasher.' "[45] His choices as experts were Rossiter Raymond and John Hays Hammond, the latter, said Clement, a man who "has a thorough knowledge of mining, & is very bright and makes a splendid witness." The selection of experts he pushed hard, in order to have a preliminary examination of the mine completed at an early date to enable the lawyers to organize their case.[46] And later, when Hammond was hesitant about spending time testifying in the remote Coeur d'Alene, Clement suggested having his family "up on a 'Summer Vacation' at some point within easy reach of the mines."[47]

Clement was also concerned with finding numerous nontechnical witnesses favorable to the company, for, as he put it, "we must overwhelm them with testimony—that is the only thing counts with these juries."[48] He would pay his witnesses well, and at the same time would use his influence in an effort to gain the appointment of a judge who would be sympathetic, but in addition expressed his fear of the appointment of "another newfledged judge—weak as a rag & and knowing nothing about mining cases as it usually happens."[49] Meanwhile, he was exercising his ingenuity to make sure that the jury was not rigged by the opposition.[50] Just to be on the safe side, he

44. Victor Clement to Simeon Reed, Wardner, September 4, 1889, Reed MSS.
45. Clement to Reed, Wardner, September 16, 1889, ibid.
46. Ibid.
47. Clement to Reed, Wardner, May 8, 1890, ibid.
48. Clement to Reed, Wardner, September 6, 1889, ibid.
49. Clement to Reed, Wardner, August 27 & October 16, 1889, ibid.
50. Clement to Reed, Wardner, October 27, 1889, ibid.

hired Pinkerton detectives to infiltrate the enemy's camp [51] and lined up newspaper support, personally reading proof on material that went into the *Wardner News*.[52]

All the while, Clement carried on his regular managerial duties and advised his superiors as to whether they ought to depend wholly on the outcome of litigation or whether they should purchase strategic interests at reasonable prices to weaken the antagonist's case.[53] He conferred regularly with the company's battery of lawyers, lamenting that he found it necessary to keep them "poked up," lest the company get "souped" in court.[54] When new litigation arose, he fully expected to be sent to Boise, some 350 miles away, to testify, but warned "I for my part shall decline any responsibility in the issue, unless I have my own way in the manner & methods to be adopted." [55] He repeatedly urged speeded-up development on the property, for "the more work we do in the Sullivan the more complex will be our defense if our adversaries get access to our works." [56] The task of laying the groundwork for the expert witnesses was both arduous and distractive and certainly not a welcome part of the mine manager's duties. "I cannot give this matter much attention," Clement complained, "as it absorbs my time, gives me nervous prostration which unsettles me for other business—and matters here do not improve by my being absent." [57]

Possibly Clement did more of the preliminary work than most mine managers, for he was exceptionally able and was given the broadest of responsibilities. The lawyers, of course, ordinarily determined the course of attack or defense, but both the resident manager and certainly the expert witnesses, im-

51. Clement to Reed, Wardner, October 16, 1889, & W. A. Pinkerton to Clement, St. Paul, Minnesota, November 5, 1889, ibid.

52. Clement to Reed, Wardner, February 16 & 20, 1890, ibid.

53. Clement to Reed, Wardner, March 1, 1890, ibid.

54. Clement to Reed, Wardner, May 8 & 11, 1890, ibid.

55. Clement to Reed, Wardner, March 20, 1890, ibid.

56. Clement to Reed, Wardner, May 15, 1891, ibid.

57. Clement to Reed, Wardner, May 11, 1890, ibid. In 1889, Clement had also complained of litigation, calling the loss of time "exceedingly annoying" and noting that he was "excessively tired of it." Clement to Reed, Wardner, December 23, 1889, ibid.

ported for the occasion, would figure prominently in the planning. Indeed, it was said that Clarence King, as the expert, directed both the legal and the technical ends of his cases and that Rossiter Raymond sat at the elbow of his attorney and coached him as to the proper questions to ask, much to the distress of rival experts who protested bitterly but unsuccessfully.[58]

Although at least one engineer believed that attendance at the apex suits in Denver "was really a delightful kind of vacation for the large companies were very liberal in their arrangements," [59] the evidence indicates that in general legal experting was hard work, with long hours and tremendous personal pressures. An engineer in 1910 insisted that most of his profession regarded court work as unpleasant, except for a small group made up of those of bellicose nature who enjoyed heated exchanges with the cross-examining attorney or those who regarded the work as an opportunity to practice mental discipline under strenuous conditions.[60] However that may be, men followed the pursuit and seemed to flourish in it, although there was no question but that it was difficult work.

Except in rare occasions in lesser suits,[61] the expert witness expected to spend considerable time underground becoming familiar with subsurface conditions and ore structure. One expert in the Kennedy Extension-Argonaut (California) case of 1910 spent nearly three months familiarizing himself with the Argonaut and even worked as a miner there for another four months, though perhaps the forty days spent by a fellow expert in the suit was more typical.[62] Letters written in the late 1890s by Nathaniel Shaler, who sometimes took time off from his Harvard teaching duties to testify in mining controversies,

58. Samuel F. Emmons, "Clarence King—Geologist," in *Clarence King Memoirs,* pp. 279–80; *EMJ, 27* (January 18, 1879), 38.

59. Henry E. Wood, "I Remember," Wood MSS, Box 3.

60. *Mining World, 32* (March 26, 1910), 655–56.

61. Occasionally, an engineer who had examined a mine for other purposes was asked to submit written testimony in a case without re-examining the property and without appearing in court. See direct interrogatories of R. W. Raymond on Coldstream mine 1873(?), Adelbert & Raymond MSS.

62. *EMJ, 96* (November 29, 1913), 1016.

indicate that not all his time was spent in scintillating conversation with other experts around the dinner table at the McDermott Hotel in Butte. "The question is a difficult one, but I hope to compass it without undue strain," he wrote. "I have had a busy day in the mine and on the surface. I shall be very glad to escape from this very tiresome place." [63] Again, from Montana, he complained: "I am pretty well through with the blessed undergound, with its dirty business, and am now doing the surface, trying to extract information from the . . . dust which wraps this wealth in. The place is a dry hell." [64]

Frank L. Sizer, an engineer whose career spanned more than half a century in the mining West, was a "regular" in apex suits, and his diaries give considerable insight into the role of the expert. Sizer was one of a number giving testimony for F. Augustus Heinze at Butte in 1900, the case opening on March 20. But Sizer had already spent at least a month in preparation: this involved conferences with lawyers and a half dozen other mining experts, a trip to Denver to consult top legal advisers, numerous trips into the underground workings of the mine, and much labor on maps, charts, models, and in the writings of the standard geological authorities. "Evening work on maps to 12 o'clock" was not an unusual entry in Sizer's journal.[65] A typical day included his spending the morning at the office, the afternoon in the mine, and the evening in conference.[66] Even after a hard day of cross-examination on the stand (during which the opposing lawyer "thought he scored a point on me"), Sizer spent the evening working in his office.[67] When testimony ended on April 20, Sizer could note: "first evening I have spent at home in many weeks." [68] But the next afternoon and night he was helping the lawyers re-

63. Shaler, *Autobiography,* p. 339.
64. Ibid., p. 398.
65. Entry for March 16, 1900, Sizer diary.
66. Entries for March 14 & 23, 1900, ibid.
67. "I was not altogether satisfied with myself—but was placed in some hard situations." Entry for April 11, 1900, ibid
68. Entry for April 20, 1900. ibid.

view testimony. On April 22, he heard one of the rival experts give a sermon: "Went to church & heard Dr. Raymond preach on Pilate and the 'Temple Ring'. Interesting but not strictly accurate." Later in the day, he was again going over evidence.[69] Even while the attorneys presented their closing arguments, Sizer was busy abstracting testimony for them.[70]

In another Butte case, of 1903, Sizer commented of William A. Clark, who had cross-examined him: "He was sharp & pressed me hard on some points." [71] He also admitted that he had been up late, discussing the case informally and that he had "Spent two hours rehearsing." [72] This was probably not unusual, for the testimony on both sides was very carefully staged for maximum effect on jury and/or judge. Verbal testimony was not in itself deemed adequate, and elaborate maps, charts, and detailed glass or wooden models, showing tunnels, ore chambers, shafts, and even the surrounding formations—and costing thousands of dollars—were presented by way of graphic evidence. Some of the scale models used in the Butte litigation were reported to have cost as much as $25,000 each; [73] and at one of the last of the large apex suits in 1915 (the Butte & Superior *v.* the Elm Orlu), attended, incidentally, by the seniors of the Montana School of Mines, more than 250 exhibits—plans, sections, sketches, and models—were introduced and the testimony ultimately filled seventeen volumes.[74]

Time on the witness stand was not always pleasantly spent,

69. Entries for April 21 & 22, 1900, ibid.

70. Entry for April 23, 1900, ibid. This decision went against Heinze. For Sizer, "it was the first case I had ever lost." Entry for June 3, 1900, ibid.

71. Entry for December 29, 1903, ibid.

72. Entry for December 27, 1903, ibid.

73. C. B. Glasscock, *The War of the Copper Kings,* p. 272. In the Colusa Parrot *v.* Anaconda case in 1901, maps and models cost the Anaconda side $35,324 to prepare. *EMJ, 69* (June 16, 1900), 720. For other references to the use of models in court see entries for January 11 & April 18, 1900, Sizer diary; Thomas, *Silhouettes,* p. 75; *EMJ, 24* (June 2, 1877), 337, *25* (January 19, 1878), 44, *27* (February 8, 1879), 87, *96* (October 25, 1913), 795; Hyman *v.* Wheeler et al., *Federal Reporter, 29* (1886), 352.

74. *M&SP, 111* (November 13, 1915), 755; Walter W. Lytzen, "The Elm Orlu *v.* Butte & Superior Apex Litigation," *M&SP, 119* (September 13, 1919), 368.

and local editors were fond of pointing out the therapeutic value of cross-examination for "them yeller-legged book miners," whose burden of conceit was greater than that of their knowledge:

A wonderful man is the mining expert, a wonderful man is he,
The earth to him is an open book, there is nothing he cannot see

. . .

He can to the fathom, the foot and the inch, tell what your ground contains,
Map out the ore chutes, the feeders, the spurs, dips, angles and crossing veins;
With book and pencil he'll write you down all that's in and out of sight,
With learned words of ponderous sound he'll demonstrate he's right.

. . .

And you would judge from his confident tone and learning from near and far
That he had heard the fiat go forth, and the song of the morning star;
That he had communed with primal man as well as the primal beast,
Had seen the world in its plastic state and the infant sun in the east.

But alas! sometimes he gets into court and on the witness stand,
And his learning then is a broken staff within a palsied hand,
And the way the lawyers then go for him is pitiful to see,
Mixing him up till he cannot tell a mocking bird from a bee.[75]

Although the seasoned veterans were no doubt hard to shake, the witness often earned his fee on the stand. Mary

75. From "The Mining Expert," *Silver Standard,* September 10, 1887.

Hallock Foote, wife of a mining engineer testifying in the case, described a Boise mining suit in 1892 as "something like a Bull Fight":

> Mr. Heyburn, the lawyer who cross-examined, is very clever. I wanted to see how it would be, in case I went there and Arthur is on the stand. He is on the Last Chance side, and will be cross-examined by Arthur Brown of Salt Lake—a clever but brutal lawyer: one of the court-room gladiators.
>
> It was painful for Mr. Loring, because there was a crookedness he was trying to hide and it had to come out. It was very dramatic and intensely exciting: the court-room filled with a typical mining crowd from the North, witnesses and owners and friends of the two mines. It is a famous case: and the interest of all the mining men in the community centers just now in this court-room. The judge is a "wool-head" but the lawyers are far above the average, and make the examinations a severe mental exercise for all concerned.[76]

The uneducated and the unwary fell easy prey. Witness old Tim Huzzy, an expert in an early case, who when asked to identify and describe a geological chart introduced as an exhibit could only look long and hard, scratch his head, and mutter "Them's the Lakes of Killarney."[77] In a Comstock suit of the 1860s, William Stewart is supposed to have trapped and confused an expert by asking how many degrees there were in a circle; he received the reply that the number varied according to the size.[78] Even the suave, polished specialists could be tripped up as shrewd cross-examining attorneys badgered unmercifully, attempting to rattle the witness or catch

76. Mary H. Foote to Helena Gilder, Boise, January 17, 1892, M. H. Foote MSS.

77. Newton, *Yellow Gold of Cripple Creek,* p. 110.

78. This story was recounted as early as June 1864, and the episode occurred in the Gould & Curry *v.* North Potosi litigation. See unidentified writer to editor, Virginia City, June 16, 1864, undated clipping, Hayes Scrapbooks, *10.*

him in a contradiction.[79] Undoubtedly, Winfield Scott Keyes, a Freiberg graduate of broad western experience, found the Anaconda-Colusa Parrot litigation at Butte something of "a severe mental exercise," as Mrs. Foote put it, although he appeared completely nonplussed when he took a firm stand on a passage from Kemp's book on ore deposits, only to have the opposing lawyer read from a later edition in which Kemp had modified his position.[80] Probably even more embarrassed was a professor of geology from the University of Minnesota, whose testimony in another case did not even coincide with what he had written in a recent book of his own.[81] In the fast legal company engaged to handle important litigation, even a good witness might look bad and a mediocre one prove a decided asset for the opposition. Frank Sizer mentions an expert for the other side in a British Columbia case in 1907 "who was really a good witness for us," and Sizer's attorney told the opposing lawyer, "now if you are not going to send a check to that man, I will." [82]

Sometimes it would be found before a case commenced that an expert's testimony would not strengthen his employer's cause; if so, he usually received fees for his preliminary work, or might even be retained on the payroll in some other capacity,[83] though this was not always true. During the 1860s, William P. Blake had been called to Virginia City by William Stewart to testify in favor of the several lode theory, as he had previously done in another case. On inspecting the mine, Blake found that the separate fissures came together in a vein at a lower level and informed Stewart that he would have to reverse his earlier position. Stewart denounced him, refused to

79. Frank L. Sizer to editor, San Francisco, January 19, 1921, *M&SP, 122* (January 29, 1921), 150.

80. Rickard, *History of American Mining,* p. 364.

81. Lytzen, "Elm Orlu *v.* Butte & Superior Apex Litigation," *M&SP, 119* (September 13, 1919), 368.

82. Entry for April 11, 1907, Sizer diary.

83. Ellsworth Daggett was hired in 1896 by the Butte & Boston Company to aid in their suit against the Lexington Company. After eight or ten days' work, when it was obvious that Daggett's testimony would not help the Butte & Boston, he was switched to another case. Ellsworth Daggett to James D. Hague, Butte, February 13, 1896, Hague MSS, Box 12.

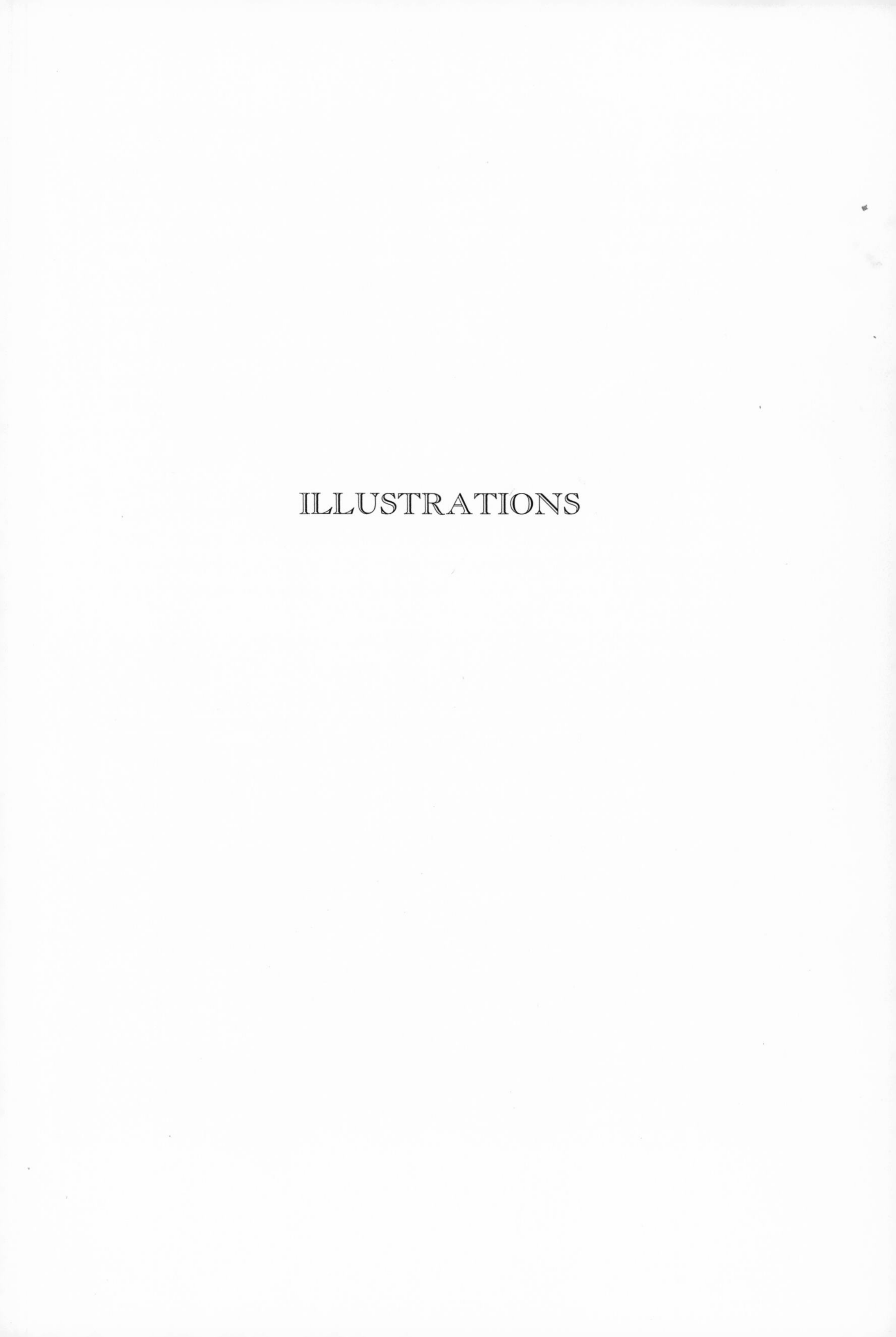

ILLUSTRATIONS

An early mine expert in California. From Walter Colton, Three Years in California *(New York, 1850), opposite p. 244.*

Class of 1889, Columbia School of Mines, at Central Mine, Colorado (?), July 1888. From Columbiana Collection, Columbia University Libraries.

Class of 1904, Columbia School of Mines, sketching timbering at Durant Mine, Aspen, Colorado, 1903. From Columbiana Collection, Columbia University Libraries.

Employees of the Savage Mine on the Comstock Lode, 1865; General Superintendent Sam Curtis seated at center. From the Bancroft Library.

Alexis Janin, 1868. From the Henry E. Huntington Library and Art Gallery.

Louis Janin. From the Henry E. Huntington Library and Art Gallery.

Charles Henry Janin. From the Henry E. Huntington Library and Art Gallery.

Joshua Clayton (?). From the Bancroft Library.

Charles Butters, 1901. From the Bancroft Library.

David W. Brunton. Courtesy William Ainsworth & Sons, Inc., Denver.

Robert Brewster Stanton at the Hoskaninni Placers, 1897. From the New York Public Library.

Dredge under construction at the Hoskaninni Placers, 1900. From the New York Public Library.

George T. Coffey (in fur coat, dark glasses) and unidentified companions. From the Bancroft Library.

John Farish in Mexico. From the Frederick G. Farish Collection, Archives and Western History Department, University of Wyoming.

Mining experts in Sonora, ca. 1901. From the Frederick G. Farish Collection, Archives and Western History Department, University of Wyoming.

James D. Hague (center) and unidentified companions at Deadwood, South Dakota. From the Henry E. Huntington Library and Art Gallery.

John Hays Hammond enters politics in South Africa. From University Club menu (Denver, May 1909), courtesy Henry Carlisle.

John Hays Hammond turns down a government position. From University Club menu (Denver, May 1909), courtesy Henry Carlisle.

John Hays Hammond prospecting in Mexico. From University Club menu (Denver, May 1909), courtesy Henry Carlisle.

Brunton pocket transit. From William Ainsworth & Sons, Inc., Denver.

Lamb's Formula on what it takes to be an engineer. From Mark Lamb, "Who Is a Mining Engineer?" EMJ, 94 *(November 16, 1912), p. 941.*

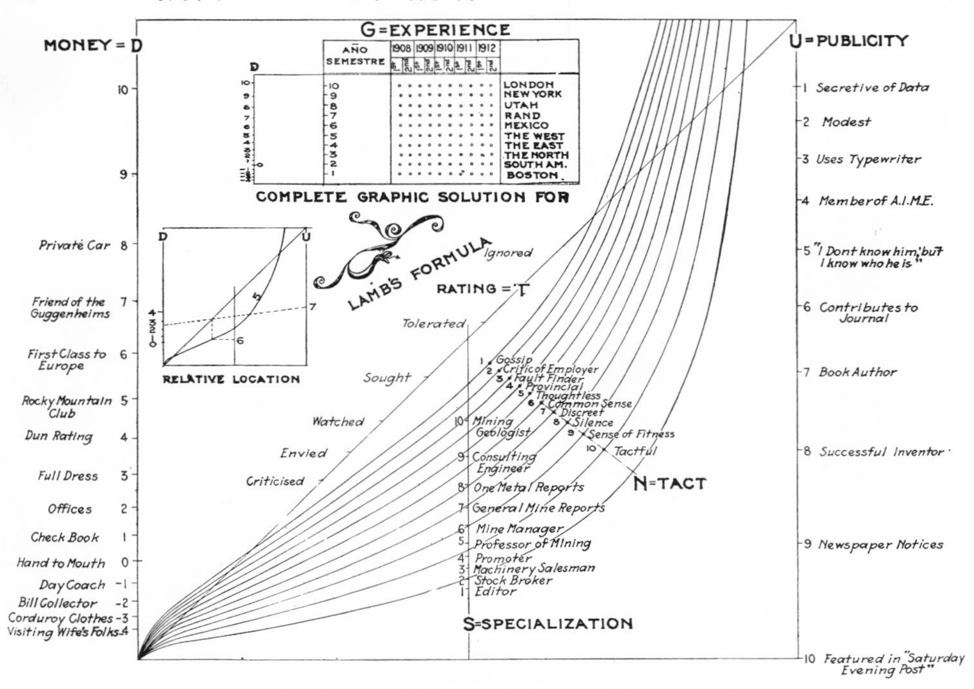

pay his expenses, and "told him to go and sell himself to the other side"—which Blake promptly did, an act that gave him a black eye in the mining profession and that also later prevented his appointment as U.S. commissioner of mineral statistics.[84]

Blake at least revised his opinions before he took the witness stand. Clarence King once changed his point of view completely on the stand, admitting that his prior testimony had been based on a superficial knowledge and that he knew more the second time.[85] An able and intelligent Colorado engineer, Walter Devereux, was put on the stand by his own company to retract a report he had made earlier. Thus forced to explain his reversal of opinions under cross-examination and "the ingenious distortion of the smartest attorneys in Colorado, and to the judgment of a suspicious jury," when he "ought to have been out of the State and unfortunately inaccessible," Devereux was denounced from one end of the state to the other, and thereafter no lawyer dared call him as an expert witness.[86] His mistake was not in changing an opinion, but in doing so in court without warning. Rossiter Raymond's advice on this point was sound:

> If you have changed your mind, say so in a technical journal or a professional paper, where the confession of a former error or incomplete knowledge will be merely a contribution to science—not for the first time on the witness-stand, where it brings you another fee, and helps a new client.[87]

84. Rossiter W. Raymond to Louis Janin, Lakewood, New Jersey, March 31, 1909, Charles Janin MSS, Box 5. William Ashburner, a contemporary engineer with no love for Blake, had a different version of the affair: Blake had examined the North Potosi, giving the owners an affidavit on which they failed to get an injunction. Later, he had informed the opposition, the Savage, that he had left a loophole in the affidavit and would like to testify on behalf of the Savage. William Ashburner to William Brewer, Virginia City, April 9, 1864, Ashburner MSS.

85. Goodwin, *As I Remember Them,* p. 172; Richmond *v.* Eureka, p. 566.

86. Rossiter W. Raymond to Louis Janin, Lakewood, March 31, 1909, Charles Janin MSS, Box 5.

87. Ibid.

It was generally acknowledged that court "experting"—at least for the real professionals—was profitable and that litigation was expensive. Individual fees varied, of course, depending on the reputation of the expert, the value of the property in question, and the laws of supply and demand. From the Comstock in 1864, Josiah D. Whitney lamented that his conscience prevented him from receiving any pay for testifying, though "some of the Experts get large fees for their services." [88] Seven or eight years later, Clarence King was hired at $100 a day to testify in the famous Dives-Pelican suit in Colorado,[89] and in 1888, Joshua Clayton, one of the "grand old men" of the profession, received $2,500 for his aid in preparing a case for the Montana Company, Ltd.[90] Two years later, when Winfield Scott Keyes, then at the very peak of his reputation, was approached by the Bunker Hill & Sullivan people, he stated his terms as $1,500 as retainer and fee for making a preliminary examination; then, if he found he could testify favorably for the company, he would charge an additional $3,500, "together with the usually allowed expenses of trip &c." [91]

Obviously, if there were a number of top flight lawyers and experts, the cost of litigation might run high. It was reported that in a suit settled in the Helena district court in 1893 a total of ten geologists and engineers were employed and that their fees alone amounted to more than $100,000.[92] Another case, heard in Utah more than twenty years later, took more than 100 days at an estimated cost of $2,500 a day, including

88. Josiah D. Whitney to William Brewer, Virginia City, April 18, 1864, Brewer-Whitney MSS, Box 1.

89. Wilkins, *Clarence King,* p. 191.

90. Entry for June 13, 1888, Clayton diary, 1888, Clayton MSS (Calif.), 9.

91. Winfield S. Keyes to Victor Clement, San Francisco, February 4, 1890. Clement suggested postponing the appointment, with the possibility of substituting the California state mineralogist, at a fee of only $1,000. Clement to Simeon Reed, Wardner, March 1, 1890, Reed MSS. In 1914, Charles Janin informed a potential client that his fee for examining property and making a deposition would be $1,000, but if he had to prepare for cross-examination his charges would be $3,000, with expenses additional in either case. Charles Janin to Louis Marshall, San Francisco, August 24, 1914, copy, Charles Janin MSS, Box 5.

92. This was the case of the St. Louis Mining Company *v.* the Montana Company, Ltd., *M&SP, 66* (June 3, 1893), 340.

model construction, exploratory work in the mines, and fees for lawyers and experts.[93] Local citizens, including the ladies, often found the courtroom scenes fascinating,[94] but there is no denying that apex litigation was expensive entertainment.

Despite the hard work, the experts themselves no doubt enjoyed themselves when not in court. The confusing complex of apex litigation in the two decades after 1890 brought a steady parade of sparkling engineers and geologists to Butte, some for Anaconda, some against. Among them was Clarence King, very much the dandy, his valet brushing his overalls clean and creasing his trousers at the end of each day. King delighted in matching wits with others, and his insatiable curiosity drove him to great lengths. One of his colleagues remembered him

> as going into a mine at early morning, taking his lunch with him; coming out late in the afternoon; bathing and dressing for dinner; then, aroused by casual table-talk, putting on his underground clothes again, and spending the greater part of the night in the mine, just to "settle the point"—though the point was not perceptibly pertinent to the immediate case in which he was engaged.[95]

Evening often brought both cooperating and competing experts together around the dinner table for delightful and engrossing conversation running the gamut from the latest scientific theory to philosophy, religion, or art. Rossiter Raymond discussed the bible and played many a chess game in such company, and Nathaniel Shaler found time to write serious literature.[96] Lawyers, experts, and company officials, on both sides of an important Utah case, forgot for a time their parti-

93. *M&SP, 109* (December 27, 1919), 911.

94. Young George Parsons attended court sessions in Arizona in 1881–1882 and recorded in his diary such expressions as "Interesting—some fun too," "Much interest manifested," and "large audience this afternoon including many ladies." Entries for December 13, 1881, & January 16 & 24, 1882, Parsons, *Journal,* pp. 280, 285, 286. See also entry for January 10, 1900, Sizer diary; John Huber to Allen Curtis, Silver Springs, Nevada, October 27, 1868, Manhattan Silver Mining Company MSS.

95. Rossiter W. Raymond, in *Clarence King Memoirs,* pp. 361–62.

96. In his hotel at Butte, Shaler wrote the second part ("The Rival Queens") of his *Elizabeth of England* series. Shaler, *Autobiography,* p. 339.

san positions in the courtroom, gathering in December 1920 to enjoy themselves at a special dinner, complete with toasts (probably nonalcoholic) and rustic verse hand-wrought for the occasion.[97]

Almost from the beginning, the use of paid mining experts in the courtroom came in for criticism, both from members of the calling themselves and from those who viewed it from apart. Even the courts conceded the shortcomings of the arrangement, but admitted expert testmony as being the best available under the circumstances.[98] It was widely acknowledged that even "honest partisans," as Rossiter Raymond called the best of the experts, could not be totally objective and scientific. Early in his career Raymond remarked:

> There is nothing that disturbs the coolness of professional observers like the knowledge that money is dependent upon the views they take of some knotty questions. Men who, if left to themselves, would modestly compare their opinions with those of their brethren, become as feed witnesses upon the stand, dictatorial, didactic, and obstinate. They defend their own half-formed theories like paid advocates. . . . People often wonder "why the geologists can't agree." Because, gentlemen, you pay them too well to magnify and perpetuate their disagreements.[99]

Naturally litigants sought to choose their experts according to reputation and to bolster whatever geological interpretation was deemed most advantageous in building a particular case. On the basis of previous testimony, the client had some idea what he was getting when he hired a technical witness. Moreover, reputable geologists and engineers often refused to com-

97. See *EMJ, 109* (January 10, 1920), 61.

98. Kahn *v.* Old Telegraph Mining Company et al., *Utah Reports, 2* (1877), 174.

99. *EMJ, 7* (August 17, 1869), 105. For other comments on the "unscientific" nature of experting, see Augustus Locke, "The Two Professions of Mining Geology," *M&SP, 120* (January 31, 1920), 161–62; "Geologists in the Courts," *M&SP, 120* (April 24, 1920), 593–94.

mit themselves fully to a client until they had visited the property in question and inspected the geological structure to make certain they could honestly testify as their employer desired.[100]

Witnesses were not usually dishonest, although occasionally one was apprehended overstepping the bounds of probity in flagrant terms. In an Aspen case of 1886 (Hyman *v.* Wheeler et al.), an expert for the defense brazenly acknowledged that an earlier report of his which contradicted his present testimony was neither true nor sincere, but had been written to sell property, a confession he justified with the excuse that "all was fair in mining" and that brought an excoriation from the presiding judge, Moses Hallett.[101] In a Denver suit a few years later, the expert on one side showed a map of underground workings he had prepared to his attorneys, who saw immediately that it would harm their case and aid the opposition. "I can fix that," retorted the expert, and he promptly did so, sitting up all night making a new map, relocating the ore channel and its direction, and the next day he presented it in court and swore to its reliability.[102]

Although such outright chicanery was not common, it was in the interest of the expert witness to soft-pedal the points unfavorable to his side of the case. "There is no use beating about the bush in the matter," said the editor of the *Mining and Scientific Press* in 1915. "It cannot be gainsaid that the retained witness is prone to be biased in favor of the signature on his pay-check." [103] Although most judges, admitting that some experts colored their opinions to suit their interests and those of their employer, believed there was no more deliberate misstating of facts, as distinguished from opinions, than from

100. J. B. Hastings to James B. Hague, Rossland, British Columbia, January 3, 1899, Hague MSS, Box 11; *EMJ, 73* (January 4, 1902), 5, *110* (July 17, 1920), 103.

101. Hymen *v.* Wheeler et al., *Federal Reporter, 29* (1886), 357.

102. W. C. Wynkoop, "The Use and Abuse of Expert Testimony," *MI&R, 20* (February 17, 1898), 358.

103. *M&SP, 111* (November 13, 1915), 755. Novelist Gertrude Atherton noted in 1914: "Scientific men are normally honest, although the great fees offered to geologists frequently infuse their judgment with that malleable quality peculiar to the lawyer under the subtle influence of his brief." *Perch of the Devil,* p. 299.

other types of witnesses,[104] Judge Phillips, in the case of Cheeseman et al. *v.* Shreeve et al., thought differently in 1889, as he charged the jury:

> Some of the witnesses, unquestionably, have not told the truth; and it is for you to conclude who they are, and whom you will believe. Judging from this trial there must be something in the altitude of that mountain, or the depths of its mines, wonderfully prolific of falsifiers and orators. From the men of science, versed in geology, mineralogy, and mining engineering, to the most unlettered miner who dwells in the bowels of the mountain, many of them seem to be orators; and, immediately after taking the witness stand, they would be found on the platform, on the same plane as the judge, making a speech to the jury, with all the warmth, energy, and zeal of hired advocates.[105]

Critics also condemned the basic idea of leading technical and scientific men propounding complex and contravening theories of geological formation before judges and jurors, many of whom could not possibly follow their arguments.

> The recurrent spectacle of half a dozen first-rate men, often university professors of distinction and scientists of high repute, testifying that something is black, while an equal number of equally able and honorable men swear that it is white, is but a public scandal and a satire upon our judicial system.[106]

104. See, for example, the comments of Judge Bartch in Grand Central Mining Company *v.* Mammoth Mining Company, *Pacific Reporter, 83* (1905), 667; comments of Judge Marshall in Wall et al. *v.* U.S. Mining Company, quoted in *EMJ, 113* (June 17, 1922), 1054. Of the celebrated Dives-Pelican controversy in Colorado, R. S. Morrison, a specialist in mine litigation, said, "There were also experts who swore to impossible conditions which nature refuses to produce, but with zeal, induced by heavy compensation, declared to be geological facts." Quoted in Thomas F. Dawson, *Life and Character of Edward Oliver Wolcott, 1,* 95.

105. Cheeseman et al. *v.* Shreeve et al., *Federal Reporter, 40* (1889), 796.

106. *M&SP, 119* (December 27, 1919), 911. See also Rickard, *Retrospect* pp. 66–67.

"A mining trial by jury is a contemptible farce, especially when the 'twelve good and true men' are selected from farmers, artisans and others who have never in their lives seen a mine," charged a Leadville mine man in 1891. "The more they hear of the technical evidence of veins, walls, faults, cleavage planes, etc., the more bewildered and confused they become." [107] "Good" witnesses were hired to convince juries and judges, and a good witness need not necessarily be a good engineer, although most were. "He must be wily, positive, clever, alert, a good mental boxer, a good bluffer, and have a single-track mind and never jump the track, and that track must be the one he is paid to run on," insisted one geologist of national reputation.[108] A witness who hedged in deducing geologic fact from the evidence and who qualified his statements was no asset in the courtroom, but the man who with great self-assurance "talked when on the witness stand as if he had been consulting engineer to the Almighty," made a marked contribution to his side.[109]

Countless hours of examination and cross-examination often obscured, rather than exposed, the truth. It was predicted that the testimony of scientific witnesses in an 1877 Nevada case "will make a book as large as any on geology yet published." And, one editor added, the opinions would be "more diversified than is usual in such cases, and some of the statements made are entirely original." [110] "After eleven witnesses had been heard the color carried mentally by an impartial observer was a hazy purple," wrote an interested bystander of the Elm Orlu-Butte & Superior trial in 1919.[111] Frequently, even the judges felt the same way, epecially those eastern territorial implants with little or no experience in mining law. Nathaniel Shaler recalled that one judge at Butte admitted that he was

107. A. Hausmann to editor, Leadville, September, 1891, *EMJ, 52* (October 10, 1891), 421.

108. Josiah Spurr, "The Quaint Usages of Apex Litigation," *EMJ, 109* (June 19, 1920), 1345.

109. Rickard, *Retrospect,* pp. 66–67.

110. *M&SP, 35* (August 25, 1877), 121.

111. Lytzen, "The Elm Orlu *v.* Butte & Superior Apex Litigation," *M&SP 119* (September 13, 1919), 368.

completely at sea in a mass of geological waves, until Shaler, using very plain, simple language, went on the stand.[112] Another, Judge Farrington of the U.S. District Court in Nevada, echoed the same sentiments in 1912, bluntly labeling the system "a farce": "Here I have listened for three months to eminent scientists talking absolutely opposing views upon scientific questions, and I as a judge not trained in geology am called upon to decide which group is right."[113] Nine years later, Judge Elmo E. Bollinger of the Arizona Supreme Court touched off a mild controversy in mining circles, when in his first mining suit he rejected the geological evidence and fell back on what he called the "practical" standpoint.[114]

Part of the basic confusion stemmed from the fact that geology was not an exact science in a full sense and that the area for argument and disagreement was wide. Expert witnesses dealt not only in facts, but also in opinions, which were much more difficult to assess. Complex geological and other problems, conditions underground, the difficulties of observation, and honest differences of interpretation by reasonable men all brought forth legitimate differences of opinion that smart lawyers marshaled to their own ends. Add to this the complications wrought by juries and judges who were not technically trained, and who judged "the hands from a pinochle standpoint" when both litigants set up their cases "with poker in mind."[115]

Many engineers, including Thomas A. Rickard, steadfastly refused to have anything to do with "courtroom mining," and believed it harmed the dignity of the engineering and geological professions. Others insisted that, because of a few cases of "professional perjury" by experts on the stand, many of the best engineers refused to appear in court "because their mo-

112. Shaler, *Autobiography,* p. 338.

113. Quoted in Rickard, ed., *Interviews,* p. 106.

114. *M&SP, 122* (May 14 & June 11, 1921), 662–64, 813; Tom Reed Gold Mines *v.* United Eastern Mining Company, *Arizona Reports, 24* (1922), 277, 290.

115. For a breezy discussion of such cases set forth in the terminology of the cardroom, see J. G. Murphy to editor, Seattle, *EMJ, 112* (December 24, 1921), 1004.

tives will be sure to be impugned and their honesty decried."[116] Though this may have been true of some, the fact remains that the "usual bunch"—the more or less "regulars" in courtroom experting—were extremely able, effective men, certainly among the elite of technical people anywhere in the United States.

Critics did more than object; many had positive suggestions for improving handling of the apex question. A common suggestion, advocated frequently by judges as well as engineers, was the appointment of nonpartisan experts to be paid by the court, rather than the litigants. Judge Farrington believed a special board, composed of one lawyer, one geologist, and one mining engineer, could provide adequate guidance for the court.[117] Some advocated that the court itself appoint all the expert witnesses and that their fees be fixed by law or come out of the public treasury; others suggested that each contestant select one expert, then agree by mutual consent on a third, the decision of the three specialists to be final, with all expenses to be paid out of a common fund advanced by the litigants.[118] Another proposal was for a third set of expert witnesses employed by the judge, with those of the contestants, a measure that some thought would only create confusion by giving the judge "one more color to choose from."[119]

From time to time, efforts were made to put into practice some of these proposed remedies, but none went very far. In 1904, principals in the case between the Morning Glory and the Mary McKinney mining companies in Colorado agreed to abide by the findings of a board of referees made up of three professional mining men, and in 1918,[120] in a suit heard at

116. Wynkoop, "The Use and Abuse of Expert Testimony," *MI&R,* 20 (February 17, 1898), 358.

117. Quoted in Rickard, ed., *Interviews,* p. 106.

118. See *EMJ,* 33 (June 17, 1882), 313, *52* (October 10, 1891), 421, *78* (September 8, 1904), 381, *109* (June 19, 1920), 1345; Wynkoop, "The Use and Abuse of Expert Testimony," *MI&R, 20* (February 17, 1898), 359; *M&SP, 119* (December 27, 1919), 911; Rickard, *Retrospect,* p. 67.

119. John J. Presley to editor, Kellogg, Idaho, June 28, 1920, *EMJ, 121* (July 17, 1920), 82.

120. *EMJ, 78* (December 29, 1904), 1041.

Tonopah, contestants agreed to choose two neutral engineers, at their own expense, to advise the court, a departure editors hoped would set a precedent for the future, despite the fact that in this particular case the "neutral" testimony was strongly on one side "and seems to some partisan." [121] Earlier, the editor of *Mining Industry and Review,* a Denver publication, had spearheaded a movement in Colorado legally to require that all expert witnesses be appointed by the court and their fees charged as part of the costs in the case. Such a law was enacted by the Colorado legislature in 1891, but was vetoed by Governor John L. Routt after prominent attorneys insisted that it would be unsafe to put such powers in the hands of judges who were often prejudiced.[122]

Many engineers and geologists were convinced that the fault lay in the apex law itself and that its repeal would go far toward eliminating litigation, although reliable reports in 1917 indicated that numerically, though not in importance, apex controversies made up only a small proportion of all mining cases during the previous fifty years.[123] One critic condemned both the law and its framers:

> Such a piece of erroneous legislation . . . could have originated only in the brain of an individual of limited

121. *M&SP, 116* (February 16, 1918), 215.

122. The bill restricted the number of expert witnesses in mining cases to a total of ten, requiring these to be men "of good repute, of sufficient learning and ability, not of kin to either party and not interested in the result of the suit, action or proceeding." The court was to select the experts, choosing three each from a list of six nominated by each litigant and selecting four more on its own. Each witness was to receive not more than $25 per day for his work in examining the property and in court, and was to be paid from a court fund deposited by the parties at law. Governor Routt, though approving the principle that the number of witnesses be limited and their compensation fixed, vetoed the measure because of the "dangerous power" he believed it bestowed on the judiciary. *Senate Journal of the General Assembly of the State of Colorado,* 8th Sess. 1891, pp. 340–41, 1250; *MI&R, 20* (September 2, 1897), 92.

123. *Mining and Scientific Press* reported in 1917 that out of 5,808 cases arising in the previous half century, only 115, or 1.9 percent, were under the apex provision. *M&SP, 114* (June 2, 1917), 755. On the other hand, in 1904, Rossiter Raymond urged Gifford Pinchot, as a member of President Roosevelt's commission on the public lands, to consider "the burning question of the 'extralateral right,' which underlies 99.9 percent of the mining litigation in the region subjected to this curse, from which the rest of the world has been free so long." Raymond, "The Mineral Land Laws," *EMJ, 77* (June 16, 1904), 958.

> mental capacity. How such a man could obtain a hearing in a council of law-framers is difficult to imagine. . . . The Law of the Apex . . . begotten in bland self-complacent ignorance by a group of opulent mechanics.[124]

Even many of those who reaped the benefits of high court fees were quick to attack the law as the root of much evil. Rossiter W. Raymond called the apex "a great curse," "an absurd law," and urged Congress to reject this "malign influence."[125] Richard Rothwell urged a real mining law, "not a chaos of confusion": "Banish that elusive thing, the apex, to the limbo of vanity, where it belongs."[126] "Even the rule of the land of poco mañana is infinitely better," said another.[127] Horace V. Winchell, even while employed in various suits, was one of the leaders of the movement to revise the law and wrote both serious prose and home-grown doggerel on "this hoary, antiquated, trouble-breeding license to trespass on a neighbor's property":

> The law of the apex is such a huge joke,
> I hope its defenders will go up in smoke.
> Courts have said that the law is for prospectors made,
> I've observed he's the man who is least often paid.
> For lawyers and experts it's all very well
> But to pay them the miner must rustle [like] h--l.[128]

As early as the 1870s, mining engineers were suggesting doing away with that portion of the law that entitled locators of a vein that apexed on their claim to follow it in depth beyond the sidelines. This extralateral provision, it was con-

124. Quoted in Shamel, *Mining Mineral and Geological Law*, p. 262.

125. *EMJ, 42* (July 3, 1886), 2, *52* (October 10, 1891), 420, *76* (September 26, 1903), 458.

126. *EMJ, 42* (October 9, 1886), 259.

127. Edward A. Belcher to editor, San Francisco, July 25, 1904, *EMJ, 78* (August 4, 1904), 173.

128. From Horace V. Winchell, "The ABC of Mining Law," *EMJ, 109* (January 10, 1920), 61; Winchell remarks, "Meeting of the Mining and Metallurgical Society of America," *Senate Document*, no. 233, 64th Cong., 1st Sess. (1915–16), 57. See also his "A Discussion of Mining Law," *EMJ, 93* (March 9, 1912), 493–97, and his "The Future of Mining," *M&M, 4* (January 1923), 12.

tended, should be eliminated in favor of a square (or perhaps parallelogram-shaped) location with mining rights confined within the four side bounds. Evidence would tend to bear out Horace Winchell's contention of 1922 that "an overwhelming majority of the mining men of the United States" favored such a change,[129] although such important figures as Chester W. Purington, Courtenay DeKalb, and Charles K. Leith argued in favor of the status quo.[130]

But despite the endorsement of prominent judges and engineers, the recommendation of the President's Public Lands Commission in 1879, and the support of congressmen who introduced measures in both the House and Senate,[131] the extralateral right remained. An International Mining Congress meeting in Denver in 1897 created a committee to suggest revision of the mineral statutes and, at its subsequent convention in Salt Lake City the following year, adopted a resolution urging Congress to do away with extralateral rights.[132] President Roosevelt was criticized in 1903 for failure to include a mining engineer on his commission on the public lands, but that body was urged to consider apex modification.[133] However, Roosevelt later appointed a special commission of engineers and geologists for the express purpose of recommending

129. Quoted in Rickard, ed., *Interviews*, 514.

130. Shamel, *Mining Mineral and Geological Law*, p. 273; Charles K. Leith, *Economic Aspects of Geology*, pp. 352–54.

131. "Report of the Public Lands Commission," *House Executive Document* no. 46, 46th Cong., 2d Sess.. (1879–80), xxxviii–xxxix. A bill introduced by Representative Symes of Colorado in 1886 would have eliminated extralateral rights in favor of a parallelogram location. Bills offered by William Stewart of Nevada in 1888, 1893, and 1894 would have modified the apex law but not abolished it. One submitted by Senator Kearns of Utah in 1902 would have established rectangular claims without extralateral rights, one introduced in the House by W. H. Douglas of New York in 1903 would have abolished extralateral rights gradually, and one by Senator Smoot of Utah in 1912 would have brought an abrupt end. *EMJ, 41* (May 8 & June 26, 1886), 343, 460, *48* (August 3, 1889), 112–13, *58* (August 4, 1894), 98, *74* (August 2 & September 6, 1902), 137, 301, *75* (February 7, 1903), 210, *93* (April 13, 1912), 728; "Report of the Commission to Amend the General Mining Laws," *House Report*, no. 639, 63d Cong., 2d Sess. (1913–14), 2, 5.

132. *EMJ, 64* (October 30, 1897), 512, *66* (July 16, 1898), 64.

133. Raymond, "The Mineral Land Laws," *EMJ, 77* (June 16, 1904), 958; Charles W. Goodale, "United States Mining Laws," *EMJ, 79* (February 2, 1905), 222.

changes in the law, and this body, comprised of John Hays Hammond, James Douglas, Daniel M. Barringer, Walter B. Devereux, and William B. Mather reported unanimously in favor of eliminating extralateral rights and substituting ten-acre square locations.[134] But there was influential opposition,[135] and the struggle continued. The Mining and Metallurgical Society gave its support and thoroughly aired the question in its 1915 meeting;[136] in his annual report for 1915, the secretary of the interior displayed a marked sympathy: "the old code is so elaborate and complicated that the best of brains cannot tell what the law is. The truth seems to be that between mining engineers and mining lawyers the rules of the game have been refined into obscurity."[137] Ultimately a committee, headed by Walter R. Ingalls, reporting to the U.S. Bureau of Mines, drafted a law for the abolition of extralateral rights in future locations, and this was introduced in the House by Representative S. S. Arentz of Nevada on July 12, 1921.[138] But even here, opposition, especially from western quarters, proved impossible to overcome. Reluctance to discard a time-honored mechanism, no matter how confusing and obsolete, kept the apex law on the books.[139]

By this time, the last of the major apex suits were coming to an end.[140] Even much earlier, some mining men had discovered that the way to circumvent litigation was by means of

134. F. F. Sharpless, "Revision of the Mining Law," *M&M, 3* (January 1922), 19. One member of the commission wrote another soon after their appointment that the abolition of the extralateral right should be the first question considered. Daniel M. Barringer to James Douglas, May 15, 1906), Letterbook L, Barringer MSS.

135. John Church to James D. Hague, New York, July 29, 1908, Hague MSS, Box 11.

136. Sharpless, "Revision of the Mining Law," *M&M, 3* (January, 1922), 20. See "Meeting of the Mining and Metallurgical Society of America." *Senate Document* 233, 64 Cong., 1 Sess. (1915–1916).

137. Quoted in *M&SP, 112* (January 1, 1916), 2.

138. Other members of the committee included Walter Douglas, J. Parke Channing, J. R. Finlay, John Hays Hammond, Hennen Jennings, and Louis Ricketts, all prominent engineers of much western experience.

139. See Percy Barber editorial, *M&M, 2* (December 1921), 1–2; Sharpless, "Revision of the Mining Law," *M&M, 3* (January, 1922), 21.

140. Interview with Ira B. Joralemon, San Francisco, April 6, 1964.

agreements with owners of adjoining locations. In 1881, William E. Church of the Detroit Copper Mining Company made such an arrangement with owners of a number of claims nearby, all parties pledging themselves to confine their operations in depth within their end and sidelines.[141] Such contracts became common in the Arizona copper regions, in Leadville after it became apparent that lawsuits were ruinous and at both Tonopah and Goldfield in Nevada.[142] Josiah Spurr, editor of *The Engineering and Mining Journal,* though recognizing the strength of human greed, preached private agreement as a solution for the "Apex muddle" in 1921, when he urged: "Boycott the apex law! Render it a dead letter. Make your vertical boundary agreements. Do it now! It will be a patriotic act, and a profitable one." [143] Spurr's plea coincided with the widespread movement for consolidation of mining property already under way, and this, rather than congressional reform or personal self-denial, ultimately provided the answer to this no longer new problem.

Meanwhile, for more than half a century, much time, effort and capital had been expended in the vexatious apex litigations, described so graphically in Ros' Raymond's inimitable style:

> A dozen lawyers on a side,
> And eminent experts multiplied;
> Maps of the biggest and the best,
> And models till you couldn't rest
> Samples of rock and vein formation,
> And assays showing "mineralization,"
> And theories of that or this,
> And revelation of "genesis,"

141. See J. Parke Channing remarks, "Meeting of the Mining and Metallurgical Society of America," *Senate Document* 233, 64 Cong., 1 Sess. (1915–1916), pp. 40–41; James Douglas, "A Remedy for the Law of the Apex," *EMJ, 84* (November 23, 1907), 975.

142. Ibid.; Josiah Spurr, "The Solution of the Apex Muddle," *EMJ, 111* (February 5, 1921), 254; Joralemon, *Romantic Copper,* p. 214.

143. Spurr, "The Solution of the Apex Muddle," *EMJ, 91* (February, 5, 1921), 254.

And summings-up of sound and fury.
No matter now which party lost—
It took the mine to pay the cost;
And all the famous fight who saw
Beheld, with mingled pride and awe,
What science breeds when crossed with law.[144]

In retrospect, can it be said that this expensive, at times unscientific, apparatus that was for so long a part of the western legal tradition was solely negative and nonproductive? Did the expert witness make contributions that balanced off his personal gains? And what of the long-term implications for the mineral industry itself?

There is no doubt that geologists and mining engineers added much to their store of general and theoretical knowledge as a result of their courtroom experience. "Experting" forced them to keep abreast of the latest scientific writing and to think in broad terms. Testifying in cases as far afield as the Comstock, Leadville, or Butte, and making in effect comparative studies, they often published their findings in professional journals or transactions, and despite partisanship, helped swell the body of scientific literature pertaining to the mineral field.

There is no doubt, too, that expert testimony, contradictory though it often was, aided the courts in interpreting the law. It was Rossiter Raymond's concept of a "mineralized zone" as a legal lode, advanced in the Eureka-Richmond dispute of 1877, for example, that paved the way for the court's acceptance of a definition, in a legal sense, of a vein that was much broader than the scientific usage of the term. For, said Justice Field, in the Eureka-Richmond decision, the mining laws

> were not drawn by geologists or for geologists. They were not framed in the interests of science and consequently with scientific accuracy in the use of terms. They were framed for the protection of miners in the claims

144. From "Lawyers and Experts," quoted in Rickard, ed., *Rossiter Worthington Raymond,* p. 95.

which they had located and developed, and should receive such a construction as will carry out this purpose.[145]

This was the spirit that in general dominated the courts throughout the era. Using scientific testimony, despite much sound and fury and wasted effort, the judiciary somehow did manage to find some guidelines. The result was a compromise: neither a purely scientific nor a purely legal point of view prevailed in the interpretation of the mineral laws.

No doubt the use of expert geologists and mining engineers in litigation sometimes led to the discovery, directly or indirectly, of new ore deposits.[146] Often in apex litigation, thousands of dollars were spent by the opposing parties developing underground works to prove the geological theories around which their cases were to be built. Anaconda Copper spent $189,020 for such development work while preparing for its suit against the Colusa-Parrot at the turn of the century.[147] In the Elm Orlu-Butte & Superior litigation, before it was finally resolved, ten raises, each from 500 to 1,200 feet long, and six miles of special drifts were run.[148]

It was at Butte that David W. Brunton, Anaconda's consulting engineer and a veteran of the Colorado legal wars, had urged a geological study of the disputed area. Horace. V. Winchell was hired and a staff organized, including young Reno Sales, to commence a systematic mapping. At first, this work was closely linked with the litigation: detailed surveys, pushed in order to prove or explain theories expounded in court, taught much about the underground features, and unsuspected ore was found in many of the "litigation drifts" during the early round of lawsuits. A report of the U.S. Geological Survey made in 1897 was "largely modified" in a new report

145. Eureka Consolidated Mining Company *v.* Richmond Mining Company, *Sawyer, 4* (1877), 302, *Federal Cases, 8* (1895), 819. See also *EMJ, 24* (August 25, 1877), 145; address by George W. Maynard, *Rossiter Worthington Raymond* (n.p., 1910), p. 57.

146. Shamel, *Mining, Mineral and Geological Law,* p. 4.

147. *EMJ, 69* (June 16, 1900), 720.

148. Arthur B. Parsons, "Mining Litigation and Common Sense," *M&SP, 122* (June 11, 1921), 815.

of 1903, "especially as to the extent and character of the faulting of veins," mainly because of the active investigation of the subterranean structure in the hunt for legal evidence. "Thus was Science helped by the most utilitarian form of commercial enterprise," said Thomas A. Rickard somewhat begrudgingly.[149] Most experts agreed that new ore found during the course of litigation work at Butte was significant.[150]

Most agreed also that the continuation of geological study was of profound import. "The Butte litigation had started a great new branch of applied science," wrote one engineer later.[151] When consolidation terminated the legal activity in 1906, Anaconda expanded its geological work on a district-wide basis, projecting known patterns of veins and faults into unopened ground, and Butte became the classic example of mineral zoning and mapping. From its "Anaconda School," many a geologist or engineer went out to apply its principles elsewhere in the mineral world.[152]

This is not to say that these same results were necessarily duplicated in other areas, although the by-product of litigation over silver-lead deposits at Tintic, Utah, in 1885, was the discovery of large important ore bodies by both contestants during their development work preparatory to the hearing of the

149. Thomas A. Rickard, "The Butte Report," *EMJ, 75* (April 18, 1903), 586.

150. For positive statements on ore discovery, see Horace V. Winchell to editor, Minneapolis, *EMJ, 110* (July 17, 1920), 103; D. W. Brunton interview, Rickard, ed., *Interviews*, pp. 79–80; *M&SP, 122* (May 29, 1921), 751; Isaac F. Marcosson, *Anaconda,* p. 134. Walter R. Ingalls registered disgust at the waste and discredit to the industry that came from the litigation, but admitted that the contributions to geological knowledge and the discovery of new ore deposits gave Butte a new lease on life. Ingalls, "The Montana Settlement," *EMJ, 81* (February 17, 1906), 329.

151. Joralemon, *Romantic Copper,* p. 111.

152. Anaconda developed a special mapping system, using vellum in such a way that "On each individual sheet is platted not only the working but the geology and as you can read through about two or three sheets of vellum it is easy to see the relation of the workings, faults, and orebodies to each other." D. W. Brunton interview, *M&SP, 122* (May 28, 1921), 751. See also Fred T. Greene to editor, Butte, April 15, 1911, *EMJ, 81* (April 29, 1911), 845–46; Joralemon, *Romantic Copper,* pp. 110–11; Read, *Development of Mineral Industry Education,* p. 15; Arthur B. Parsons, ed., *Seventy-Five Years of Progress in the Mineral Industry*, pp. 14–15.

case, and Jonathan Bourne, Portland capitalist, was able to use the geological knowledge developed in a 1908 apex suit to take profits of $1 million from a faulted portion of an ore body within the vertical bounds of his Ontario mine.[153] But these and certainly Butte were exceptions, rather than the general rule.[154] And if Anaconda was the first of the major concerns to establish its own geology department, it had before it the examples of district studies made by the U.S. Geological Survey. Yet, its great mass of litigation at the turn of the century focused attention on the geologist and soon led to consolidation, which in turn not only made for greater operational efficiency, but curbed future lawsuits and also permitted the scientific investigation of an entire district under a single auspices.

153. *EMJ, 40* (November 7, 1885), 317; *M&SP, 120* (May 29, 1920), 794.

154. For comments on the waste and futility of litigation as regards the net discovery of ore, see *Statistics of Mines and Mining in the States and Territories West of the Rocky Mountains; being the Sixth Annual Report of Rossiter W. Raymond* (1874), p. 515; *EMJ, 67* (May 13, 1899), 571; *M&SP, 99* (December 27, 1919), 911, *122* (September 10, 1921), 360.

CHAPTER 7

"If Everyone Specialized, Where Would Be the Common Meeting Ground?"

> *Who steals our gold and silver, and copper, zinc and lead?*
> *Who takes the joy all out of life and strikes our high hopes dead?*
> *Who never wrote a schedule that to anyone else was clear?*
> *The sulphur-belching, miner-welching smelter engineer.*
>
> —Anonymous [1]

The versatility of the mining engineer was proverbial. Like the young graduate—a gem in the rough—who offered George A. Sala a "plug of terbaker" on a westbound train in 1880, the engineer was "ready for anything" from stock speculation to becoming "the conductor of a freight train." [2] A man of the world, the mining engineer could call on many talents:

> he is as much at home in some primitive, far-off, revolutionary country, with his hat for an office, as in a city skyscraper. . . . He can throw a diamond hitch, cook his food, drive a mule or a tunnel, manage an Irish foreman, sun-dodging Mexican or a Chinese coolie. He is an expert with the slide rule and with the can opener, and some ancient writer, possibly Agricola, has intimated that, in the "good old days," he understood the technique of the corkscrew. Like the country doctor, he can make an emergency repair with a jackknife, a piece of string, a scanting and the tail of a shirt. . . . He is facile in language, ranging from burro Anglo-Saxon to pigeon Spanish or Chinese.[3]

1. From "The Engineer," *EMJ, 105* (April 13, 1918), 696.
2. George Augustus Sala, *America Revisited,* pp. 390–91.
3. Charles H. MacDowell, quoted in *M&M, 1* (April 1920), 7.

Within his profession, the mining engineer was at once a generalist and a specialist, and frequently moved from one type of work to another—from mine examination to management, erection of machinery, surveying, milling, or development work, perhaps in the end settling on one particular emphasis. A war-time census of some 7,500 mining engineers, taken in 1917, indicated a wide range of experience (see Table 4).[4] Likewise, many engineers jumped from one mineral to

Table 4. Experience of Mining Engineers, 1917

TYPE OF EXPERIENCE	NUMBER
Assaying	718
Construction (building)	1,770
Construction (machinery)	735
Development work	2,325
Designing and drafting	945
Dike and levee construction	98
Drainage and pumping	563
Dredging	115
Erection of machinery	1,050
Explosives (mine and quarry)	863
Hydraulic mining	173
Metallurgy (all kinds)	2,920
Ore concentration	1,038
Prospecting (with boring machines)	840
Quarrying	165
Steam-shovel mining	330
Surveying, mine or topographic	1,935
Tunnel and shaft work under compressed air	83
Underground mining	7,500
Miscellaneous *	1,615

* Includes 467 consulting engineers, 312 geologists, 121 professors of mining, metallurgy, or geology, 129 who held administrative executive posts in mining companies, and an unspecified number of inspectors, accountants, editors, sales or purchasing agents and efficiency, fuel, or safety engineers. The figures represent the entire country, and, because many men had experience in more than one category, their total is more than 7,500.

SOURCE: Albert Fay, "Preparedness Census of Mining Engineers, Metallurgists and Chemists," U.S. Bureau of Mines, *Technical Paper,* no. 179 (Washington, D.C., 1917), pp. 8–9.

4. How many mining engineers there were in the country at any given time is difficult to determine. Samuel Christy estimated in 1893 that the mineral

another. Fred Bradley, for example, first made his reputation in the gold quartz mines of California, then moved to the silver-lead ores of Coeur d'Alene, tried gold dredging, and was connected with the production of tungsten. The 1917 census showed that the mining engineer's experience by the type of mineral was almost as varied as the type of work he performed (see Table 5).

Table 5. Mining Engineers' Experience by Minerals, 1917 *

Coal	1,795
Copper	1,758
Gold and silver	1,840
Iron	842
Lead and zinc	1,384
Manganese	37
Mercury	35
Molybdenum	20
Nickel	18
Petroleum	1,045
Phosphate	27
Pyrite	19
Rare metals	55
Tin	34
Tungsten	57
Other metals	303

* These figures represent the entire country, with some engineers listed under several headings.

SOURCE: Albert Fay, "Preparedness Census of Mining Engineers, Metallurgists and Chemists," U.S. Bureau of Mines, *Technical Paper,* no. 179 (Washington, D.C., 1917), p. 9.

Editors, commencement speakers, and textbook writers invariably extolled the virtues of broad training and experience,

industry employed *"not over 6,000 persons who may be said to require technical training as engineers."* Samuel B. Christy, "The Growth of American Mining Schools and Their Relation to the Mining Industry," AIME *Trans., 23* (1893), 452. The Twelfth Census (1900), the first to list the profession, showed 2,908 mining engineers, but included an additional 6,034 under the heading of "Surveyors" and 8,887 more as "Chemists, Assayers and Metallurgists." *Occupations at the Twelfth Census,* p. 7.

and certainly a majority of the old-time engineers were by necessity general practitioners of a most adaptable sort. But, as one speaker told the American Mining Congress in 1907, "Specialization has produced as remarkable results in mining as in other technical branches."[5] Advancing technology, large-scale operations, and increasung complexity prompted a gradual, if inexorable, trend in the direction of particularization, a movement that bothered some members of the profession. "If everyone specialized, where would be the common meeting ground?" asked one critic in 1915.[6]

Those who focused on mine examination, management, or litigation experting were specialists of a kind. Even as early as the 1880s, it was being acknowledged that specialists were beginning to make inroads, and soon even such top-level veterans as Rossiter Raymond and Thomas A. Rickard were being criticized by their colleagues for venturing into specializations not their own.[7] The process was slow and never complete, but gradually the generalist—"the man who could walk through a mine and tell you all about it when he came out"—gave way to the specialist who handled one phase of an examination or process of treatment.[8] Ultimately, in considering the basic question of whether a mine would pay or not, a consulting engineer came to augment his own opinion with reports and estimates from electrical, mechanical, and metallurgical engineers.[9] Engineering partnerships began to recognize this growing need for specialization and to align themselves, as did the San Francisco firm of Benjamin, Hunt, and Meredith in 1899, to include one mining engineer, one mechanical and hydraulic engineer, and one electrical engineer.[10] Leading professional

5. Milnor Roberts, "What Can the Profession Really Expect from the Mining School Graduate?" *Report of the Proceedings of the American Mining Congress,* Tenth Annual Session, Joplin, Missouri, November 11–16, 1907, p. 176.

6. Charles F. Willis, "The Need for Better Mining Education," *Report of the Proceedings of the American Mining Congress,* Eighteenth Annual Session, San Francisco, California, September 20–22, 1915, p. 152.

7. Robert Peele to Eben Olcott, New York, October 26, 1894, Olcott MSS (N.Y.); William S. Potter, "A Present Need in the Engineering Profession," AIME *Trans., 17* (1888–89), 383.

8. *Pacific Miner, 16* (June 1910), 205.

9. *EMJ, 116* (December 29, 1923), 1104.

10. *M&SP, 78* (May 6, 1899), 500.

schools turned out an increasing number of petroleum, metallurgical, fuel, or geological engineers—specialists all—along with mining engineers proper. Less than half the 1927 graduating class at the Colorado School of Mines were in mining engineering; the remainder were in the more particularized fields.[11]

In practice, no well-defined boundary was drawn between the fields of work of the mining engineer and the metallurgist. Most trained engineers received some instruction in ore milling and in metallurgy, and most at some time in their careers had first-hand experience in these areas. Sometimes, the engineer's functions ended with the delivery of the ore to a custom reduction works; in other cases, the mining company's plant included a concentrating mill, amalgamating or cyaniding works, or even a smelter.[12] But some engineers, from the beginning, devoted their full energies to the basic questions of separating mineral from ore and of extracting metal from miners. The Columbia School of Mines had recognized this emphasis as early as 1873, when it offered a degree in metallurgy; by 1887, it offered one in metallurgical engineering, and other technical institutions were following the lead.[13]

Certainly the period from 1849 to 1932 saw great advances being made in the means of working ores—especially complex ores and low-grade deposits—and in these advances lay one key to the continuing development of the mineral industry itself. Innovation followed countless experiments, many of them

11. Of 64 graduates, 30 were mining engineers; the rest were in petroleum or geological or metallurgical engineering. *Colorado School of Mines Magazine, 17* (July 1927), 4, 11.

12. Peele, ed., *Mining Engineers' Handbook, 1,* iii. The term "reduction mill" is a general one used to designate a plant designed to reduce ore—that is, to separate metals from ore—by whatever method, be it the use of chlorine, cyanide, smelting, or otherwise, The concentrating mill was any plant that used air, water, specific gravity, or mechanical means to eliminate much of the worthless ore content, leaving a concentrate for final processing. In a cyanide works, gold would be extracted from concentrates, tailings, or crushed ore by means of a dilute solution of potassium cyanide. Gold dissolved in this solution was subsequently deposited on metallic zinc or separated by other means. "Smelting" implies converting ore to molten state by means of heat and chemicals and separating the metallic content by specific gravity. Fay, "Glossary," pp. 175, 176, 203, 561, 627.

13. *EMJ, 93* (November 4, 1911), 877.

fantastic and wholly unscientific, which drained both capital and confidence. But, at the same time, technically trained and experienced men were able to draw on information and observations from all parts of the globe to evolve and adapt such important processes as Washoe pan amalgamation, chlorination, cyanidation, and flotation—all vital to western mining.

Early California techniques were simple. Placer gold or even free-milling quartz posed few problems, although the stamp mill was an inefficient mechanism that lost a substantial percentage of gold in the tailings. But even in California in the 1850s, ores were being encountered in which gold was in combination with slight amounts of sulphur or base metals and not so readily removed. Hence, mining men constantly sought new means of working such ores or techniques to decrease the loss of precious metal in the tailings. One onlooker wrote Joshua Clayton from California in 1859: "There are 6 or 8 other plans in vogue here—by different parties,—*all better than the other, and sure to take out all the gold!*" [14] About this time, the Plattner process was introduced at Grass Valley by G. F. Deetken, who placed fine roasted ores in wood or lead-lined tanks, allowed chlorine gas to penetrate for several days, then leached out the gold with water. This was the first practical use of chlorination in the West, but when improved and modified, it would for years remain one of the standard methods. The next step was to speed up the process by using revolving, lead-lined barrels. Next came the washing in the revolving barrel, and eventually the technique was to dump the entire charge into a separate filter vessel, with washing done under pressure.[15]

Most silver along the eastern edge of the Sierra Nevada was combined with gold, with very little base metal, and could be treated with California and Mexican methods. Almarin Paul and others modified and expanded the *patio* and *cazo* processes

14. Redick McKee to Joshua Clayton, San Francisco, June 6, 1895, Joshua E. Clayton MSS (Calif.), *1*.

15. Rossiter W. Raymond, *Statistics of Mines and Mining In the States and Territories West of the Rocky Mountains* (1875), p. 110; *EMJ, 60* (September 21, 1895), 270.

into what was known as Washoe pan amalgamation, which ground silver ore in pans or tubs with the addition of mercury and sometimes blue vitriol and salt or other chemicals. Even so, in 1862, it was estimated that one third of the ore value was being thrown away with the tailings and slimes on the Comstock.[16] This problem was attacked by the Janin brothers, among others. At the Mexican Mill in 1864, Louis Janin sought to apply a new process but was thwarted by the manager ("The new Supt. of the Mex. Mill is an ass," said Janin.), so he took charge of the Gould & Curry, where he reported in 1865 "everything is going on swimmingly." The Gould & Curry turned to pan amalgamation, but it also built the first mill designed to work tailings in 1866. At the same time, Janin pushed his experiments and, with several others, acquired some 100,000 tons of tailings and slimes, and in 1867 began to work them in a specially built plant, only to have a spring freshet sweep down the canyon and carry away much of their rich ore mud and put an end to the enterprise. But the Janins discovered the proper method of amalgamating tailings, and various Comstock mills then began to process their own. Louis found the work monotonous, however, and soon left metallurgy to his brother Alexis, who was regarded as the country's leading authority on ore amalgamation at the time of his death in 1897.[17]

16. *AJM, 1* (June 30, 1866), 216; San Francisco *Bulletin,* June 12, 1860; A. D. Hodges, "Amalgamation at the Comstock Lode, Nevada," AIME *Trans., 19* (1890–91), 197. In the simple *patio* process, crushed silver ore was mixed with salt and copper sulphate by the treading action of horses or oxen and amalgamated with mercury. In the *cazo* approach, amalgamation was achieved by boiling the finely mashed ore with mercury in a copper-bottomed vat, or *cazo,* with salt and additional mercury added. Both were slow and cumbersome, but the Washoe pan process adapted them for large-scale use. Tailings were the leavings after the ore that was most readily removed was taken out and slimes were a thin ore mud, often the product of wet crushing valuable ore. Fay, "Glossary," pp. 141, 492, 623, 671.

17. Hodges, "Amalgamation at the Comstock Lode, Nevada," AIME *Trans., 19* (1890–91), pp. 211–13, 226; *EMJ, 52* (November 21, 1891), 599, *63* (January 23, 1897), 95; Louis Janin to Louis Janin, Sr., Virginia City, July 30, 1864, November 20, 1864, & February 8, 1865, & Louis Janin to Juliet C. Janin, Virginia City, October 5, 1864, & February 19 & March 20, 1865, Janin family MSS, Box 10; Alexis Janin to Juliet C. Janin, Galena, Nevada, July 18, 1871,

But the problem of treating the refractory ores found in central and eastern Nevada, Utah, Montana, and Colorado was every bit as perplexing as the loss of metal in tailings. Silver is a complicated, deceptive metal. Frequently, it, or gold as well, was in combination with sulpur and base metals and stubbornly resisted early attempts at separation. Many mining men attributed the slump in the mineral industry of the mid-1860s to an inability to work these rebellious sulfuret ores.

> Our German fathers, working mines,
> First exorcised the devil;
> While we affirm that sulphurets
> Are the sole cause of evil.[18]

"In no other department of metallurgy have our inventors made such a brave show, with so little success," wrote one knowing editor in 1870.[19]

Considerable progress was already being made. Already, Nathaniel P. Hill, with the aid of a German metallurgist, had established a smelter at Blackhawk, Colorado. Soon Hill would combine his talent with that of Richard Pearce, a Cornishman of considerable experience in tin and copper, to build a new plant that would successfully adapt Swansea processes to Colorado with considerable success.[20]

And already a number of remarkable European-born, German-trained engineers were at work in the West, combining the best of the Old World processes with the products of their own fertile imaginations to produce a working milling and smelting technology second to none. Guido Küstel was one. Born in Austria and trained at Frieberg, Küstel was in California in 1851. He improved barrel amalgamation and introduced it into the Santa Rita mines of Arizona in 1858, and he was one of the pioneers in the use of chloride and lixiviation

---

Janin family MSS, Box 5; Louis Janin to James D. Hague, November 9 & 27, 1891, Hague MSS, Box 12.

18. Unidentified clipping, 1867, Hayes Scrapbooks, 4.
19. *EMJ, 10* (December 13, 1870), 377.
20. *DAB, 14,* 353–54; *EMJ, 37* (June 14, 1884), 438.

—the separation of a soluble from an insoluble by washing with a solvent. "In his special field," wrote one editor, "he has probably, more than any other man, influenced and shaped progress." [21]

For a time, Küstel's junior partner was Ottokar Hofmann, another Freiberg-educated Austro-Hungarian, who carried chlorination into Sonora and worked out a number of improvements in lixiviation.[22] Küstel was also in partnership for a number of years with Eugene N. Riotte, a Prussian with a Freiberg background, who was one of the early pioneers in Nevada smelting and who also invented a quicksilver furnace of some importance.[23]

Riotte was forced to give up his work in Nevada, where he was associated with Carl A. Stetefeldt in 1870, because he contracted "a fearful salivation" while roasting tailings. Stetefeldt had come from his native Germany via Clausthal in 1863. Soon after, he patented an improvement on the Gerstenhofer roasting furnace and, in 1869, constructed a unit for an English company at Reno. The Stetefeldt furnace for roasting and chloridizing silver ore marked a definite advance in handling sulphide ores, although critics scoffed at its originality: "I reflect that Stet is not modest: that he stole the Gerst. idea & now has tried to act independently in the chloridining idea," complained one engineer. But Stetefeldt's equipment was widely adopted; he did much to improve his own and others' processes—especially the Russell lixiviation method; and his book on the lixiviation of silver ores became standard after 1888.[24]

21. Brunckow, *Report . . . of the Sonora Exploring & Mining Co.,* pp. 8, 12–13; *M&SP, 8* (April 16, 1864), 248, *17* (November 7, 1868), 297, *45* (August 19, 1882), 120; *EMJ, 39* (September 16, 1882), 145.

22. *M&SP, 17* (November 7, 1868), 297; *EMJ, 52* (August 8, 1891), 161.

23. *EMJ, 51* (May 16, 1891), 583

24. Amos Bowman to Allen Curtis, December 2, 1869, Manhattan Silver Mining Company MSS; *DAB, 17,* 595; *EMJ, 10* (July 12, 1870), 21; *61* (March 28, 1898), 300; *M&SP, 95* (August 10, 1907), 158; Rossiter W. Raymond, *Statistics of Mines and Mining in the States and Territories West of the Rocky Mountains* (1870), pp. 125–26. The basic Russell process was one of chloridizing and roasting, then leaching with water to remove base metals. Sodium hyposulphite and cuprous-sodium hyposulphite were used for leaching

Others of the German school would include Albert Arents, who, at Eureka, Nevada, introduced a revolutionary rectangular furnace equipped with his patented "leadwell," or "siphon-tap," an arrangement by which molten lead could be drawn off without disturbing the operation of the furnace. This set the tone for such smelting equipment down to comparatively recent times.[25] August Wilhelm Raht came to America in 1867 fresh from Freiberg and pioneered in silver-lead smelting practice all over the West, not to mention Australia; eventually he became the metallurgical troubleshooter for the Guggenheims and was well known for his bessemerizing of copper matte and for his work in developing the "gum-drop" method of sampling silver-lead bullion.[26] Yet another, Frederic Anton Eilers, introduced a more precise system of metallurgical accounting; built the Eilers Smelter at Pueblo in 1883, a plant that gained a reputation as "a veritable metallurgical training school", and was known as a leader among those who "changed lead smelting from a rule-of-thumb affair to an exact science of working out the theory and practice of slag formation on an accurate chemical basis." [27]

This is not to say that German-born and educated engineers dominated western metallurgy, but certainly they were among the most important brilliant advance guard that greatly influenced the field in the late nineteenth century. No one would minimize the contributions, however, of such men as Edward D. Peters, who left an indelible imprint on the early copper and nickel industries and ended his career as professor of metallurgy at Harvard and MIT.[28] Nor can the Irishman Philip Argall be ignored. Argall came to Leadville in 1887 with mining experience on three continents. During the 1890s, he saw

---

silver, which was then precipitated by the action of sodium sulphide. Carl A. Stetefeldt, "Russell's Improved Process for the Lixiviation of Silver-Ores in Its Practical Application," AIME *Trans., 15* (1886–87), 355–81; Fay, "Glossary," pp. 492, 584.

25. *DAB, 1,* 343–44; *Who's Who in Mining,* p. 2.

26. *DAB, 15,* 326; *MI&R, 20* (June 30, 1898), 571.

27. *DAB, 6,* 63–64; *Who's Who in Mining,* p. 29.

28. *DAB, 14,* 504; *EMJ, 38* (September 13, 1884), 169, *103* (March 3, 1917), 380, *106* (August 31, 1918), 424.

the potential in the MacArthur-Forest cyanide process, a Scottish innovation, then alternatively proclaimed as a "humbug" and an American idea.[29] Argall designed, built, and managed a large-scale cyanide plant at Cripple Creek, which by 1895 was handling 3,000 tons of ore per month, and which resulted in the lowering of the cost of treating ore from $15 a ton of $3.50 in 1898 and to $1.38 in 1913.[30]

Another who worked out improvements and was instrumental in applying the cyanide process successfully was Charles W. Merrill, a Berkeley graduate, who started his career with Alexis Janin, but by 1893 was on his own. "Cyanide Charlie," as he was known behind his back, built a large plant at the Standard Consolidated mine at Bodie, another on the Arizona desert, and in 1899 was awarded a ten-year contract to install equipment for his advanced version of the process at the Homestake mine in South Dakota, his profits to include a share of the gold saved. This work being completed in 1908, Merrill organized the Merrill Company at San Francisco to spread his equipment and methods (he took out twenty-five metallurgical patents during his career) and met with phenomenal success.[31]

Mention must also be made of Charles Butters, who had worked with chlorination during the 1880s, but was converted to cyanide in South Africa during the 1890s. On his return, he built the first plant in Mexico to treat slimes by agitation and cyanidation, and he installed another on the Comstock to reclaim tailings, but lost nearly half a million dollars before perfecting the process. Butters was also one of the first in the country to utilize oil flotation. Although his detractors referred to him as "that second edition of Barnum & Bailey, that pseudo-metallurgist" and he had the reputation of being a

29. This was a basic process of recovering gold by leaching pulped ore with a solution of potassium cyanide and then with water, precipitating the gold on zinc or aluminum. A Newark resident named Simpson was supposed to have patented the idea and Louis Janin, Jr., claimed to have experimented with it before MacArthur and Forest. *EMJ, 50* (December 13, 1890), 685, *53* (May 21, 1892), 540; Fay, "Glossary," p. 411.

30. *DAB, 1,* 344–45; Rickard, ed., *Interviews,* pp. 9–33.

31. Ryder, *Merrill,* pp. 2, 24–41, 55–58; Bunje, *Careers,* p. 10.

"grinding employer," Butters nonetheless has to be regarded as one of the leaders in his field.[32]

So also was Louis D. Ricketts, who gained his professional standing primarily in the field of Southwestern copper. At Morenci, he built a plant with concentrator, crushers, graders, washing and shaking machinery connected by conveyor belts —the first time such an arrangement had been tried. At Globe, in 1904, he designed and built for Phelps Dodge a new concentrator and smelter that returned more profit from 4 percent ores than previously had been made on 12 percent ores. At Ajo, he developed a leaching process, which made possible the working of low grade ores that previously had not responded to concentration; while at Inspiration, he combined leaching with flotation on a large scale, and, with the use of block-caving mining techniques, ore of 1.226 percent became profitable.[33]

Obviously, the history of milling, smelting, and metallurgical techniques is a story in itself, involving thousands of individual contributors, both trained and untrained. Suffice it to say that many mining engineers were part of that story, either as specialists or as general practitioners who at one time or another had occasion to work in ore or mineral processing. It is a telling point that in the three years prior to 1918, fully two-thirds of the papers published in the AIME *Transactions* dealt with metallurgical practice. In that same year, this organization again took up the question, first raised in 1876 and intermittently thereafter, of whether to change the name of the society more in keeping with its members' interest and emphasis. Hence the change to American Institute of Mining and

32. Frederick C. Roberts to Walter Bradley, San Francisco, June 24, 1909, Walter W. Bradley MSS, Box 1; J. S. Wallace to H. E. West February 28, 1912, copy, Bradley MSS; Rickard, ed., *Interviews*, pp. 119–24; Bunje et al., *Careers*, pp. 5–6. Oil flotation took advantage of the principles of surface tension and colloid chemistry to separate mineral from pulverized ore by floating it on the surface of oil while the residue sank to the bottom of a settling tank. Fay, "Glossary," p. 277.

33. Bimson, *Ricketts*, pp. 15–23. Block caving involved mining from the top down in successive layers of considerable thickness. Each block was undercut over most of its bottom area and its pillars blasted out. Fay, "Glossary," p. 87.

Metallurgical Engineers in 1919 was a recognition of reality.[34]

In addition, there were other forms of specialization. Engineers often concentrated on a single metal. Henry W. Gould, for example, was known as "a world authority" on the mining and metallurgy of quicksilver; [35] F. L. Bartlett was a zinc expert; Edwin Chase specialized in manganese; Arthur Thatcher, Eugene N. Engelhardt, and Henry Collins, in lead or silver-lead ores. James T. Beard and Lyndon K. Armstrong were known for their expertise in coal mining,[36] whereas Edward D. Peters, according to his card in 1893, was one of many who "Attends exclusively Copper mining and Smelting." [37]

The vast bulk, of course, were experienced in underground techniques. Some, such as Victor M. Braschi, focused on a single facet—"Machine rock drilling in all its branches." [38] Others specialized in a particular approach to mining. Hamilton Smith and George T. Coffey were primarily known for their work in hydraulic mining before restrictive legislation brought hydraulicking to its knees in California after 1884.[39] Following the 1890s, dredging would be an increasingly important specialization. In this field, the pioneer was Robert H. Postlethwaite, a New Zealander who introduced New Zealand dredge techniques into California in late 1895 or early 1896 and who would remain as one of the leading dredge designers in the country, first for the Risdon Iron Works, then the Union Iron Works, even after the two merged in 1911.[40] As dredging flourished in California, it migrated into other areas of the West and the Far North. An intricate business, it involved highly expensive equipment, new techniques of testing

34. *M&SP, 116* (May 25, 1918), 729.

35. Wolfe, ed., *Men of California,* p. 67.

36. *M&SP, 116* (June 8, 1918), 806; *M&M, 13* (November 1932), 501–02; *Who's Who in Engineering, 1922–23,* p. 64; *Who's Who in Mining,* p. 6; Rickard, ed., *Interviews,* pp. 484–86; *Lists of the Alumni of Columbia . . . 1892,* p. 28.

37. "The Mineral Industry Advertiser," in Rothwell, ed., *Mineral Industry 1893, 2,* 46.

38. Ibid., *2,* 43.

39. *DAB, 17,* 273; *M&M, 10* (July 1929), 249–50.

40. *Who's Who in Engineering, 1922–23,* p. 1009; *EMJ, 54* (December 11, 1897), 699, *61* (June 29, 1901), 823.

and sampling ground, and the large-scale working of low grade material—all of which required a high degree of professional skill to accomplish. Thus, there quickly emerged a substantial body of dredge engineers who both designed machinery and supervised its installation and operation. Thomas Barbour of San Francisco specialized in dredge construction; Robert E. Cranston, O. B. Perry, and Charles H. Munro were among the leading consultants in dredge engineering, and often depended on such specialists in the field as Fred L. Morris or William H. Lanagan.[41] Lanagan, like J. P. Hutchins, became an expert in dredging in cold northern climates, where distinct adaptations had to be made, using steam for thawing as well as for power.[42]

The development of low-grade copper techniques early in the twentieth century also created a group of specialists steeped in the new methodology. These men—the Jacklings, Ricketts, Requas, and Channings in the vanguard—supervised the ingenious "block-caving" techniques, the open-cut methods using steam or electric shovels to strip away the overburden and dump raw ore into railroad cars, and they devised and perfected new metallurgical processes that enabled them to work profitably ores that averaged as low as 1 per cent copper.[43]

By the early 1900s, another specialty had emerged—the geological engineer. Almost from the beginning, such trained geologists as Clarence King, Richard Penrose, and Josiah Spurr had followed the mining engineering profession, either wandering in and out or going over to it completely. And from the beginning, some working knowledge of geology was essential to effective engineering work. In some instances, as in the Tombstone district of Arizona, the Bendigo Reefs in Aus-

41. *Who's Who in Engineering,* p. 98; *M&M, 13* (September 1932), 425; *EMJ, 82* (November 10, 1906), 889, *118* (July 19, 1924), 84.

42. W. H. Lanagan, "Notes on Winter Trip from Nome through Council, Candlee, Kugruk, Inmachuk, Kongarox Districts," April 1910, copy, Morris MSS, Box 1.

43. Parsons, *The Porphyry Coppers,* pp. 10–11; James K. Finch, *The Story of Engineering,* pp. 314–21; Parsons, ed., *Seventy-Five Years of Progress,* p. 511.

tralia, and the Witwatersrand in South Africa, the engineer with a knowledge of geology was able to locate ore bodies with considerable precision.[44] But only later did this specialization take on a separate being, falling somewhere between the mining engineer and the geologist per se.

The geological engineer was concerned with the broader question of locating or developing ore by studying geological formations. His function was

> to read the story of the rocks, to note their promise and the probability of fulfillment, and to determine what possibilities the classified facts of geologic science may bring to light beyond those bare realities which have been touched with the hand, measured with tape and transit, or proven with the diamond drill.[45]

Anaconda set the pace, but large firms followed the lead with staffs of their own to conduct geological work on a large scale and provide an accurate mapping or sectioning of strata vital to sinking or raising shafts from several levels at once and essential in the development of systematic mining as it emerged in the twentieth century.[46] By 1906, when Walter Ingalls of the *Engineering and Mining Journal* editorialized on the trend, the geological specialist was already important in major operations, though not for small ones:

> The big mining company now requires a mining geologist to study the evidence presented as a guide toward the discovery of ore deposits; a mining engineer to extract the ore; a mill engineer to extract the minerals from the ore; a metallurgical engineer to extract the metals from the mineral; and finally a mechanical engineer to build the various plants and maintain them in good running order.[47]

44. Forbes Rickard to editor, Denver, April 14, 1903, *EMJ, 75* (April 25, 1903), 625–26; T. A. Rickard, "The Bendigo Gold-Field," AIME *Trans., 20* (1891), 463–544; John A. Church, "The Tombstone, Arizona, Mining District," AIME *Trans., 33* (1903), 3–37.

45. *EMJ, 80* (December 16, 1905), 1125.

46. See pages 228–29, this book.

47. *EMJ, 81* (May 19, 1906), 956.

But the geological engineer had to win acceptance; at first he was regarded by the mining engineer "with the same collegiate upper-class contempt which the miner had accorded to him." [48] By 1920, the geological engineer had firmly established himself, and was even specializing more narrowly in such fields as coal, petroleum, metalliferous deposits, and nonmetallic deposits. Were it not for the unwieldiness of the title, probably the AIME officially would have become the American Institute of Mining, Metallurgical, and Geological Engineers in 1919 when it changed its name to give a better indication of the scope of its membership.[49]

Although "efficiency" was the byword of the modern engineer, discussion of the "scientific management" ideas of Frederick Winslow Taylor are curiously absent from the columns of mining publications until the era of World War I. But consciously or unconsciously, most mining engineers worked toward the same goals as Taylor and his followers, and here and there in the twentieth century would appear one of the profession, such as George T. Harley, who made his real reputation as an efficiency engineer, advising on standardized practices, cost reduction, and more effective use of human resources.[50]

During the years after 1900, there was an increasing awareness, both corporate and governmental, of the need for safety measures and programs in the mineral industry, and, especially with the establishment of the federal Bureau of Mines in 1910, a few skilled men appeared who might be termed safety engineers. Some of these, such as Daniel Barrington and C. L. Colburn, came from the ranks of ordinary mining engineers and worked in conjunction with the Bureau of Mines. Others were affiliated with such organizations as the California Metal Producers Association, and some, of whom Charles W. Good-

48. *M&SP, 107* (December 13, 1913), 921; "The Geologist as Engineer," *EMJ, 109* (February 7, 1920), 374.

49. Ibid.

50. *Who's Who in Engineering, 1922–23,* p. 558; Glenville A. Collins, "Efficiency-Engineering Applied to Mining," AIME *Trans., 43* (1912), 649–62.

ale was an excellent example, worked with private corporations, in Goodale's case, Anaconda Copper.[51]

But despite increasing specialization, it should be reiterated that the general practitioner, though constantly losing ground, was never supplanted. So long as there were small companies, there continued a need. But the trend was clear. And the independent engineer, long the backbone of the profession, gradually came to be retained by the large corporation, often as a specialist.

Mining engineers sometimes served as agents or technical advisers to the mine machinery industry, sometimes on a part-time, sometimes on a full-time basis. William Bredemeyer advertised his services in 1875 as including the normal functions of the consulting engineer, U.S. mineral surveyor, and representative for the Humboldt Company, manufacturers of mining and concentrating equipment.[52] John A. Church had the agency for Paddock's Ore Dressing Machinery in 1879, and Joshua Clayton a few years later sought the exclusive franchise to handle the Rappid Ore Concentrator for the Utah, Idaho, and Montana district.[53] John M. Adams, the first graduate of Columbia Mines, was for years the western agent for the Frue vanner, a widely used ore separating machine.[54] Daniel M. Barringer and Richard Penrose handled the Jenish Patent Ball Mill, a German-built pulverizing mill, in the 1890s. Although they actively promoted the machine, by the end of the decade they preferred not to advertise it, "owing to the trouble that we had with the manufacturers." [55] During the

51. *Who's Who in Engineering, 1922–23,* p. 560; *EMJ, 104* (September 8, 1917), 448, *112* (October 8, 1921), 576; *M&M, 3* (February 1922), 18.

52. *EMJ, 20* (November 6, 1875), 465.

53. *M&SP, 39* (October 18, 1879), 250; Joshua Clayton to Rappid Ore Concentrator Company, Salt Lake City, September 2, 1884, Clayton to Levi Newcomb, Salt Lake City, September 14, 1884, and Clayton to Sperry, April 25, 1884, copies, Letterbook 1883–84, Clayton MSS (Calif.), 4.

54. *SMQ, 5* (January 1884), 179, *14* (April 1893), 190.

55. Daniel M. Barringer to Woodworth Wethered, December 13, 1894, Barringer to Herman Loehnert, January 21, 1895), and Barringer to Charles W. Jones, October 14, 1895, copies, Letterbook A, Barringer MSS; Barringer to Richard P. Rothwell, December 1, 1899, copy, Letterbook D, ibid.

first fifteen years or so of the twentieth century, Charles Janin supplemented his normal engineering work by selling dredges on a commission basis to British firms operating in Siberia or Malaysia.[56]

Others did part-time consulting to important mine equipment manufacturers, served as ore buyers for reduction works[57] or made the mining machinery field more than a mere sideline. At one point in his career, after his pregnant wife had vetoed his going to West Africa, William Kett was employed in the Spokane office of Fraser & Chalmers, one of the nation's leading producers of mine and mill equipment. But the depression of 1893 closed down this office. Kett then helped build a mill in Oregon and took charge until the ore petered out and left him unemployed. After working for a time as a sauerkraut salesman in Chicago, he finally went to work for the Gates Iron Works at Butte, selling mine machinery until he could get back in the main stream of mining engineering.[58] Walter McDermott was for years full time with Fraser & Chalmers, and John W. Young, after a brief engineering career, subsequently became affiliated with the same concern as a dealer in machinery and, eventually, after its reorganization as Allis Chalmers Company, became vice president.[59] Young Thomas Rickard, the son of Reuben Rickard and a Berkeley-trained engineer, practiced in Nevada for a year, then became interested in Park, Lacey & Company, San Francisco manufacturers of mine equipment. In 1901, Rickard and others bought out the firm and created the still-existing concern, Herran, Rickard & McCone.[60]

Nor were mining engineers confined merely to the basic

56. Charles Janin to Edgar Rickard, San Francisco, May 19, 1916, copy Charles Janin MSS, Box 6.

57. Both John Hays Hammond and Frederick Corning were consultants to the Union Iron Works in San Francisco; William Weston was Cripple Creek ore buyer for the Metallic Extraction Company in 1895. John Hays Hammond to Samuel Barlow, New York, July 8, 1884, Barlow MSS, Box 165; *EMJ, 37* (December 27, 1884), 421; *MI&R, 16* October 10, 1895), 141.

58. Kett, *Autobiography,* pp. 62–74.

59. *EMJ, 93* (January 27, 1912), 232, *113* (April 29, 1922), 733.

60. *EMJ, 91* (April 15, 1911), 778.

functions of their profession. Even in mid-career or later, many strayed into interrelated or even unconnected fields: education, journalism, railroads, irrigation, or government work. And they moved back into mining as readily as they shifted out.

Many crossed the line to the Halls of Ivy as professors or administrators in technical schools throughout the country. Robert Peele, Ross E. Browne, William J. Sharwood, and Courtenay DeKalb were among those who successfully combined teaching and active engineering at different times in their careers.[61] Thomas A. Rickard sought, but did not receive, a professorship at Berkeley, but Frank Probert, with a wide range of copper experience, did become dean of the University of California School of Mines in 1916, and Theodore J. Hoover, though overshadowed by his more illustrious brother, eventually held a comparable post at Stanford.[62]

Theodore B. Comstock was the founder and first director of the Arizona School of Mines (1891–95) and was president of the University of Arizona during part of this period, but stepped aside in 1895 to open a private practice in Prescott.[63] His successor as director was William P. Blake, a man with a long, active career both as an engineer and an academician, who continued his consulting on the side, despite protests charging a misuse of state funds.[64] Regis Chauvenet was for nearly twenty years president of the Colorado School of Mines, but resigned in 1902 to open a Denver office after the school's trustees had intervened in questions of student discipline.[65]

Technical journalism was another field that attracted many

61. *Who's Who in Engineering, 1922–23,* p. 979; *M&SP, 53* (July 3, 1886), 5; *Royal School of Mines Register,* p. 173; *Who's Who in Mining,* p. 25.

62. Charles Janin to Chester W. Purington, San Francisco, March 27, 1915), copy, Charles Janin MSS, Box 5; Harper, ed., *Who's Who on the Pacific Coast,* p. 463; Winfield S. Downes, ed., *Who's Who in Engineering, 1931,* p. 638.

63. *EMJ, 60* (October 12, 1895), 351, *100* (August 7, 1915), 243.

64. *EMJ, 79* (May 28, 1910), 1099; *M&SP, 76* (March 19, 1898), 313.

65. *EMJ, 82* (March 22, 1902), 422; "Report of the Committee of Inquiry on the Colorado School of Mines," *Bulletin of the American Association of University Professors, 6* (May 1920), 20–21; *Who's Who in Mining,* p. 17.

mining engineers. Such periodicals as the *Engineering and Mining Journal* and *Mining and Scientific Press* were edited almost invariably by trained engineers, Rossiter Raymond, Thomas A. Rickard, Courtenay DeKalb, Percy S. Barbour, and Richard Rothwell among them. Other engineers served from time to time as assistant editors or as western correspondents for the major journal.[66] Or they founded journals of their own: Theodore F. Van Wagenen was editor and publisher of the *Mining Review,* a Colorado periodical, during the 1870s; W. C. Wynkoop was the proprietor of *Mining Industry and Tradesman,* also in Denver; and Theodore B. Comstock had tried an eight-page monthly, the *San Juan Expositor,* during the previous decade.[67] Several mining engineers became editors and publishers of local newspapers, and one, Amos Bowman, "a scientific adventurer of the Bliss type," did some of the work on Hubert Howe Bancroft's monumental history of the West.[68]

Mining engineers were employed regularly by railroad companies;[69] a number came to be connected with western irrigation projects;[70] and some built breakwaters and habor fa-

66. Among these were Claude T. Rice, George J. Young, Francis L. Vinton, Louis Janin, Jr., and Theodore F. Van Wagenen. *EMJ, 18* (July 18, 1874), 36, *25* (February 16, 1878), 107, *87* (March 13 & May 22, 1909), 572, 1058; Frank H. Probert, "Mining Alumni, An Appraisal and Appreciation," *The California Monthly, 18* (April 1925), 450.

67. *Helena Herald,* August 31, 1875 & October 23, 1876; *EMJ, 52* (October 3, 1891), 391, *29* (January 31, 1880), 83.

68. Charles W. Haskell left engineering in 1883 to edit the *Mesa County Democrat* at Grand Junction, Colorado. Allen T. Bird became editor and proprietor of the *Nogales Oasis,* Nogales, Arizona, early in the twentieth century. Charles W. Haskell dictation, Grand Junction, January 18, 1887; *Pacific Miner, 10* (November 1906), 8. For Bowman, see Hubert Howe Bancroft, *The Works of Hubert Howe Bancroft, 39,* 272, 540–41; John M. Duncan to Allen Curtis, San Francisco, October 2, 1868, copy, Manhattan Silver Mining Company MSS.

69. Among others employed by railroad companies were A. F. Wuensch (Union Pacific, 1888), John Hays Hammond (Central & Southern Pacific, 1890, Professor Bibikov (Atlantic & Pacific), W. S. Ward (Denver & Gulf, 1896, George Milliken (Northern Pacific, 1891), and Horace V. Winchell (Great Northern, 1906). *EMJ, 46* (September 8, 1888), 199, *50* (July 5, 1890), 13, *52* (November 7, 1891), 535, *81* (May 5, 1906), 868; *MI&R, 17* (May 7 & 21, 1896), 520, 547.

70. Arthur D. Foote devoted more than a decade of his life to an abortive irrigation project in Idaho. See Foote, *The Idaho Mining and Irrigation Com-*

cilities—one, George M. Hill, became California state harbor commissioner in 1912.[71] Some went from mining into highway construction, and one, at least, helped build the Panama Canal.[72] Cities hired them as municipal engineers,[73] and states and territories employed them as official assayers, geologists, mine inspectors, or tax consultants.[74] The federal government employed them in a multitude of capacities: as members of scientific expeditions; as assayers or superintendents of the mints; as mineral examiners for the Land Office; as inspectors or safety engineers with the Bureau of Mines; as tax evaluators; or in one case, simply to build a cold storage plant in Manila.[75]

And a number of mining engineers wandered even further afield, entering pursuits as diversified as law, cemetery management, advertising, automobile tires, and movie direction.[76]

---

pany (New York, 1884); Mary H. Foote to Helena Gilder, Boise, April 30, 1884), M. H. Foote MSS; Arthur D. Foote to James D. Hague, Boise, March 5, 1882 Hague MSS, Box 11. William Blake was in charge of building a dam across the Hassayampa River for the Walnut Grove Water Storage Company in 1886. *EMJ, 42* (August 7, 1886), 99; Pierre A. Humbert did the planning for the Folsom dam, canal, and powerhouse in California. *M&SP, 63* (September 12, 1891), 168.

71. *EMJ, 94* (July 20, 1912), 130; *M&M, 3* (November 1922), 60.

72. *EMJ, 97* (April 11, 1914), 779; *Who's Who on the Pacific Coast,* p. 7; *Salt Lake Mining Review, 18* (April 30, 1916), 34.

73. George W. Riter was city manager of Salt Lake City, 1904–06; Alexander Leggett held the same position at Butte. *Mining World, 28* (June 20, 1908), 1002; *EMJ, 102* (September 2, 1916), 435.

74. Among others, Edward D. Peters served as territorial assayer of Colorado, Thomas A. Rickard as Colorado state geologist, Louis D. Ricketts as geologist of Wyoming, Wayne Darlington as Idaho state engineer, George C. Swallow as Montana inspector of mines, W. H. Storms as state mineralogist of California, and J. R. Finlay was hired by New Mexico to appraise mines in that state for tax purposes. Rickard, *Retrospect,* p. 65; *Who's Who in Mining,* p. 79; *EMJ, 34* (November 11, 1882), 254, *78* (November 17, 1904), 805, *95* (February 1, 1913), 297 *91* (May 21, 1921), 879; *M&SP, 109* (October 10, 1914), 579, *121* (October 30, 1920), 611; *Helena Herald,* July 12, 1889.

75. Byron Defenbach, *Idaho* (New York, 1933), *2,* 129; *M&SP, 84* (May 10, 1902), 264, *116* (April 27, 1918), 597; *EMJ, 29* (June 5, 1880), 383, *68* (July 29, 1899), 126, *72* (December 14, 1901), 796, *92* (July 1, 1911), 33, *98* (September 26 & November 21, 1914), 588, 932, *109* (February 28, 1920), 431, 583; *M&M, 3* (September 1922), 51, *4* (August & November 1923), 431, 578, *16* (March 1935), 164–65; *The Mines Magazine, 54* (April 1964), 47; *Who's Who in Mining,* pp. 77, 104; Len Shoemaker, *Pioneers of the Roaring Fork,* pp. 116, 122.

76. John H. Boalt, a Freiberg graduate, drifted into law practice after several years in mining, but his engineering background served him well in

Undoubtedly, this defection from the main core of professional work was common to some extent in most fields of endeavor, and how many mining engineers remained in their chosen calling is purely conjectural.

---

intricate mineral cases. J. C. Bates, ed., *History of the Bench and Bar of California,* p. 191. For other examples of the engineer turned lawyer, see Harper, ed., *Who's Who on the Pacific Coast,* p. 600; *EMJ, 77* (June 23, 1904), 1018. Hubert Eaton went from mining to manager of the Forest Lawn cemetery in Los Angeles, which under his supervision became one of the fanciest in the country. St. Johns, *First Step Up,* 90–96, 295. Roger L. Wensley left mining in 1920 to go into advertising in New York City; Carroll D. Galvin went from copper to dredging to promotion of Mexican rubber plantations and finally automobile tires; Louis R. Ball, Colorado Mines, 1900, served in the army, practiced as an engineer, and in 1916 was a movie director in Hollywood. *EMJ, 96* (September 6, 1913), 469, *109* (May 22, 1920), 1180; *Colorado School of Mines Magazine, 6* (June 1916), 138.

CHAPTER 8

"We Are Not Doing This for Glory, or Mere Philanthropy"

There are mines that make us happy,
There are mines that make us blue,
There are mines that steal away the tear-drops
As the sunbeams steal away the dew.
There are mines that have lost the ore chute faulted,
When the ore's forever lost to view,
But the mines that fill my heart with sunshine,
Are the mines that I sold to you.

—Anonymous [1]

Mining engineers, as Walter Ingalls told the war-conscious Missouri Mines class of 1916, could be divided into a number of different categories. Those who managed mines and metallurgical works, he said,

> are the subalterns, captains and colonels of our army. Fewer in number are those who advise about the development of mines, build metallurgical works, devise new metallurgical processes. They are the staff-officers. Many of them are great scientists, whose work is often inadequately requited. Finally there are the engineers in whom the business instinct is highly developed—men like Jackling, Hoover, Hammond, Bradley—who are our generals.[2]

What Ingalls left unsaid was that the vast bulk of mining engineers desired to emulate the "generals" of the profession and to amass a fortune through financial interests in mining, as

1. To the tune of "Smiles," *M&M, 1* (April 1920), 6.
2. Walter R. Ingalls, "The Business of Mining," *M&SP, 113* (August 19 1916), 277.

apart from their technical earnings. As one wrote in 1894, "The successful mining men are not the poor devils who toil for soulless corporations, but the schemers who place the properties, or who get in on leases."[3] Few, indeed, were not at some time in their careers involved in some way with the business side of the industry—as mine vendors, company promoters, investors or owners of property, or as corporate executives. From such endeavors emerged the affluent, although most aspirations were not assuaged. Yet desire died hard.

The young engineer particularly kept ever alert for promising property to be picked up cheaply on the side. Many were like H. J. Jory, a California graduate, who disliked the idea of being an employee all his life and who yearned to "hear the pounding of my own stamps," even though he expected "to be busted many times" before he was through.[4]

> Every engineer has visions of some day getting hold of "something for himself." Thus the mining districts, and every conceivable spot, to the ends of the world, are constantly being scoured by an army of keen-eyed, clear-brained, carefully-trained men, each looking for something good, and looking for himself, which guarantees that his heart is in his work.[5]

Engineers sometimes grub-staked the conventional prospector, either in the search for virgin ground or for property worth leasing, and as early as the 1850s, Joshua Clayton was advocating the organization of formal prospecting parties under a trained engineer or geologist to comb systematically the California mining districts for new possibilities,[6] a pro-

3. William Akers to Wilbur E. Sanders, Denver, February 13, 1894, Wilbur Edgerton Sanders MSS.

4. H. J. Jory to Samuel B. Christy, Kennet, California, February 18, 1891, Christy MSS.

5. "S.T." to editor, San Francisco, July 22, 1907, *M&SP, 95* (August 17 1907), 206.

6. Joshua E. Clayton to editor, *Union Democrat,* copy, Joshua E. Clayton MSS (Yale). See also Gressley, ed., *Bostonians and Bullion,* pp. 105, 112–13; Henry Curtis Morris, *Desert Gold and Total Prospecting,* p. 20.

posal that would become reality with the advent of the exploration company late in the nineteenth century.

Throughout the entire period, mining engineers themselves carried on prospecting work, often in their spare time or in slack seasons. Arthur Thatcher, Columbia, 1877, built up a good local consulting practice during the Tombstone boom, but what he made "experting," he said, "we usually blew into prospecting, so we just about kept even."[7] Mary Hallock Foote mentioned a "young man of fine education, good family & prospects as an Engineer" in early Leadville, who abandoned his career "because he isn't making money fast enough and here he is 'prospecting,' living in a log cabin with a 'partner'—cooking his own meals, wearing a dark flannel shirt and concealing the gentleman as much as a gentleman can be concealed by such 'lendings.' "[8]

Prospecting could be hard, unrewarding work, from which men often returned "with renewed energy and a fair stock of good nature, if nothing more."[9] During the Goldfield boom, Ernest Locke, an English engineer of considerable experience, spent several months probing the wilds of Nevada using horses and pack mules, but found nothing except some low-grade cinnabar deposits. The miner who accompanied him later assessed the effort: "As a prospecting venture the trip to the ridge was not fruitful, but as a mountain climbing feat, with four-footed animals, it was a grand success."[10]

Depression periods particularly turned engineers in the direction of prospecting. Following the panic of 1893, Raphael Pumpelly, just back from Europe with little hope of a good professional job, spent several years prospecting with his son-in-law. During five years of toil, which he found "very interesting, generally disappointing, and often very rough," he

7. Quoted in *M&SP, 124* (February 25, 1922), 254.

8. Mary H. Foote to Helena Gilder, Leadville, May 28, 1879, M. H. Foote MSS.

9. Ross E. Browne to Samuel B. Christy, Iowa Hill, California, October 8, 1885, Christy MSS.

10. Crampton, *Deep Enough,* pp. 73–74; see also Morris, *Mining West,* pp. 37–38.

discovered but three promising properties. In two of these instances, the financial backers withdrew support, one because of the Venezuela boundary controversy; the third property, which showed a thin vein of rich ore, became a cropper because "high graders" stole most of the product.[11]

On the eve of World War I, the profession seemed overcrowded. Mining schools were turning out graduates at a rapid pace, the war in Europe had dislocated mining activity and had closed down several areas, large corporations were consolidating their holdings under their own engineering staffs, and the Mexican Revolution had displaced a "small army of competent engineers." With the employment market at home apparently glutted, there was increased activity by technical men to prospect and develop their own mines.[12] Some of this was individual effort by such men as Henry Sanderson in Colorado, an engineer described as "wiry and a veritable billy goat for climbing," who came to enjoy the distinction in his day of having "patented more mining claims than any other engineer in the state."[13] Some was also done by groups of engineers, ten or a dozen strong, banding together in mutual associations.[14]

But few were successful. Perhaps the most imaginative and eventually the most fruitful were the attempts to locate low-grade copper deposits, after the techniques for working such ores had been successfully demonstrated. Not only in the Southwest did these explorations bear fruit, but far-sighted engineers, led by Seeley Mudd, Phillip Wiseman, and others, sent a man to comb the Mediterranean in search of slag heaps remaining from days of antiquity that might be profitably worked. The results, found on the island of Cyprus, ultimately provided the basis for the operation of the Cyprus Mines Corporation.[15]

11. Pumpelly, *My Reminiscences, 2,* 691–92.
12. *EMJ, 100* (November 20, 1915), 856, *99* (May 1, 1915), 789; *Mining and Engineering World, 40* (May 23, 1914), 957–58.
13. Otis A. King, *Gray Gold* (Denver, 1959), pp. 52–53, 55–56.
14. *M&SP, 110* (January 16, 1915), 96.
15. Lavender, *Story of Cyprus Mines Corporation,* pp. 46–80.

Another avenue open to ambitious mining engineers with a little capital was the leasing of property, both mines and mills. Normally, leasing was on a royalty basis, but operating funds were required, and, frequently, several engineers would combine their resources. But even with supposed technical know-how, leasing was risky, and more than one enterprising young man lost both his labor and his investment when the undertaking failed. With friends, Louis Ricketts leased and became manager of a mine in Colorado early in his career. Striking rich ore, the group built an expensive mill without bothering to ascertain the extent of the deposit. When the ore petered out, the project went "belly up," but Ricketts learned a valuable lesson.[16] An "impecunious friend" induced William Kett, Phillip Wiseman, and Charles W. Goodale—all engineers—to lease the Keystone gold mine, near Georgetown, Montana, in the summer of 1895. The undertaking was a failure, and Goodale, the only one of the group with any money, had to assume full responsibility, though Kett, at least, ultimately repaid his share with interest.[17]

But young engineers were optimists. When Charles Munro arranged a lease on dredging ground in 1910, he anticipated a return of his investment of $20,000 in one year—and "90% besides." "This looks like easy money," he said.[18] And so it was for a fortunate few. Henry Morris was one of several engineers who took about $300,000 out of a leased mine near Reno in six months.[19] Lewis A. Hayden, an MIT graduate and scion of a wealthy Denver family, once borrowed $5,000 from his father to lease part of the Elkton mine at Cripple Creek. At Christmas time, he handed his father a check for $100,000, his half of the Elkton lease profits.[20]

The experience of young Robert Livermore in Colorado was mixed. With several others, he leased part of the Indepen-

16. Bimson, *Ricketts,* pp. 13.

17. Kett, *Autobiography,* pp. 84. See also *M&M, 17* (December 1936), 602.

18. Charles H. Munro to Fred L. Morris San Francisco, December 14, 1910, Morris MSS, Box 2.

19. Morris, *Desert Gold,* pp. 23–24.

20. Newton, *Yellow Gold of Cripple Creek,* p. 108.

dence mine early in the twentieth century. But it was "probably the worst and most unpromising block in the mine, far beyond our means to develop," so the group gave it up and managed to obtain a better lease on the same property. Livermore put about $5,000 and a great deal of hard work into the enterprise, which produced some $50,000, to be split four ways. His other efforts at Cripple Creek were less rewarding: one lease was flooded out; another pinched out. And even the lease on the Independence, a steady producer, was lost. When Livermore approached the superintendent for renewal, the latter tactfully suggested that in addition to royalties—already 60 percent—an under-the-table bribe would be required. Livermore declined, and the lease lapsed. "Looking back on it," said Livermore, "I think I was too noble." [21]

Over the years, enough engineers were successful at leasing to keep the tradition alive and popular. And a number of men, such as Benjamin B. Lawrence, who reopened the old Dives and Pelican in Colorado, and Arthur Macy who revived the Silver King in Arizona and the Standard Consolidated at Bodie, made their professional reputations by reopening abandoned properties, usually on a profit-sharing basis.[22]

Frequently, an engineer recognized potential profits to be gleaned more indirectly from mining. In 1870, Joshua Clayton noted that there was money to be made by buying up claims along the proposed wire tramway route of the Eberhardt Company at White Pine.[23] Benjamin Lawrence believed a "good thing" could be made of purchasing and reorganizing Burlingame's assay business in Denver, adding a testing works designed by himself and Franklin Guiterman.[24] Ernest Wiltsee once put up funds to fight a case through the courts on behalf of a taxi-driver–prospector named Scott, and recovered 50,000 shares of Round Mountain Consolidated Mining Com-

21. Gressley, ed., *Bostonians and Bullion,* pp. 105, 106, 108, 111–12, 116–17.

22. *M&M, 2* (June 1921), 25; Rickard, ed., *Interviews,* pp. 260–61.

23. Joshua Clayton to James Clough, Treasure City, January 21, 1870, copy, Clayton MSS (Calif.), 2.

24. Benjamin B. Lawrence to Eben Olcott, Denver, February 27, 1894, Olcott MSS (N.Y.).

pany—enough when sold, to buy his wife an expensive new automobile and settle her in a Parisian apartment.[25]

Many a mining engineer supplemented his income, and a few managed to build fortunes from the sale of mining property or the promotion of mining companies. At least one segment of the profession believed that one of the functions of the engineer was to bring buyer and seller closer together, and to do so he had to cultivate and retain close ties with capital. Thus, when John Hays Hammond returned from Africa in 1896 with grandiose plans for working with Fred Bradley to place large western mines in London, his advice to Bradley was to broaden his circle of contacts with capitalists—"to strengthen himself with the various cliques." [26]

Long before this, editors had noted a shift in mining finance and speculation. "The tide of mining matters is still rising in the East and ebbing in the West," said one in the late 1870s. "The good mines and the mining experts are all coming to New York, and the wolves of the Comstock will have only each other to devour in San Francisco." [27] This transition never came fully to pass and the San Francisco area would always remain one of the major mining and mining engineering centers, but New York did indeed become the center for financial activity, and many engineers who—like Hammond—were engaged in mine sales, promotion, or financing would early move their offices to Manhattan to be near the source of capital. As Eben Olcott wrote in 1895 on the virtues of New York for the mining engineer: "The heaviest capitalists are here, and once in a while they get the mining fever bad." [28]

Hammond was one of those who argued that the engineer actively should undertake to promote mining property in order

25. Wiltsee, "Reminiscences," pp. 228–29.

26. John Hays Hammond to Frederick Bradley, London, October 9, 1896, Hammond to N. H. Harris, London, October 9, 1896, & Hammond to James Houghteling, December 17, 1896, copies, Letterbook 3, Hammond MSS.

27. *EMJ, 28* (September 27, 1879), 217.

28. Hammond, *Autobiography, 1,* p. 145; Eben Olcott to W. E. Newberry September 6, 1895, copy, Letterbook 23, Olcott MSS (N.Y.). See also Frank L. Sizer, "Professional Status of Mining Engineers," *EMJ, 106* (December 29, 1923), 1104; *M&SP, 111* (October 23, 1915), 620.

that it be done honestly and effectively, and from the 1880s he acted on this premise. A minority of the profession, however, disagreed, insisting that promotion created a bad public image and was outside the province of the technical man. For every successful engineer turned promoter, contended Thomas A. Rickard, "a hundred wreck their careers in their eagerness to drive short-cuts to wealth." [29]

But mere words could not blunt the business instinct. Even some of those who expressed a distaste for the crass commercialism of their colleagues were sometimes not averse to a little commercialism themselves. Thus, a young assistant engineer at a mine in Washington could in one breath condemn his immediate superior for his promotional activities and in the next report he was seeking placer ground in Oregon, "as I have a bunch of tenderfeet in Mass. who want me to find them some dry placers, but I will give them something good or nothing." [30] Eben Olcott criticized his friends for peddling mines at unreasonably high prices, castigated others for their tendency "to incline a little more to promoting than to strictly professional Mining Engineering," and expressed himself of the belief that "mining sales must be made outside of the Engineer's ranks." [31] At the same time, Olcott made a number of efforts to sell property and raise capital. In 1887, he took an option on the Polar and Accidental at Silver Cliff—"one of the best properties to buy that I ever reported on"—and sought to promote its sale both in New York and London. A few years later, he was actively attempting to place a Gunnison placer mine belonging to D. D. Fowler and was dickering with Londoners to whom he sought to sell a promising quicksilver

29. John Hays Hammond, "Professional Ethics," AIME *Trans., 39* (1908), 626; *M&SP, 47* (October 20, 1883), 241, *66* (May 27, 1893), 823; Hammond to George W. Trustlow, December 2, 1895, copy, Letterbook 2, Hammond MSS; Rickard, "Some Aspects of Mining Finance," *EMJ, 76* (November 28, 1903), 803. See also *EMJ, 45* (March 3, 1888), 158, *45* (April 28, 1888), 310, *79* (May 4, 1905), 859; *M&M, 6* (January 1925), 2; Brereton, *Reminiscences,* p. 34.

30. S. M. McClintock to George Louderback, Buston, Washington, October 16, 1904, Louderback MSS.

31. Eben Olcott to Robert Peele, June 5, 1895, & Olcott to William B. Parsons, February 28, 1896, copies, Letterbook 23, Olcott MSS (N.Y.).

property in Oregon. Subsequently, he would seek financial backing for a mill scheme presented by Thomas B. Stearns, an engineer-friend who had left consulting to manufacture mining machinery.[32]

Sometimes, the engineer played a role in promotion as an aide to one who was more professional in the business. A promoter named Zellerbach brought in Melville Attwood in 1871 to help place western property in England, and George Daly promised to go east in 1879 on behalf of several Nevada mine owners. Daly planned to address the Bullion Club in New York and assured his clients that "you may be sure I will place all properties you are interested in in as favorable a light as possible." [33] For this, the engineer expected a sales commission, just as he did if he introduced an ultimate buyer to an owner. Fred Morris was to receive half the promoter's commission if he introduced someone who purchased the Alvarado mine in Arizona.[34] Robert Bell, an engineer who was also Idaho state mine inspector, set a 10 percent commission for himself, although "for political reasons," it would have to be in his wife's name.[35]

A few mine vendors owned their property outright, but most simply "bonded" it—that is, took an option to buy at an agreed-on price before some specified date. Some options ran for as long as a year, but those of three or four months' duration were not uncommon, for mine owners hesitated to tie up their property if it had real promise, just as they were often

32. Olcott to J. C. McCreary, July 7, 1887, Olcott to W. D. King, August 9, 1887, & Olcott to Burrage & Taylor, New York, September 6 & 12, 1887, copies, Letterbook 12, Olcott MSS (N.Y.); Olcott to D. D. Fowler, February 8, 1889, Olcott to J. B. Austin, October 19, 1889, and Olcott to G. W. Larkman, October 19, 1889), copies, Letterbook 16, ibid.; Olcott to Thomas B. Stearns, New York, September 2, 1894, copy, Letterbook 22, ibid.

33. George S. Roberts to Asbury Harpending, San Francisco, September 29, 1871, Harpending MSS, Box 1; George Daly to Henry Yerington, Aurora, Nevada, October 9, 1879, Yerington MSS, Box 1.

34. Byron O. Pickard to Fred Morris, Phoenix, October 22, 1915 & February 22, 1916, Morris MSS, Box 7; see also Morris to Norman Smith, June 30, 1916, copy, ibid., Box 1.

35. Robert N. Bell to Fred Morris, Boise, February 26 & June 23, 1913, ibid., Box 7.

skeptical of bonding to known speculators.[36] Before his option expired, the promoter tried his best to find a buyer at a price that would provide a margin of profit. This margin varied: one Nevada engineer was accused of offering a Goldfield mine for which he had paid $32,000 for a cool $1 million; Joshua Clayton, a much more modest man, expected 10 percent profit on an Idaho property and believed that the one-fifth interest he received in a Butte mine for concluding its sale was "a pretty good months work." [37] But in general, the margin was all the traffic would bear, for as Clayton noted, ". . . we are not doing this for glory, or mere philanthropy." [38]

Either a mine could be sold outright to an individual or a small group or the vendor could undertake the task of forming a company to acquire it. Either approach was laborious and the chances of success risky. Clarence King, who was promoting Mexican property, remarked in 1881: "I have wasted all my Autumn placing Salido, and never saw such up-hill work before. The market is as dead as Julius Caesar;—in fact there is *no* market." [39] John B. Farish lamented in 1894 that his sale of the Victor mine had fallen through in New York because he was unable to find subscribers, though he had worked and was still working hard at the task.[40]

The diaries kept by Joshua Clayton would underscore the experience of King and Farish. Mine promotion was time consuming, difficult, and frustrating work. Early in 1881, Clayton was in Louisville with Governor Murray to work up support for the Moulton mine. He distributed a prospectus to the governor's friends, explained the project to them, and patiently set about the task of prying open their pocketbooks. Typical

36. Entries for January 9 & 10 & February 25, 1880, Clayton diary, 1880, Clayton MSS (Calif.), 7; Joshua Clayton to W. F. Noble, August 24, 1884, copy, Letterbook 1883–84, ibid., 4.

37. George Graham Rice, *My Adventures with Your Money,* p. 50; entry for April 16, 1881, Clayton diary, 1881, Clayton MSS (Calif.), 8; Clayton to F. Phillips, February 13, 1883, copy, Letterbook 1883–84, ibid., 4.

38. Clayton to H. A. Monroe Butler-Johnstone, Salt Lake City, April 24, 1883, copy, Letterbook 1883–84, ibid., 4.

39. Clarence King to L. Gilson, December 10, 1881, copy, Letterbook 4, King MSS.

40. John B. Farish to Eben Olcott, Denver, August 7, 1894, Olcott MSS (N.Y.).

entries in his diary for this period read: "Still working up Moulton matters—slow conservative people, can't hurry them"; "Got subscription started at last, with fair prospects"; "It is slow work but one needs close and patient attention to make it go"; "Drumming on the same string. Slow work." After several weeks of this, with disappointing results, Clayton left Louisville, leaving another promoter to take over the job on a commission basis. Even his trip home on the B&O was frustrating: "Rough road and very bad whiskey," he recorded.[41]

Nor was the outright sale of mining property to individuals who already possessed capital any less difficult and discouraging. Few capitalists were willing to plunge into a mining venture single-handedly, and those who were had to be convinced, which was not easy, as W. H. Lanagan discovered in 1916, when he tried to interest capitalists in the La Paz property near Yuma. San Francisco mining men declined the proposition, but Tom Aiken, a successful Alaska operator, agreed to examine the ground with the idea of acquiring it if impressed. First Lanagan had to iron out difficulties with one of the owners, who wanted an additional cash payment before any sampling was done. Surreptitious sampling was a possibility, acknowledged Lanagan, were it not for the owner's reputation "of being quick on the trigger."[42] But before Aiken arrived, this problem had been amicably resolved. Accompanied by his wife, Aiken visited the mine in 120 degree weather and was impressed neither with the climate nor the mining potential of the quartzite area. His proposition rejected, Lanagan was left to lament:

> I now know how all the poor devils felt whose pet placer propositions I have turned down in the last ten years. Mine was turned down all right. Aiken not only

41. See entries for January 12–25 & February 15 & 25, 1881, Clayton diary, 1881, Clayton MSS (Calif.), 8.

42. W. H. Lanagan to Fred L. Morris, Los Angeles, July 17, 1916, Barstow, July 21, 1916, & Bouse, Arizona, July 23, 1916, Morris MSS, Box 1; Morris to Lanagan, San Francisco, July 21, 1916, copy, ibid., Box 1.

> failed to show any enthusiasm but left here with the aggrieved air of a man who thinks he has been taken on a wild goose chase. . . . I couldn't interest him in history, geology, sources of enrichment, options, yardage, or anything else. What he wanted to see was gold, and he couldn't find enough to make any impression. . . . Tom quit cold and said the whole country was bunk. . . .
>
> It was a lovely party. Everything that could possibly go wrong went wrong. Mrs. A kept tab of the pans and made nice little sarcastic remarks about the conspicuous absence of gold. This of course improved my disposition a lot. . . . Now that it is all over I am rather glad that we didn't find any gold, for she would have suspected me of salting it if we had.[43]

Whatever the problems, most mining engineers at one time or another ventured into sales and promotional waters, sometimes as slack season work,[44] more often as a sideline activity whenever opportunity presented itself. Some undoubtedly put more emphasis on this than on their engineering work proper. Most—Phillip Deidesheimer, William P. Blake, Lionel Nettre, Charles H. Munro, George Maynard, and Seeley Mudd among them—generally operated as individuals, but a few organized more formal agencies for the same purpose. Alexander Del Mar, a versatile engineer who was at one time director of the U.S. Bureau of Statistics, established a concern in London to market western mines in England.[45] Edward Bates Dorsey, a man with experience both on the Comstock and in Chile, was listed at the end of 1879 as president, director, and engineer for the U.S. Mining Investment Company, a New York firm organized to deal in "ONLY Dividend-Paying Mines," according to its advertising.[46] But perhaps the best known promotional agency of this type was the Exploration

43. Lanagan to Morris, Dome, Arizona, August 2, 1916, ibid., Box 1.

44. Erich J. Schrader to George Louderback, Yerington, Nevada, January 5 1914, Louderback MSS.

45. *M&SP, 43* (October 1, 1881), 218; *DAB, 5,* 225–26.

46. *M&SP, 17* (November 28, 1868), 345, *40* (February 21, 1880), 128; *Mining Record* (New York), *6* (December 20, 1879), 528.

Company, Ltd., founded in London in 1886 with Rothschild-backing by two American engineers, Hamilton Smith and Edmund DeCrano. From a modest beginning ("The said clients, however, are bashful—not to say backward—& we may get in the poor-house before they come with proper fees"),[47] the Exploration Company blossomed into a highly successful endeavor that helped sell and find financing for such important mines as the Alaska Treadwell, the Tomboy, and Anaconda Copper. When DeCrano died in 1895, Smith brought in Henry C. Perkins as a partner and moved the main offices to New York. The concern had a global emphasis, but it retained western representatives and worked hand in hand with such engineers as Henry Bratnober and Alfred Wartenweiler in the promotion of undertakings in the Idaho, Montana, and Colorado regions especially.[48]

Most engineer-promoters were engaged in honest and sincere endeavors. There were undoubtedly exceptions, but not many fell into the category of promoters described by Joshua Clayton in 1883:

> The idea of all that class of men is to *work the public,* first, then work the mine. If it turns out well they will again *work* the public, until the cream is skimmed off. Then again, *work the public.* They don't care a damn for "legitimate mining," any further than to use it as a means of robbing the public.[49]

Clayton himself was an example of the scrupulous engineer-promoter. In 1871, when his coworker in London tried to place a number of claims on the market without clear title, Clayton insisted that they be withdrawn immediately. He condemned the greed of promoters on both sides of the Atlantic and damned the misleading reports and "flaming prospectuses

47. Hamilton Smith to James D. Hague, London, February 18, 1886, Hague MSS, Box 12.

48. *EMJ, 70* (July 14, 1900), 34; *M&SP, 105* (December 21, 1912, 812; *DAB 17,* 273.

49. Joshua Clayton to H. H. Mason, Salt Lake City, June 5, 1883, copy, Letterbook 1883–84, Clayton MSS (Calif.), 4.

&c." used to sell overpriced properties. What he wished to further, he said, was "a *Square, straight forward mining enterprise* to be established on practical business principles and to be conducted purely as a legitimate business. . . . I intend to see that my parties get *good* mines at *fair* prices." [50]

Clayton was a man with unexcelled reputation for probity and fair play—and he was never a financial success. Like other engineers caught up in the promotional side of the industry, he believed in the undertakings he pushed and often had funds of his own invested in them. Louis Janin, one of the original promoters of Adolph Sutro's great Comstock tunnel scheme, was a stockholder from the beginning.[51] Though early in his career, James D. Hague wrote a friend that he would "try to get some good mining scheme in shape to do something with in the East while I am here," his efforts were usually confined to mines in which he already held an interest, and they were usually concerned with raising additional operating capital. Thus, in the early 1870s when he sought £40,000 in London, it was as working capital for a mine in which he already had an investment, as did Samuel Emmons, Rossiter W. Raymond, and George Opdyke, the mayor of New York City.[52] Nearly twenty years later, with Clarence King, Hague sought unsuccessfully to find a purchaser for the De Lamar mine in Idaho,[53] subsequently one of the few British-owned properties in that state to pay consistent dividends.

Daniel M. Barringer and Richard Penrose were as much promoters as they were practicing engineers, but they almost

50. Clayton to Henry Hughs, Treasure City, May 13, 1871, copy, Clayton MSS (Calif.), 2, Clayton to W. W. Lowe, Salt Lake City, February 22, 1873, copy, Letterbook 1873, ibid., 3; Clayton to H. A. Monroe Butler-Johnstone, Salt Lake City, April 24, 1883, and Clayton to James Moriorty, December 27, 1884, copies, Letterbook 1883–84, ibid., 4.

51. Constitution and By-Laws, Sutro Tunnel Company, August 1865, handwritten copy, Sutro MSS, Box 20; Louis Janin to Adolph Sutro, Virginia City, February 1, 1866, ibid., Box 6.

52. James D. Hague to Samuel Emmons (San Francisco, June 24, 1875), copy, letterbook 6, Hague MSS; notations of November 9 & 11, 1870, Hague to John Taylor, London, March 13, 1871, & Hague to John H. Bird, London, May 10, 1871, copies, Letterbook 3, ibid.

53. James D. Hague to R. J. Wilson, New York, May 15, 1890, & Hague to Edmund DeCrano, August 19, 1890, copies, Letterbook 13, ibid.

invariably had capital of their own tied up in the enterprises they pushed. In 1895, the two were attempting to float an Arizona gold or silver proposition, and Barringer advised his partner to proceed carefully:

> In talking with people in Chicago you might throw out hints that you *may* have a good thing to propose to them next summer, but that we are not now absolutely certain as to the intrinsic value of the properties, and therefore do not want to say anything until we are. *It will excite their curiosity, and they will be ready and anxious to hear all about it when we are in a position to speak to them. I am doing the same thing here.*[54]

During negotiations for the property, Barringer suggested that one of their friends and a potential investor, Count James Pourtales, be taken to visit the mine, but Barringer's lawyer brother, Lewin, who was also in the undertaking, vetoed the idea:

> Keep him informed, but ask him to keep the knowledge confidential. But, if you take down there to Arizona such a *big sputtering German Count,* as Pourtales is, the Irish Mexican greasers that you are dealing with will think they have heaven and earth combined, and will jump the option and leave you in the lurch. So in dealing with these people, you have to deal with them differently from what you would with an honest man (the honest man is a rarity anyway, especially in the southwest).[55]

In addition to the Common-Wealth Mining and Milling Company, in which Pourtales did invest, Barringer also raised capital from friends, including Cyrus McCormick, for an iron mining venture in Mexico and for a concern conceived to locate and work commercially a giant meteor believed imbedded in a crater in Arizona. He also helped float the Shasta Dredg-

54. Daniel M. Barringer to Richard Penrose, February 7, 1895, copy, Letterbook A, Barringer MSS.

55. Lewin Barringer to Daniel M. Barringer, February 11, 1896, copy, ibid.

ing Company, a firm in which he persuaded novelist Owen Wister to invest $10,000.[56]

Undoubtedly, a fair number of engineers speculated in the shares of mining companies, although some agreed with J. Ross Browne that "Mining speculations, like transactions in horse-flesh, have a tendency to blunt the moral perspectives." [57] "Assuredly we believe that the engineer should have ideals higher than those of a hodcarrier, a ward politician, or a bucket-shop broker," insisted Thomas A. Rickard, always a spokesman for a more ethical approach to engineering. Personal speculation in mining shares "saps the foundations of professional integrity." "In Nevada today," wrote Rickard in 1906, "there is many a promising young mining engineer who is being undone by the gambling fever, who makes his office a bucket shop and his technical training a lure for the ignorant." Rickard particularly deplored the practice of the engineer taking or buying stock in a company with which he was associated, but recognized its prevalence:

> It is a brutal fact that if dabbling in stocks on the strength of information obtained professionally is warrant for depriving an engineer of his degree, then a majority of the financially successful men would have to go without that luminous tail.[58]

But many, such as John Hays Hammond, dissented and continued to speculate as they saw fit, though such men as Hennen Jennings argued that no engineer should have hidden holdings in property on which he reported nor should he buy on margin.[59] Some insisted that a competent engineer attached to a reputable stock brokerage firm would help curb wild speculation and add stability in the industry; Josiah Spurr, editor of

56. Daniel M. Barringer to Cyrus McCormick, December 14, 1903, copy Letterbook I, Barringer to Owen Wister, September 11, 1905, copy, Letterbook K, & list of bondholders, Shasta Dredging Company, Letterbook N, ibid.

57. Browne, *Adventures in the Apache Country,* p. 528.

58. "Engineers and Ideals," *M&SP, 94* (June 15, 1907), 737; "The Engineer as a Financier," *M&SP, 97* (October 17, 1908), 509; "Gambling in Mining Shares," *M&SP, 93* (December 15, 1906), 701.

59. Rickard, ed., *Interviews,* p. 250.

the *Engineering and Mining Journal* in 1924, urged the development of a new branch of the profession, in which groups of mining engineers with integrity and business sense would organize to examine, approve, and offer for sale stock in meritorious mine ventures.[60]

Many were speculators on a small scale: they were like the engineer who was installing a mill at Mineral Hill, Nevada, who wrote a stock operator friend in 1906 "to see if there was anything in the stock line that you could put me on to so I could clean up enough plunks to make some Christmas presents on without going down into me jeans." [61] At the other extreme, a few went into the market in a big way. Edmund DeCrano paid $30,000 for a seat on the San Francisco Stock and Exchange Board in the spring of 1875. Richard H. Rickard, a prominent engineer with holdings in the Lake Superior and Colorado districts, was one of the organizers and treasurer of the Mining Exchange in New York City prior to his death in 1885. After twenty-eight years in Montana, A. J. Seligman moved to the East Coast, where he was a member of the New York Stock Exchange for about a quarter of a century.[62]

Fortunes were made in share speculation, and some engineers were willing to prostitute themselves by capitalizing on "inside" knowledge. George Daly was not averse to passing along confidential information to help friends or even to playing the market in shares of the companies he managed. Pierre Humbert, manager of the Bullion, a Comstock property, was accused in 1880 of "bulling" stock of that company by spreading deceptive reports of ore discoveries.[63]

60. Francis C. Nicholas, "Mining Stocks and Mining Engineers," *Mining World, 27* (September 7, 1907), 395–96; Glenville A. Collins, "Who Is Your Engineer," *Mining and Engineering World, 45* (September 16, 1916), 505; Josiah Spurr, "Bunk for Morons," *EMJ, 117* (June 21, 1924), 995.

61. Edward B. Jones to Henry J. Amigo, Mineral Hill, November 15, 1906, Goldfield (Nevada) Mining Companies MSS, Box 1.

62. King, *History of the San Francisco Stock and Exchange Board,* p. 56; *EMJ, 39* (February 21, 1885), 117; *M&M, 16* (June 1935), 280.

63. George Daly to Henry M. Yerington, Aurora, Nevada, September 13, 1878, Yerington MSS, Box 1; John F. Cassell to Asbury Harpending, San Francisco, May 6, 1880, Harpending MSS, Box 1; *Virginia Evening Chronicle,* June 28, 1880.

No doubt a number of engineers made handsome speculative profits. Robert Livermore once netted a tidy $12,000 on $2,500 put into Esperanza stock.[64] But many were much less successful. Ferdinand Baron Von Richthofen was reported to have lost "a good deal of money" in quartz stock gambles during his brief professional career in the West in the 1860s. Samuel T. Curtis, a prominent manager on the Comstock and later in Mexico, reputedly made and lost several fortunes in mining speculations, but died penniless in 1907.[65] Henry Janin also had the reputation of being something of a speculator—one who "always wants to get in at bed-rock," but who was "not a stayer." Yet, in the long run, he seems not to have profited greatly. "I am proud to say that as a speculator, I never did anybody any good," he wrote in 1899.[66]

Probably the majority of engineers were stockholders in mining companies to one extent or another, not always for speculative purposes, except insofar as mining itself was a speculation. Often they received shares in exchange for inspection, promotion, or even managerial work, and frequently they invested cash of their own, for as professionals they could not help but consider mining a legitimate business enterprise. Still, most realized that in the nineteenth century, at least, more capital went into the ground than came out of it.

Many were engaged in mine investment in a small and probably never very rewarding way. Joshua Clayton always had a few mining interests and occasionally made a few hundred dollars, but seldom more.[67] During their Leadville days, Arthur Foote's wife could "glory in the temptations which surround Arthur and which are not even temptations to him." Mining investments were too great a risk, she insisted. "The only way in which *I* consider them safe is as furnishing income to resi-

64. Gressley, ed., *Bostonians and Bullion,* p. 126.

65. Josiah Whitney to William Brewer, San Francisco, December 23, 1865, Brewer-Whitney MSS, Box 2; *EMJ, 83* (January 26, 1907), 203.

66. *Pacific Coast Annual Mining Review* (1878), p. 51; Henry Janin to James Hague, London, April 4, 1899, Hague MSS, Box 12.

67. Entries for April 14 & 16, 1877, Clayton diary, 1877, Clayton MSS (Calif.), 5; Share Certificate, McDonald Silver Mining Company (Nevada), Hague MSS, Box 10.

dent employees & managers."[68] Later in his career, when on a firmer financial basis, Foote did invest on a limited scale. In 1904, he was president of a power company he had started on the Middle Yuba, and records indicate that, in the 1913–30 period, he put money into a number of small California and Nevada mining enterprises, such as the Mugwump Mines Company, which, he said in 1920, "has kept me very poor."[69]

By his own figures, Robert B. Stanton invested $2,169.40 in the Golden Gate Concentrator Company between 1884 and 1890 and listed it as "All a total loss." Stanton also figured a personal deficit of $8,180 on the Hoskannini placer endeavor of the 1896–1901 era, including $5,180 in stock and $3,000 in salary due.[70] When William Leete, another engineer, was seeking a job in 1897, he explained that he was in dire need after losing every thing in an unfortunate mining venture. "I put all I had into it and it left me 'broke,' " he said.[71]

Both Louis Janin and his brother Henry invested in mines and mills early in their career, with but dubious results. Louis put some $7,000 for himself "and others" into Reese River property in 1863, and at the same time anticipated substantial profits from a mill enterprise—"a tremendous undertaking" —in which he was interested. But three years later he reported that business in Virginia City was at a standstill, with no water to operate the mill.[72] In 1869, Henry Janin regretted his inability to help his father financially, having put all his ready cash into Louis' mill. "I am not in a position to assist anyone not even my creditor," he wrote philosophically. "A creditor is not a bad thing to have, as he takes more interest in one's life

68. Mary H. Foote to Helena Gilder, Leadville, July 10, 1880, M. H. Foote MSS.

69. Mary H. Foote to Helena Gilder, Grass Valley, August 1904, ibid.; Arthur D. Foote to George B. Agnew, July 9, 1920 & January 14, 1921, copies, A. D. Foote MSS.

70. Financial summary, undated field book 30, pp. 170, 171, Stanton MSS.

71. William M. Leete to Samuel Christy, Spokane, March 18, 1897, Christy MSS.

72. Louis Janin to Juliet C. Janin, Empire City, September 11, 1863, & Virginia City, September 2, 1866, Janin family MSS, Box 10; Louis Janin to Edward Janin, San Francisco, October 6, 1863, ibid., Box 9.

& doings than a mere friend would." [73] But better days were ahead. In 1878, Henry was one of the directors of the Golden Terra Mining Company in Dakota and of the Paradise Valley Mining Company in Nevada; two years later, he had $30,000 to invest in the Membres Mining Company in New Mexico.[74]

James D. Hague enjoyed mixed success. In 1873, he was president of the Twin River Mining Company, on whose property he had attempted to raise working capital several years before. But the mine proved unprofitable, and in the fall of 1874 Hague was forced to write one of his fellow shareholders:

> I am very sorry to say however that the mine is dead. The funeral was held at noon, Sept 8, when the Company became the owner, at the delinquent sale, of more than half its stock. The remaining effects of the deceased, in the shape of machinery at Ophir Canon are now for sale.[75]

Hague was also a shareholder in the Ruby Gold Gravel Mining Company, in the Excelsior Water and Mining Company, and in the Tomboy Gold Mine Company, Ltd., the latter a strong Colorado dividend-payer. In 1897, he was president and a major owner of the Mount Pleasant Gold Mines Company, a concern that floundered for lack of funds for further development.[76] He missed a fine opportunity when he failed to pick up a $100,000 option on the Empire at Grass Valley, a property that eventually produced $15 million in gold, but he took a fortune from neighboring mines, including the North Star, one of the most consistent of the California deep-level producers.[77]

Ernest Wiltsee made "a considerable profit" from the

73. Henry Janin to Louis Janin, Sr., Virginia City, April 7, 1869, ibid., Box 6.

74. *Black Hills Daily Pioneer,* January 31, 1878; *M&SP, 37* (September 14, 1878), 166; James D. Hague to Henry Janin, New York, June 4, 1880, copy Letterbook 4, Hague MSS.

75. James D. Hague to Rossiter W. Raymond, San Francisco, September 11, 1874, copy, Letterbook 6, ibid.

76. See miscellaneous share certificates, ibid., Box 10; William P. Bonbright & Co. to Hague, New York, February 15, 1908, ibid., Box 11; Hague to Charles G. White, New York, September 23, 1897, copy, Letterbook 21, ibid.

77. Wiltsee, "Reminiscences," pp. 14–15; *DAB, 6,* 87.

W.Y.O.D. mine at Grass Valley before moving on to larger money in South Africa at the turn of the century. On his return, he invested $25,000 in claims on the Harquahala Desert, north of Yuma, but lost it all because of a flaw in title.[78] But subsequent investments in dredge mining and petroleum put Wiltsee in the wealthy but eccentric bracket.

The Bunker Hill & Sullivan was an engineer-oriented concern, not only in terms of operation, but also in terms of investment. After he became manager in 1887, Victor Clement advised greater capital outlay to develop the property, and owner Simeon Reed commissioned John Hays Hammond to find $300,000 for this purpose. Half of this came from D. O. Mills of California, some from a Chicago group headed by James H. Houghteling, and some from various mining engineers around the country. Christopher Corning, who had examined the mines, put in $25,000, and Eben Olcott, a friend of Hammond, disposed of at least 2,000 shares to associates, keeping 500 himself, for he considered them "a gilt edged investment." [79] Victor Clement also had an interest. When the reorganization occurred, Reed offered him $20,000 in shares at cost. Clement hesitated, "owing to an aversion I have towards mining speculation" and because he had been previously "unfortunate in this class of investments." But in June 1888, convinced the mine was good, he purchased the stock. Besides, he pointed out, "Being placed here upon the footing of an interested party, the Company will feel better served." [80] Ultimately, Clement built his holdings to 36,301 shares, which his wife, after his death, sold for over $900,000. Hammond, who had received 2,500 paid up shares for his role in organizing the company, also acquired additional blocks of stock and, by September 1892, owned 32,962 shares.[81]

By the turn of the century, the more perceptive engineers

78. Wiltsee, "Reminiscences," pp. 17, 157–59.

79. Thomas A. Rickard, "The Bunker Hill Enterprise—IX," *M&SP, 120* (May 15, 1920), 709; Eben Olcott to S. S. Palmer, March 2, 1892, and Olcott to E. Scofield, March 8, 1892, copies, Letterbook 18, Olcott MSS (N.Y.).

80. Victor Clement to Simeon Reed, Wardner, June 14, 1888, and Reed to Clement, June 19, 1888, Reed MSS.

81. Rickard, "The Bunker Hill Enterprise," p. 709; Cloman, *I'd Live It Over,* p. 228.

were well aware of the growing tendency of trained, experienced mining engineers to become presidents of mining companies. John A. Church, for example, admitted that "the movement has not spread very far as yet," but he recognized it as an optimistic omen for the future. John Hays Hammond, another who successfully bridged the gap between engineering and corporate management, echoed the same sentiments. "This is the era of the engineer," he proclaimed—an era in which the engineer was destined to "invade the sphere now monopolized by the employer and capitalist and eventually become, in fact himself, the master." [82] By 1909, Herbert Hoover correctly gauged the direction of evolution taking place within the profession. It was a shift from the advisory to the executive capacity: "The mining engineer is no longer the technician who concocts reports and blue prints. It is demanded of him that he devise and finance, construct and manage the works which he advises." Thus, the "bridge of engineering," said Hoover, was based on two "piers"—one of technical knowledge, the other of "commercial experience and executive ability." Greater capital requirements made more common each year the use of mining engineers as business executives, and this, Hoover believed, was a distinctly American development.[83]

Thus, although mining engineers served as presidents or directors of companies throughout the entire period, the tendency became more pronounced during the late nineteenth century, with the engineer assuming the characteristics of the modern executive. And not a few would have to be counted among the captains of industry. Some achieved their eminence in milling, rather than mining proper. David W. Brunton, for example, combined with F. M. Taylor to build and operate a successful mill in early Leadville. In 1889, he constructed a public sampling works at Aspen, which he expanded into a

82. John A. Church to editor, New York, May 2, 1904, *EMJ, 77* (May 12, 1904), 755; John Hays Hammond, "The Engineer, His Scope and Qualifications," quoted in *Northwest Mining Journal, 3* (April 1907), 50.

83. Hoover, *Principles,* pp. 185–86, 188–91.

chain of similar plants flung throughout Colorado, Utah, and Nevada, all the while continuing his consulting practice.[84] Henry E. Wood came to Leadville in 1878, fresh from Yale, and established an assay office and laboratory, which he ultimately moved to Denver, adding an ore testing works in 1898. Wood consulted on the side, and in 1909 patented and began to employ the Wood ore flotation process, specializing in the concentration of molybdenite. By World War I, Wood and his sons owned and operated the largest molybdenite mine in Canada, and during the postwar slum, he turned successfully to petroleum.[85] Another Coloradan, James B. Grant, a former Confederate soldier who had studied at Freiberg, early opened a smelter at Leadville with financing from an uncle, and eventually consolidated this with others to create one of the most important smelting empires in the West.[86]

J. Parke Channing was an excellent example of the new breed of engineer-executive of the twentieth century. Channing was instrumental in forming the General Development Company in 1906 and the Miami Copper Company two years later, and he served as vice president and director in both; in addition, he was an official of Seneca Copper, Kerr Lake Mining Company, and Tennessee Copper and Chemical.[87] Harvard graduate Benjamin B. Thayer began his career as foreman at Anaconda in Montana. After a varied experience there and in the Southwest, he became assistant to the president of Amalgamated Copper, and later vice president, then president of Anaconda Copper. He also served as president of the AIME, and at the time of his death in 1933 was listed as an official in twenty-seven different mining, railroad, and utility companies.[88]

Also numbered among the "greats" would be James Doug-

84. *EMJ, 87* (February 27, 1909), 458.

85. Henry E. Wood to Max Mailhouse, Los Angeles, April 5, 1921, Wood MSS, Box 2.

86. Griswold, *Carbonate Camp,* pp. 66–67; *EMJ, 192* (November 18, 1911), 1004.

87. Rickard, ed., *Interviews,* pp. 163–64; *Who's Who in Engineering, 1922–23,* p. 253.

88. *M&M, 14* (March 1933), 165.

las, Mark L. Requa, and Daniel C. Jackling. Douglas, who would serve as president of the AIME, was early affiliated with the Copper Queen Company in Arizona, a concern that was subsequently incorporated into Phelps Dodge Company, of which he was president.[89] Requa, the son of an old Comstock superintendent, in 1903 organized the White Pine Copper Company at Ely, Nevada, utilizing low grade techniques pioneered in Utah and Arizona, and subsequently consolidated his holdings to form the Nevada Consolidated Copper Company, a Guggenheim-oriented enterprise.[90] Jackling, the most visionary of the three, was a millionaire by the time he was forty, primarily as a result of his pioneering work in handling low-grade porphyry ores and the development of such concerns as Utah Copper, Ray Consolidated, and Chino Copper. Jackling was also affiliated with numerous other enterprises, at least a few of which did not prove as profitable as his copper investments. In 1918, he referred to Alaska Gold as "my imbecile child," expressing regret that friends had lost money in it. "We nursed and tended it as faithfully as we did our successful porphyry copper mines," he said, "but this child just didn't have the stuff." [91]

Frederick W. Bradley has been cited as "a good example of the dominance of the engineer in the control of the operations and policy of a large mining undertaking." [92] In 1923, Bradley was listed as having been president of sixteen different mining and smelting companies in the West and Alaska, not to mention an assortment of other concerns.[93] These ranged from the Bunker Hill & Sullivan, of which he was president from 1897 to his death, to dredge enterprises in California and low-grade copper in Nevada. With Requa, he was interested

89. *EMJ, 106* (July 6, 1918), 18–19; *DAB, 5,* 396–97.

90. Schmidt, "Early Days," pp. 1–4; Russell R. Elliott, *Nevada's Twentieth-Century Mining Boom,* pp. 178–79; L. K. Requa to author, Salt Lake City, May 11, 1964.

91. Wolfe, ed., *Men of California,* p. 101; Rickard, ed. *Interviews,* pp. 193, 210–20; *Who Was Who in America, 1951–1960, 3,* 440–41; unidentified clipping, stamped February 26, 1918, Bradley MSS.

92. Rickard, *History of American Mining,* p. 324.

93. *Who's Who in Engineering, 1922–23,* p. 179.

in property at Stites, Idaho; he was a heavy investor in, and for a time president of, the Alaska Juneau Gold Company, a firm in which Bernard Baruch was one of the strong financial backers.[94] Originally, he held the controlling interest in the Atolia Mining Company in the Randsburg district of California, but by 1917 his holdings had dropped to 41 percent. Baruch, Requa, and J. H. Mackenzie, another engineer, were also stockholders, but in the interest of more efficient operation, Bradley managed to squeeze both Requa and Mackenzie off the board of directors and in 1923 assumed managerial responsibility.[95] With Atolia, Alaska Juneau, Bunker Hill & Sullivan, and other holdings, Bradley managed to keep busy. "Between times," he said, "I still keep occupied in 'chasing' up what appear to be attractive mining propositions; but I am not giving time to anything of this kind that is not worthwhile on its face."[96]

The list might easily be extended if desired. Herbert Hoover, John Hays Hammond, and Hennen Jennings are often regarded as the very embodiment of the successful "new" engineers—the "generals" in Walter Ingalls' modern industrial "army." In their careers, as in those of the Jacklings, Bradleys, and Requas, was symbolized the passing of the old technical man, who gave way before the great advance of science and the growing complexity of the mineral industry. Now came increasing specialization at the operating level, but a broader background at the top, requiring, in the viewpoint of Hoover, not only a wide knowledge of the fields of engineering, but an understanding of economics and the humanities, "and in addition to all this, engineering sense, executive ability, business experience, and financial insight."[97]

94. J. S. Wallace to Fred Bradley, San Francisco, August 23, 1907, & Bradley to Bernard Baruch, Kellogg, August 3, 1921, Bradley MSS.

95. Baruch to Bradley, Washington D.C., December 21, 1917, Bradley to Baruch, San Francisco, December 28, 1917, n.p., June 15, 1920 & March 20, 1923, ibid.

96. Bradley to Baruch, June 15, 1920, ibid.

97. Hoover, *Principles,* p. 186.

CHAPTER 9

"No Spot upon the Globe Is Too Remote"

You can talk about the traveling men who roam,
Of the sailor boys who never have a home;
But the scouting engineer
has no chance to pound his ear
As he travels round this little ball of loam.

—Anonymous [1]

The quarter of a century or so following the mid-1890s has been labeled "The Elizabethan Age of gold-mining" by the British engineer-author, J. H. Curle. The era was "short and brilliant, like its prototype in dramatic literature," said Curle, and included the great discoveries in the Transvaal, Western Australia, the Yukon, and Colorado and Nevada:

> The unapproachable Rand was Shakespear; Kalgoorlie, so richly veined, doomed to so early death, was Marlowe; Cripple Creek, high above the world, yet vitally of it, was Sir Philip Sydney; the gold dredges—the intrusion into imaginative mining of realism—were Ben Johnson; the Klondyke, of a lesser calibre, was Massinger; and the adjacent Nevada camps of Goldfield and Tonopah were Beaumont and Fletcher.[2]

This "Elizabethan Age of gold-mining" coincided with a comparable "Golden Age" of American mining engineers in foreign countries according to Herbert Hoover,[3] for from 1890 to the beginning of World War I, the services of the trained American engineer were in demand in all parts of the

1. From "The Scout Engineer," *EMJ, 105* (April 20, 1918), 765.
2. J. H. Curle, *This World of Ours* (New York, 1921), p. 43.
3. Hoover, *Memoirs,* p. 1, 116.

world at unprecedented premiums. If the engineer had always been a nomad, now he became the traveler par excellence, his horizons vastly expanded as new mineral regions cried out for the most advanced technology.

As early as the middle 1880s, American editors lauded the work of American engineers in Latin America, Australia, and as far off as China, Japan, and India. In 1886, Richard Rothwell of the *Engineering and Mining Journal* could write with all sincerity:

> Every year brings a wider recognition of the fact that, in mining and in practical metallurgy, our American engineers are the most successful and economical in the world, and their services are in request in nearly every country of the world. The difficult and unusual conditions under which they have been forced to carry on their work have developed an ingenuity and fertility of resources, that, guided as they now are by very thorough scientific training, have made the American mining engineer and metallurgist the most successful in the world.[4]

British periodicals noted, with some misgiving, the replacement of English engineers in the 1890s, especially by the versatile, self-reliant Californians, and as far away as Calcutta, *Indian Engineering* commented favorably on the superiority of American engineers over the British, who were less modern, less adaptable, and less tactful.[5] An imaginative speaker at the Colorado School of Mines in 1904 saw the American mining engineer as "the Jason of to-day" leading "The Quest of the Golden Fleece:"

> No spot upon the globe is too remote or too difficult of access for him to reach. He is rapidly Americanizing Mexico, conquering it not as Cortez did, but with those more peaceful arts which not only enrich the people of that country, but are awakening them from the sleep of

4. *EMJ, 41* (January 2, 1886), 1; see also *EMJ, 39* (June 13, 1885), 402.

5. London *Mining Journal,* cited in *M&SP, 73* (October 10, 1896), 294; *Indian Engineering,* cited in *EMJ, 63* (May 1, 1897), 422.

serfdom and ignorance in which they have so long dwelt. He has invaded Peru, but only to extract from the heart of her towering mountains that which was hidden from the rapacity of Pizarro. He has sailed to Japan, to Korea, to Brazil, and to Australia to search for, to examine, and to work the mineral resources of each country. He carried the war into Africa, and after developing the greatest gold district known to the world, rebelled against an ignorant and rapacious government, and, under a new and independent flag, brought on a war that involved nations, and called more than 300,000 men into the field of battle. Nay, he has even invaded the realm of the Great White Czar himself, and now quietly explores the mines in territories where a few years ago no foreigner was permitted to intrude. Far away in the great Northwest under the twilight of an Arctic sky, over 2,600 miles of snow and ice, he has followed his dogs and sledge with the same old tireless energy and indomitable courage which have ever been the marks of those who seek the Golden Fleece.[6]

Not all commentators of this period were so eloquent, but they made the same point. The editor of one technical journal noted in 1904 that the major part of one man's work in the office was to change the addresses of engineer-subscribers, "who, in some capacity of their work, are rambling the earth over."[7] The personal column of a single issue of the *Engineering and Mining Journal* shows American engineers in, en route to, or coming from Alaska, London, Cornwall, Mexico, West Africa, South Africa, Kalgoorlie, Asia Minor, China, Tibet, Chile, Newfoundland, and Nicaragua.[8]

Several factors account for the demand for American engineers abroad during the "Golden Age." By the 1890s, technical schools in the United States were deemed the best in the

6. Harry Hugh Lee, "The Quest of the Golden Fleece," *EMJ, 77* (June 23, 1904), 998.

7. "Mining Engineer a Nomad," *M&SP, 88* (April 16, 1904), 256.

8. *EMJ, 78* (September 8, 1904), 404.

world. Beyond that, the Americans could boast of a variety of experience, especially in deep-level mines. The American engineer had a reputation for versatility and adaptability. The preference for Californians in managerial posts outside the country was explained by one onlooker in 1902 as a result of "the rather American habit of getting out of the ruts, and applying in one place ideas gathered from another place." [9] J. H. Curle believed the American possessed an intangible "mining instinct, that knowledge of values, and that grasp of the whole mining horizon" that most of Curle's own countrymen could not match.[10]

Foreign posts paid higher salaries and stipends, an incentive in itself. But beyond this, American engineers were attracted abroad by other considerations. In a distant, undeveloped land, the young engineer might "be a pioneer, and distinguish himself by overcoming great difficulties, earning fame and fortune." [11] Certainly, the careers of such men as Herbert Hoover, John Hays Hammond, and Charles Butters would bear this out. When Eben Olcott contemplated a professional assignment in Mexico in 1881, he believed he could "do more towards building my reputation there than in the U.S."

> I hate to change but it seems necessary for an M.E. to do so pretty often. In one sense this is very valuable & desirable for it enlarges our scope as consulting engineers when we rise to that dignity—& settle down in N.Y. with a swell office and other blessings.[12]

It is fair to say that American mining engineers blanketed the globe and that this movement was under way to a lesser extent even before the 1895–1914 era. (See Table 6.) The heaviest demands would come in Mexico, Australia, and South Africa eventually, where mining came to play a larger role in

9. Charles H. Finch, "Workings of a Manager's Mind," *M&SP, 84* (February 1, 1902), 65.

10. Quoted in *M&SP, 78* (April 30, 1904), 294.

11. *EMJ, 90* (February 18, 1905), 98.

12. Eben Olcott to Euphemia Olcott, Silverton, Colorado, July 29, 1881, Olcott MSS (Wyo.), Box 1.

national economies. But from an early date, American technical men, like those of other nationalities, were noted as visitors to out-of-the-way places. Not the least was Asia, to which both men and machines were invited as early as the 1860s.

William P. Blake, a Yale graduate and proprietor of the unsuccessful *Mining Magazine,* and geologist Raphael Pumpelly, were employed by the Japanese government in 1861 and 1862 to explore the island of Yezo and to teach American mining techniques. Apparently, either Louis or Henry Janin

Table 6. Foreign Experience of American Mining Engineers, to 1917

COUNTRY, &C.	NUMBER
Africa	74
Australasia	46
Canada	384
Central America	74
Cuba	68
Europe:	
Austria-Hungary	7
Belgium	6
Denmark	1
France	23
Great Britain	116
Germany	61
Holland	2
Italy	5
Norway-Sweden	19
Russia	25
Switzerland	8
Other	22
Not specified	101
Far East (including Philippines)	105
Greenland	3
India	11
Mexico	679
Newfoundland	7
South America	241
West Indies	17

SOURCE: Albert H. Fay, "Preparedness Census of Mining Engineers," U.S. Bureau of Mines, *Technical Paper,* no. 179 (Washington, D.C., 1917), p. 11.

was first considered and might have gone had not Blake decided to go.[13] In any event, Louis Janin did accept a Japanese appointment twelve years later at a salary of $12,000 a year, plus expenses. By July 1873, Janin was in Tokyo, but was by no means enchanted with the Land of the Rising Sun, perhaps in part because a doctor who examined him for insurance before he left San Francisco had diagnosed a heart disease and predicted he might not live out the year.[14] Janin died in 1914, after a long and vigorous career. Janin's employer was a Japanese, who in turbulent times had gained control of a number of mines and sought to expand his holdings. But the effort came to naught, and, early in 1874, Janin prepared to leave. "The game is up and I return home in April to swell the noble army of martyrs," he said. "My Company has given up all their mines, and consequently don't need my valuable services," he noted, explaining that the death of the leading figure in the enterprise, political troubles, and overcapitalization all combined to bring about the end. But it was Janin who is supposed to have convinced the Japanese government to send students to Europe and the United States to learn mineral technology.[15]

Meanwhile, some American mining machinery was being imported, and other American engineers were going to Japan. Alexis Janin, Louis' younger brother, arrived in Tokyo in the fall of 1873, en route to a professional appointment on the island of Sado. Alexis remained in Japan three years, learned the Japanese language, and considered his work there highly successful.[16] During the same period, John C. F. Randolph,

13. *DAB, 2,* 345; Pumpelly, *My Reminiscences, 1,* pp. 267, 308–15, 338–39; Josiah D. Whitney to William Brewer, San Francisco, August 13, 1861, Brewer-Whitney MSS, Box 1; Merle Curti & Kendall Birr, *Prelude to Point Four,* pp. 38–39.

14. Charles Hoffmann to Josiah Whitney, San Francisco, March 28, 1873, Hoffmann MSS; Louis Janin to Henry Janin, Tokyo, July 7, August 6, & October 30, 1873, Janin family MSS, Box 9; *Reese River Reveille,* March 27, 1873.

15. Louis Janin to Henry Janin, Tokyo, March 8, 1874, Janin family MSS Box 9; *DAB, 9,* 609.

16. *M&SP, 21* (July 2, 1870), 8, *25* (July 27, 1872), 57; James D. Hague to I. G. H. Godfrey, San Francisco, March 19, 1873, Letterbook 5, Hague MSS; Alexis Janin to James Hague, Tokyo, October 31, 1873, & Alexis Janin to Juliet C. Janin, Sado, December 12, 1875, & Paris, December 24, 1876, Janin family MSS, Box 5.

who had spent 1869–71 on government service in Germany, was in Japan in the employ of the Japanese government, as was the English-born engineer, J. H. Ernest Waters. Randolph would later become manager of a Borneo diamond enterprise, and Waters would pursue a career in Colorado, New Mexico, and Mexico, but was hired by the Chinese government in 1882 as an adviser to introduce foreign technicians and mining equipment into North China, a post he soon resigned to organize a company of English residents in Shanghai to purchase mines in Colorado.[17]

In 1886, Viceroy Li Hung Chang brought in John A. Church, a Columbia man of much western experience, to import American mining techniques into northern China. With his wife and a few Americans as mine and smelter foremen, Church would spend nearly three years at old silver mines some 150 miles north of the Great Wall, among people who had never before seen a Westerner. There, amid incredible obstacles, Church built a modern smelter, reopened and updated the mines, and put them on a producing basis. In the face of primitive Chinese metallurgical practice, it took infinite patience and resourcefulness to overcome deep-seated prejudices and points of view: because of a superstition that a sacred dragon dominated a convenient coal mine, Church had to go elsewhere for fuel; labor proved crude and difficult to handle; and marauding robber bands were a constant threat until broken up by Imperial cavalry.[18]

Other American engineers in China had comparable experiences. One was forced to cease underground work "because the provincial necromancer reported to Pekin that the mining operations disturbed the slumbers of the dragon that lies coiled around the Imperial tombs," but the payment of a fee overcame this difficulty.[19] A number ran afoul of bandits, and

17. *EMJ, 22* (August 5, 1876), v, *51* (February 28, 1891), 265, *52* (August 22, 1891), 213, *91,* (February 11, 1911), 337.

18. *DAB, 4,* 103–04; *EMJ, 103* (March 10, 1917), 428; *Mining Magazine* (New York), *11* (May 1905), 400; Ellis Clark, "Notes on the Progress of Mining in China," AIME *Trans., 19* (1890–91), 589.

19. Thomas A. Rickard, "Superstition and Mining," *EMJ, 77* (June 23, 1904), 993.

others, including Herbert Hoover and Auguste Mathez, were there at the time of the Boxer uprising in 1900.[20] Hoover, of course, was the best known of all the engineers in China at the turn of the century. Supervising a foreign capitalized coal and cement enterprise, Hoover would spend two and a half years in China, with a staff drawn from three continents, and would penetrate as far as Ulan Bator, the Mongolian capital, in search of gold.[21]

It was one of the first graduates of the Colorado School of Mines, Walter H. Wiley, who examined the gold properties in northern Korea in 1896 that were included in a concession granted to the Colbram Syndicate,[22] and other American engineers, including Hoover, were soon active in that region. The Britisher, Edward McCarthy, who had fled South Africa with his family during the Boer War and China during the Boxer difficulties, met a number of them there. In Korea, McCarthy encountered an American diamond drill expert who "seemed to have no other thought but his beloved drill, indeed he almost slept with it like the Italian organman with his organ." [23] McCarthy also became acquainted with an American engineer who had spent some time in China and who was studying both Chinese and Korean with considerable success. Subsequently McCarthy met him again in England and on the Utah deserts.

> From this little episode of my life it will be seen how small the world appears to us Mining Engineers. We meet each other in all parts of the world and take it as an everyday occurrence, and think no more about it than friends do when they meet each other in different parts of England—perhaps less.[24]

Like those in China proper, American engineers in Korea worked in a strange, sometimes hostile environment. One went

20. *M&SP, 116* (February 16, 1918), 246; *EMJ, 93* (March 30, 1912), 662; Eugene Lyons, *Herbert Hoover: A Biography,* pp. 50–52.
21. Ibid., pp. 45–55; Hoover, *Memoirs, 1,* 38, 45–46.
22. *M&M, 12* (June 1931), 294.
23. McCarthy, *Further Incidents,* pp. 126–27, 152, 158, 240.
24. Ibid., p. 247.

to prison for eighteen months for having killed a Chinese employee who had criminally assaulted his twelve-year-old daughter. Another was petitioned by his workers to hire a sorceress to preside at a feast to be prepared for the particular devils responsible for a rash of accidents at the mine. Another, who eventually died "close to the bottle," was supposed to have developed the habit during his stay in Korea.[25]

Herbert Hoover inspected tin mines in Penang, but late in 1904 became interested in ancient lead and copper mines in Burma, which he proceeded to develop profitably. Many others spent time in the Orient as inspecting engineers, consultants, or managers—among them William Ashburner, John B. Farish, Richard M. Geppert, and Arthur B. Foote. But compared with the major mining areas of the Golden Age, they were there in limited, though not unimportant, numbers, spreading American technology often under the auspices of British capital.

Closer to home, Latin America attracted American engineers by the hundreds. Intemperate climate and uncomfortable living conditions were offset by high consulting fees, managerial salaries, and the knock of opportunity. Many western engineers moved to and fro across the Mexican boundary as if that republic were merely another state, but travel to South America or the Isthmian nations was a good deal more arduous. Alexis Janin described a trip, made to Central America late in 1883 that took forty days at sea coming and going and that was hot and unhealthy all the way. Yet, two years later, Janin was off for Bolivia on professional business.[26] Eben Olcott, who made a number of such trips, was in Colombia in 1887–88 and found the experience pleasant, "but I did not make as much as I expected and I was associated a little too much with speculators." Two years later, preparing for another Latin American journey, Olcott wrote John Hays Hammond, "I am sorry to say that I am going out of the world

25. Rickard, "Superstition and Mining," *EMJ, 77* (June 23, 1904), p. 993; *EMJ, 96* (August 16, 1913), 323; *M&M, 10* (March 1929), 168.

26. Alexis Janin to Juliet C. Janin, San Francisco, February 11, 1884, & LaPaz, August 31, 1886, Janin family MSS, Box 5.

again on a long trip to remote portions of Peru. This seems to be my fate and it is not the most agreeable for a married man." [27] But wherever there was mining or even potential mining, there were mining engineers with western experience. American mining machinery went also. In 1879, the Union Iron Works in San Francisco built a large Morcan fourteen-inch suction gravel pump, patterned after those being tried on the Feather River in California, for use in South America.[28] Hardly had dredging been successfully adapted in California in the late 1890s than such men as Henry C. Granger were introducing the new equipment into Colombia's Choco district, where he had worked professionally for several years.[29]

Numerous engineers, including Alfred Wartenweiler, Ernest Wiltsee, and William Argall, made consulting trips to various parts of Central and South America, and numerous others managed mines in countries as diverse as Chile, Bolivia, Ecuador, and British Guiana.[30] A few were in the employ of Latin American governments. Russell F. Lord, Yale graduate and Civil War brigadier general, left his western engineering practice in 1886 to become chief engineer for the government of Salvador; six years later, he went to Ecuador in the employ of a mining firm, where he remained until 1897, when he returned to the United States in failing health.[31] John C. F. Randolph originally went to Colombia in 1888 on professional business, but remained for an additional year as commissioner of mining for the state of Tolima.[32]

Some, such as Charles Butters, went on their own as capitalists. When nearly seventy years of age and after a successful ore milling and reducing career in the West and on the Rand,

27. Eben Olcott to E. F. Eurich, April 20, 1888, copy, Letterbook 14, Olcott to Hammond, May 7, 1890, copy Letterbook 17, Olcott MSS (N.Y.).

28. *EMJ, 27* (March 8, 1879), 162.

29. *EMJ, 65* (April 23, 1898), 498.

30. See *EMJ, 48* (December 28, 1889), 571, *55* (April 15, 1893), 348, *60* (August 10, 1895), 131; *MI&R, 20* (July 1, 1897), 2; Ernest A. Wiltsee, "Reminiscences," pp. 262–63; Arthur F. Wendt, "The Potosí, Bolivia, Silver District," AIME *Trans., 19* (1890–91), 74.

31. *EMJ, 68* (July 22, 1889), 104.

32. *EMJ, 46* (July 14, 1888), 28, *47* (January 5, 1889), 16.

Butters purchased a mine in Nicaragua and set out to develop it personally. Not only did he encounter "every obstacle in the mining engineer's manual," he also incurred the wrath of his former bookkeeper and storekeeper-turned-bandit, Sandino, who took over the district and sent his henchmen to kill Butters. The American escaped with his life by jumping down an ore chute, but he lost the mine and his investment in it.[33]

More than one American mining engineer was robbed and killed in Latin America, and numerous others succumbed to disease. Arthur Wendt died in 1893 of disease contracted in Bolivia, where William Prince was robbed and murdered a year later. William Argall, of a prominent Colorado family of mine engineers, died of yellow fever in quarantine in New York, after returning from Bolivia in the summer of 1897. Another Coloradan, Wolcott E. Newberry, a Columbia graduate and son of Professor John Newberry, was forced to leave Ecuador in 1898 when he came down with tropical fever. En route home, he was shipwrecked on a Cuban island and was captured by the Spaniards during the Spanish-American War. Soon released, he died within a few weeks.[34] Even in the twentieth century, there might be some danger in certain areas from hostile natives. Guy N. Bjorge and William L. Taylor were captured by Motilones Indians in Venezuela while making explorations near Lake Maracaibo in 1913, but managed to escape.[35]

Of all Latin America, however, it was Mexico that attracted Americans most. Proximity, rich mines, cheap labor, and the potential for applying modern equipment and processes brought both capital and technical know-how from north of the Rio Grande, especially during the era of strong man Porfirio Díaz (1876–1911), when the country's natural resources were thrown open to ruthless foreign exploita-

33. Biographical sketch of Charles Butters, typescript, Charles & Jessie Butters Memorial Collection.

34. *EMJ, 55* (April 15, 1893) 348, *56* (October 14, 1893), 393, *65* (June 11, 1898), 708, *65* (June 25, 1898), 769; *M&SP, 68* (May 19, 1894), 307; *MI&R, 20* (July 1, 1897), 2.

35. *EMJ, 95* (March 8, 1913), 537.

tion. An estimated forty American companies were already operating in Mexico at the beginning of the period, and the number increased substantially as the nineteenth century gave way to the twentieth. The entry of the Guggenheims in the 1890s brought "big capital" and their fusion with the American Smelting and Refining Company a few years later gave American capital a solid grip on the Mexican mineral industry. It was no mere accident that the AIME held its annual meeting in Mexico City in 1901, attended by 165 American engineers, who visited mines in nearly every part of the country.[36]

Long before this, however, undeterred by the rigors of climate, the threat of savage Indians, and a high incidence of violence, countless Yankee engineers had penetrated Mexico's vast wastelands, either to examine or to manage property. Such men as James Hague, the Janins, and Eben Olcott were as familiar with Mexico as with the American West. Beginning in 1882, John Hays Hammond spent nearly two years managing a Sonora property in which the Janin brothers had an interest.[37] Edwin H. Garthwaite spent approximately one third of his twenty-four-year professional career in that country, and Richard Chism devoted his entire career to it, both as editor of *El Minero Mexicano* and as a practicing engineer.[38]

Life in Mexico posed many challenges to the American. Living conditions were usually isolated and primitive, and not everyone was as fortunate as young Robert Livermore, who in 1902 had two Chinese servants to attend to his wants—including the morning chore of filling his boots with hot water to rout any scorpions or centipedes.[39] When the Englishman Edward McCarthy arrived at his new post in Chiapas in the 1890s, wending his way through eleven drunken Mexicans lying on the road, he was greeted by the outgoing manager, "a tall lanky Brother Jonathan" with the words "Well at last you have arrived and I guess you have come to hell." Before his

36. AIME, *Trans.*, *32* (1902), cxlii, cxlvi.

37. Hammond, *Autobiography, 1*, pp. 109–14.

38. Garthwaite, "Reminiscences," pp. 24–45; *M&M, 17* (July 1936), 369; *EMJ, 52* (July 18, 1891), 78; San Francisco *Chronicle,* April 28, 1936.

39. Gressley, ed., *Bostonians and Bullion,* p. 81.

three years in Chiapas were over, McCarthy was inclined to agree.[40] A few years earlier, an American in Chihuahua had written:

> As I write I can feel the fleas playing hide & seek up the legs of my drawers, and the smell of the tough beef and onions I have eaten for supper clings to me still. It is a tough country to live in, you bet, and Uncle Sam will arrange things differently some day when he gets it—which he surely will.[41]

No doubt many engineers had the same initial reaction as Edwin Ludlow who was sent to Coahuila in 1899 with instructions to open up a large coal field and to "obtain a production of 5,000 tons per day as soon as possible." The setting, he said, was "not entirely encouraging": "He saw a cactus and mesquite desert with no trees, no houses (except a few 'jackals'), and no water; but he was told that a small spring, 2 miles away, would furnish enough for drinking."[42] Edwin Garthwaite, who took his wife with him when he went to manage a property at hot, dusty Sierra Modpa, later admitted that had he first seen the region, he would not have accepted the job.[43]

Prolonged isolation bothered not only wives, but some engineers as well, although with traditional adaptability, men often joined together to form "a kind of club where *tequila* cocktails circulated with some freedom."[44] Ralph Ingersoll, a young bachelor, recalled that after a year or so in a Mexican mining camp "I had to be very careful to call on the least prepossessing of my feminine acquaintances first and gradually work up, until I had regained the strength to say more than a

40. McCarthy, *Further Incidents,* pp. 2, 15.

41. W. H. Armstrong to Solomon Noel, Jesus Maria, Mexico, August 22, 1889, Noel family MSS.

42. Edwin Ludlow, "The Coal-Fields of Las Esperanzas, Coahuila, Mexico," AIME *Trans., 32* (1902), 143.

43. Edwin Garthwaite to Augusta Garthwaite, May 2, 1891, quoted in Augusta Garthwaite, "Addenda to Reminiscences of a Mining Engineer," p. 2.

44. Wagner, *Bullion to Books,* p. 31.

dozen words to a pretty girl without proposing to her." [45] Ingersoll also described the American colony at the El Monte de Cobra mine, where the married men rented company houses on "Pershing Drive," and the bachelors lived in a large dormitory on the hill. Too many servants, too little to do, and very narrow social circles soon created an elite, a "Four Hundred," that filled Ingersoll with disgust. Moreover, he soon tired of formal dinners and the current craze, Mah-Jongg, which he learned to hate.[46]

But most engineers took both the environment and the artificial social situation, where it existed, in stride. Their concern was their work, and this they pursued until for one reason or another they felt compelled to move on. Those of experience passed on essential information to newcomers. Neophytes were urged to go armed with a revolver and an adequate supply of boots. Knowledge of the language or the acquiring of a good interpreter was taken for granted. Because travel was often by muleback, the engineer was advised to pack his belongings in two small steamer trunks, each weighing about a hundred pounds.[47] Often, those heading for Mexico deliberately sought out old-timers who were familiar with conditions. Before an inspection trip in 1900, for example, Frank Sizer "had a long talk with Victor Clement who is familiar with Mexico and gave us many valuable hints." [48]

Travel in Mexico, especially away from the railroads, was difficult enough at best. On one occasion, traveling into Sonora, Robert Livermore had an additional handicap—the son of his client, "a New York stripling, whose chief interest throughout the journey was the baseball fortunes of the Yankees" and who went along "for educational purposes." [49]

The engineer-manager in Mexico had the special problem of handling native labor—a cheap, relatively docile labor, but

45. Ingersoll, *In and Under Mexico,* p. 218.

46. Ibid., pp. 134–38, 143–44, 147.

47. A. R. Townsend, "Some Suggestions for Travel in Northern Mexico," *EMJ, 77* (February 25, 1904), 315–17.

48. Entry for July 15, 1900, Sizer Diary.

49. Gressley, ed., *Bostonians and Bullion,* p. 165.

one that required some intimate knowledge of the Mexican laborer and his way of life. Wages were often shockingly low, and the laborer had to be trained in drilling, tramming, or mucking. In coal mines, it was not unusual for a miner to demand to see his car weighed before loading another, and, in most mines, wages were paid weekly or sometimes even daily, generally on a contract basis rather than at an hourly or per diem rate.[50]

As in the American West, engineers found themselves engaged in all sorts of subsidiary activities, ranging from road and railway construction to the building of steel-bottomed scows for hauling ore and the installation of electrical plants.[51] American machinery and processes went with them, of course. H. F. Reinhard, a German-trained expert, who had spent fifteen years in San Francisco, was credited with bringing "millions of dollars" worth of mining equipment into Chihuahua from 1880 to 1895. Ottokar Hofmann early introduced the lixiviation process of roasting and leaching silver ores, thus successfully challenging the age-old patio process in northwestern Mexico.[52] Subsequently, the use of the cyanide process, by Henry R. Batcheller in Sinaloa and by others elsewhere, enabled the profitable working of lower-grade ores and would bring a veritable mining boom. Likewise, the introduction of the standard square-set timbering by American engineers and timbermen would permit the efficient reworking of more than one ancient property.[53]

But Mexico was an unhealthy land. Engineers faced the ravages of disease in areas where medical attention was often lacking. More than one engineer served as physician and sur-

50. Ludlow, "Coal-Fields," AIME *Trans., 32* (1902), p. 144; Marvin D. Bernstein, *The Mexican Mining Industry, 1890–1950,* pp. 84–95.

51. *EMJ,* (March 8, 1913), 501.

52. *M&SP, 70* (January 26, 1895), 50; *EMJ, 52* (August 8, 1891), 161; F. H. McDowell, "American Mining Machinery in Mexico and Central America," AIME *Trans., 13* (1884–85), 408–17; Ottokar Hofmann, "Trough-Lixiviation," AIME *Trans., 16* (1887–88), 668–77.

53. James W. Malcolmson "The Sierra Mojada, Coahuila, Mexico, and Its Ore-Deposits," AIME *Trans., 32* (1902), 133; E. A. Tays & F. A. Schiertz, "The Treatment of Clay-Slimes by the Cyanide Process and Agitation," AIME *Trans., 32* (1902), 179.

geon in an emergency: Edward McCarthy once amputated the arm of a Mexican worker maimed in an explosion.[54] Many contracted typhoid or malaria, and some, including Louis Janin's youngest son, never survived.[55] Others perished as a result of blood poisoning.[56] Indians posed a problem, and, as late as 1919, two American engineers were reported killed by Yaquis.[57]

Even in the relatively stable period of Porfirio Díaz, Mexico was a violent land. Mine managers frequently exposed themselves when they sought to enforce regulations or prevent ore theft. William R. Boggs, a VPI graduate with nearly a quarter of a century of experience in Colorado and Mexico, was beaten to death near Topia in 1907 when he tried to prevent the sale of liquor on the night shift.[58] Payroll robberies were not uncommon, sometimes with the engineer-manager killed in the process.[59] During the 1880s, one American engineer was arrested for killing a Mexican who tried to steal part of the payroll. Apparently, the Mexican government compromised the situation by permitting him to enlist in the army, from which he promptly deserted. Years later, he was recognized while managing a gold mine in Chihuahua and imprisoned, but escaped to San Francisco. But the pull of Mexico was too strong: he returned and disappeared there in 1903.[60]

Beginning in 1910, Mexico underwent a period of violent revolutionary convulsion, as the result of which Díaz was ousted in 1911 and an assortment of contenders sought control of the government for the next half dozen years. Meanwhile, the country was bathed in blood, not only because of the revolutionary movement itself, but also because of the inability of

54. McCarthy, *Further Incidents* pp. 36–37.
55. *EMJ, 58* (September 15, 1894), 252.
56. *M&SP, 95* (July 18, 1907), 612.
57. *M&SP, 119* (July 12, 1919), 42.
58. First reports, later corrected, said Boggs was killed when his company was unable to meet its payroll. *M&SP, 95* (December 14, 1907), 727; *EMJ, 84* (December 7, 1907), 1081, *75* (June 20, 1908), 1256.
59. *EMJ, 84* (September 28, 1907), 606.
60. *EMJ, 51* (March 14, 1891), 330, *52* (October 3, 1891), 391, *76* (August 15, 1903), 244.

the regimes at Mexico City to suppress bandits and maintain order. It was estimated that during these upheavals 270 American mining men died violently in the country.[61]

In the movie, *Wings of the Hawk* (1953), the hero, a handsome young American engineer, who has struck it rich in Mexico, is at the same time involved in the revolution and in love with a bandit queen. This is in keeping with the romantic stereotype of the engineer, but probably far from reality. Most American engineers considered the revolutionaries as "dreamers and schemers"; they agreed with Ernest Wiltsee, who regarded Porfirio Díaz as "one of the two greatest men that Mexico has ever produced." "What a country Mexico was," Wiltsee reminisced, "and what it could be again if properly governed by educated, decent white people! Not entirely given over to Indians, aided and abetted by a Communist and Bolshevist regime." [62]

During the upheavals, from 1910 to 1918, mining operations were disrupted and American engineers found themselves, usually inadvertently, involved in domestic strife. One result was a mass exodus, which glutted the market for technical men in the United States, just at a time when World War I was disclocating mining in other areas. Those who chose to remain in Mexico, either for all or part of the period, were faced with the threat of constant peril. At least one had a nervous breakdown because of the strain, and several were imprisoned—one on charges of instructing the Zapatista rebels in the use of dynamite, another for aiding the rebels by purchasing ore stolen from a mine confiscated by the government.[63] Numerous mines were abandoned, and engineers forced to

61. Russell H. Bennett, *Quest for Ore,* p. 139. According to a Senate investigation, between 1910 and May 20, 1920, a total of 397 American civilians of all occupations were killed in Mexico and 32 were "outraged or wounded." An additional 32 were killed along the border. "Investigation of Mexican Affairs," *Senate Document,* no. 285, 66th Cong., 2d Sess. (1919–20), *2,* 3382.

62. Wiltsee, "Reminiscences," pp. 128, 168, 179.

63. *EMJ, 101* (June 17, 1916), 1089, *102* (July 22, 1916), 197, *102* (August 19, 1916), 361, *103* (February 3, 1917), 242; *M&SP, 113* (December 2, 1916), 823.

flee. Walter Douglas narrowly escaped capture by rebel forces in 1912 after a harrowing trip by railroad motorvelocipede; his father, James Douglas, was forced out of Cananea by a mob in the following year.[64] In the absence of protection, payroll and bullion robberies increased, and a number of engineers were captured and held for ransom.[65] One manager, S. F. Shaw, and his crew killed nine rebels in a running fight in Charcas early in 1913 and with a locomotive and several cars rescued fifteen persons besieged in a local church. Frank B. Harding, consultant for a company on the Sonora-Chihuahua boundary, was taken prisoner when Villistas raided the camp. Beaten, stripped, and finally released, he walked the seventy-five miles to Agua Prieta.[66] Alfred M. Hamilton, who had been in Mexico since 1906, lost all his belongings when outlaws burned his house. He saw some of his coworkers hanged on telegraph poles near the American Smelting and Refining Company smelter, and at one point he entertained the leader of the victorious rebel faction, while concealing the loser under a bed. Several times, he saved his own life by bringing out a camera and urging bandits to pose on their horses. He returned from Mexico weak from heart and lung trouble and suffering from nervous prostration to die at the age of thirty-seven.[67]

Many American engineers were killed in sporadic affrays,[68] but the most sensational episode was the so-called Santa Isabel massacre of January 10, 1916. On that day, eighteen American mining men were taken from a Mexican Northwestern train, robbed, stripped, and shot by Villistas while on their

64. *M&SP 105* (September 21, 1912), 360; *EMJ, 94* (September 21, 1912), 562, *95* (April 26, 1913), 868.

65. *M&SP, 105* (December 7, 1912), 745, *109* (November 21, 1914), 825; *EMJ, 94* (December 7, 1912), 1096, *95* (January 1, 1913), 202; *Papers Relating to the Foreign Relations of the United States, 1914,* p. 680.

66. *M&SP, 106* (March 22, April 12, 1913), 464, 563, *118* (June 14, 1919) 826.

67. *M&SP 118* (January 4, 1919), 30.

68. See *M&SP, 109* (July 11, 1914), 71; *Papers Relating to the Foreign Relations of the United States, 1914,* pp. 685–87.

way to reopen the mines of the Cusihuiriachic Mining Company.[69] Mining men and the public at large condemned this "result of a watchful waiting policy," [70] but it would take a Villista attack on the little town of Columbus, New Mexico, later in the year to bring forceful intervention by the U.S. government.

The fruits of the Mexican Revolution were to curtail American ownership of natural resources, but mining again revived, and Yankee engineers continued in great demand, although not always for American employers. As before, the republic to the south continued to provide a challenging field for the engineering profession.

Since the 1850s, Americans had had some interest in Australian mining. Californians had been among the "diggers" in New South Wales and Victoria almost from the beginning.[71] During the 1860s, Americans debated the relative merits of American and Australian equipment, and J. Mosheimer of San Francisco apparently was hired to introduce "the Hungarian process as it is now called"—a concentration process.[72] The Union Iron Works shipped some of Hendy's concentrating machinery to Melbourne as early as 1868, but not for several decades was this more than a mere trickle. When J. S. Phillips left California for Australia on professional business in 1885, he could announce that he was prepared to introduce American mining machinery to that far continent.[73] It would appear that Parke and Lacy, a San Francisco firm, was the first to export machinery on a large scale and, ultimately, during the 1890s, would have a branch office in Sydney.[74]

Meanwhile, during the 1880s and 1890s, the opening of new deposits, especially deep in New South Wales and in

69. *EMJ, 101* (January 22, 1916), 194; *M&SP, 112* (January 15, 1916), 76; *Papers Relating to the Foreign Relations of the United States, 1916,* pp. 581–83.

70. *M&SP, 112* (January 15, 1916), 76.

71. See Rodman W. Paul, "'Old Californians' in British Gold Fields," *The Huntington Library Quarterly, 17* (February 1954), 161–72.

72. San Francisco *Bulletin,* July 14, 1866, clipping, Bancroft Scrapbooks, *94, no. 1,* 3441.

73. *M&SP, 13* (September 29, 1866), 200, *16* (March 21, 1868), 177.

74. *EMJ, 68* (November 25, 1899), 634.

Western Australia made such names as Broken Hill, Kalgoorlie, and Coolgardie common utterances among American professional men. In July 1886, according to one of Australia's most perceptive historians of mining, the directors of Broken Hill Proprietary "made perhaps the most momentous decision in Australia," when they sent to the United States to find the best mine manager available. From the Comstock they hired William H. Patton at £4,000 a year—"twice the allowance the highest-paid Australian politician received." From Colorado, with its advanced metallurgy, they took young Freiberg-trained Herman Schlapp, a native of Iowa. Patton would encounter trouble when he sought to use the familiar square-set timbering at Broken Hill: lodes were wide, and heavy pressures sometimes crushed the timber; ground shifts ruined mill foundations or threw machinery out of line. Schlapp built the largest smelters in Australia to that time, but the plummeting price of lead and silver in the 1890s sorely hurt Broken Hill. Yet, Schlapp came to be regarded as "possibly the foremost metallurgist in Australia." [75]

Patton and Schlapp represented the first of an influx of experts from the American West—an influx that, says Geoffrey Blainey, tied Australia "to a new powerhouse of skills and attitudes." [76] This infusion of fresh technological blood, together with a strong flow of British capital, would make Australia an important mineral producer by the end of the nineteenth century. Among other American visitors was Edward D. Peters, with experience in Colorado silver and Montana copper, who was hired in 1893 (at a fee of £2,500, and who, while cursing the food, climate, and rough environment, gave a solid endorsement of the Broken Hill fields.[77] Louis Janin, Jr., made several trips to Australia, remaining more than two years on one of them. Reuben Rickard came out of retirement in 1895 to examine mines at Coolgardie, but died of dysentery the next year; his brother, Alfred, who had gone to Australia

75. Geoffrey Blainey, *The Rush That Never Ended,* pp. 154–58, 220.
76. Ibid., p. 252.
77. Ibid., p. 220.

to meet him, died a few months later.[78] Others went to manage properties, among them Carl F. Hesse of California and Ralph Nichols, who came from the DeLamar mines in Nevada.[79] An American engineer, who returned from Western Australia in 1897, named a number of Californians he had met there, including "a young man by the name of Hoover, a miner from the Leland Stanford University." [80] It was here that Herbert Hoover, under the auspices of Bewick, Moreing & Company of London, would make his reputation and in the process draw around him a talented coterie of engineers and metallurgists trained at Stanford, Columbia, California, and Colorado Mines.[81]

Americans were also drawn to Tasmania and New Zealand. One, Lamartine Cavaignac Trent, general manager of the North Lyell Company in Tasmania, has been described as "decisive, impatient, a striking American with his cloth cap, flat nose, and the trace of Red Indian in his complexion." But he left behind a negative image as well: the smelters he built were expensive, but poorly constructed; he was both inefficient and extravagant, using a special locomotive to carry his lunch, mail, or morning paper. After a running tiff with company directors, Trent is supposed to have departed with a salute, leaving Tasmania "to the working man and the servant girl!" [82]

Another American, Robert C. Sticht, pursued a far more successful career in Tasmania. A graduate of Brooklyn Polytechnic and Clausthal, with extensive experience in Colorado and Montana milling, Sticht went to Mount Lyell in 1895 and established pyritic smelting, in theory using the iron and sulphur in the copper ore as fuel in the furnace, thus using less coke in the smelting process. A few years later, using cold air,

78. *EMJ, 58* (December 22, 1894), 587, *60* (November 2, 1895), 423, *61* (April 4, 1896), 327, *63* (May 15, 1897), 486, *68* (July 8, 1899), 44; *M&SP, 72* (March 7, 1896), 192.

79. *EMJ, 60* (November 2, 1895), 423, *65* (January 22, 1898), 108, *67* (April 29, 1899), 506.

80. Quoted in *M&SP, 76* (January 1, 1898), 4.

81. Hoover, *Memoirs, 1*, 28–29, 37, 78.

82. Blainey, *The Rush That Never Ended*, pp. 227–28.

rather than hot, blown through his furnaces, he smelted ore with neither coal nor coke. Sticht's work was regarded as "a milestone in copper metallurgy," and at the time of his death in 1922, flags flew at half-mast and work halted for five minutes through the western part of the land in honor of "Tasmania's foremost citizen." [83]

As Blainey points out, Australia was quick to borrow and adapt, and rapidly emerge as a leader in the metallurgical field: "In innovation and adaptation no mining country matched Australia in this triumphant era of world metallurgy." It promptly adopted the cyanide process in North Queensland, beginning in 1891, and Mount Morgan developed chlorination to its highest efficiency. It was the Australians who pioneered in the practical development of the flotation process, though Americans working independently had discovered the basic techniques. Still, when the first successful froth flotation plant opened at Butte in 1911, Broken Hill had already handled 8 million tons of ore by the process.[84]

Australia-New Zealand also pioneered in dredge mining, and the first successful dredges in California during the late 1890s were patterned on those from Down Under. Within a few years, however, dredge engineers in western America would be the world's recognized leaders in this field, and eventually dredges manufactured in San Francisco would be exported to Australia and New Zealand.[85] But it was American mining and metallurgical engineers, in the important 1890 to 1910 period, who did much to modernize and to transplant western processes and equipment to the region. American engineers, indirectly at least, would provide the incentive and the model on which Australian mining schools and the Australian Institute of Mining Engineers would be patterned.

During the 1890s, Americans also began to discover Siberia. Even as early as the 1860s and 1870s, a few engineers,

83. Ibid., pp. 222, 229; *M&M, 4* (February, 1923), 99.

84. Blainey, *The Rush That Never Ended,* pp. 255, 270.

85. Lucien Eaton, "Seventy-Five Years of Progress in Metal Mining," in Parsons, ed., *Seventy-Five Years of Progress,* p. 78; interview with George Hurst, Bethlehem Steel, San Francisco, June 5, 1961.

including George Maynard, had been in Russia,[86] but now the Trans-Siberia Railroad was being pushed to completion, and British firms were able to get mineral concessions. What was more logical than to introduce the new efficient dredges being perfected in California?

Russell L. Dunn inspected Amoor River placers in 1896, receiving "one of the largest fees ever paid an American mining engineer for similar services." Charles Hoffmann was on the Amoor in 1897 and a year later made what he called "an interesting trip down the Lena to examine potential dredge property." Hoffmann was again in Siberia in 1900, as were two of his engineer sons, George and Ross, not to mention Forbes Rickard of Central City, Colorado.[87]

Early in the twentieth century, the profession began to publicize in more detail the possibilities of the Siberian fields. The English expert J. H. Curle was wary. He admitted that plenty of gold was available and that the Russian engineers had ignored alluvial deposits, but he saw serious handicaps in the form of climate, the language barrier, primitive labor methods and accounting systems, and official red tape.[88] On the other hand, W. W. Behr, in St. Petersburg at this time (1904), disagreed, contending that political problems had been overdone and that, despite outmoded labor devices, opportunties in Siberia were unlimited.[89] John Hays Hammond, who visited Russia a few years later, took the same position,[90] and, by 1910, some of the potentialities were being realized, and Americans were coming in in larger numbers.

By the following year, at least a dozen engineers with west-

86. "T. A." Rickard recalled meeting Maynard on the Pashkoff estates in the Urals, where the elder Rickard managed mines and smelters. Rickard, *Retrospect*, p. 10.

87. *M&SP, 73* (August 1, 1896), 100; *EMJ, 69* (June 2, 1900), 656, *70* (October 20, 1900), 465, Charles Hoffmann to William Brewer, London, November 19, 1898, & San Francisco, May 18, 1900, Hoffmann MSS.

88. "Gold Mining in Siberia," *The Economist* (London), *62* (August 27, 1904), 1408–09.

89. W. W. Behr to editor, St. Petersburg, October 6, 1904, *EMJ, 78* (December 8, 1904), 902.

90. *EMJ, 91* (February 4, 1911), 253.

ern experience were in Russia, including Charles Munro, J. P. Hutchins, Chester W. Purington and A. C. Perkins.[91] Purington set up offices in London to specialize in Siberian properties. He advised a colleague in San Francisco, who had expressed a desire to examine mines in Siberia: "It will pay you to get a knowledge of Russian in any case, as sooner or later the services of all the dredging engineers in California are bound to be required in Siberia." [92]

Horace H. Emrich, Colorado School of Mines (1903), went out to Kyshtim to manage a British property in 1910, but a year later was murdered.[93] Edward McCarthy visited the manager of the Spassky Copper Mines after a 525-mile trip across the steppes. The manager was a graduate of the Royal School of Mines in London, married to an American, and a naturalized citizen of the United States himself.[94] In 1912, W. H. Lanagan was manager of the Orsk Goldfields, Ltd., at Kolchan, and through Fred Morris in California, ordered dredge components to be shipped via Vladivostok. Orsk Goldfields employed a number of American technical experts under a tightly drawn contract that prescribed a standard ten-hour work day, plus "work at such times in such shifts as may be required, irrespective of Sundays or weather conditions, until the termination of the work," and a pledge not to interfere in Russian politics or religion. Apparently, Lanagan failed to please the directors of the company, for at the end of 1914 they were looking for a replacement. "At the present moment," wrote Edward Hooper, as he solicited recommendations, "the Board say they have had enough of Americans, (I know you will excuse my saying this) and are therefore look-

91. *EMJ, 92* (November 11, 1911), 958; Charles Munro to W. B. Watts, San Francisco, April 4, 1911, Morris MSS, Box 2.

92. *EMJ, 92* (July 15, 1911), 130; Charles Janin to Chester W. Purington, San Francisco, July 15, 1911, copy, Purington to Janin, "Aboard the Vlad-Mos Express," November 2, 1911, Charles Janin MSS, Box 5.

93. *EMJ, 92* (October 28, 1911), 842. Herbert Hoover was in charge of general planning for the Kyshtim estate in the Urals, and brought in a number of Butte engineers. Lyons, *Hoover,* p. 68.

94. McCarthy, *Further Incidents,* p. 327.

ing out for a Britisher who will 'fill the bill.' This, as you can imagine, is not an easy task." [95]

The coming of World War I dislocated mining in Russia somewhat. Some American engineers, such as Karl and Ross Hoffmann, left by way of Norway, but others remained and new ones arrived.[96] The Russian Revolution further demoralized the industry and gave a number of engineers a first-hand view of the political upheaval. Charles Janin arrived in Petrograd "just in time to see the first chapter of the revolution, but unfortunately not in time to see Rasputin polished off." [97] Janin joined Horace Winchell, Frederick Kett, and Ira Joralemon, who were already in Russia.[98] Subsequently, Norman C. Stines, an engineer formerly associated with Herbert Hoover and in 1917 attached to the American Embassy in Petrograd, joined with other engineers quietly to gain control of a substantial supply of platinum, vital to the American war effort. This move was successful, and F. D. Draper, also an engineer, escorted the platinum, worth $2 million, with considerable difficulty, to a bank at Yokohama.[99]

During and immediately following the revolution, engineers were an important source of information on Russia for the U.S. government. It was an American engineer who is supposed to have first suggested armed intervention in Siberia, and it was American engineers who provided much of the evidence under which that intervention was carried out. Chester W. Purington and Reuben E. Smith were among those called to Washington to work with military intelligence in 1917–18.[100]

But the new Soviet Union discouraged foreign capital, and

95. W. H. Lanagan to Charles H. Munro, Kolchan, October 19 & 20, 1912, Morris MSS, Box 1; copy of agreement, 1913, ibid., Box 9; Edward Hooper to Charles Janin, London, December 29, 1914, Charles Janin MSS, Box 5.

96. Edgar Rickard to Charles Janin, London, August 20, 1914, ibid.

97. Charles Janin to Charles W. Purington, August 21, 1917, ibid., Box 6.

98. Ira B. Joralemon to Charles Janin, Omsk, June 2, 1917, ibid.; Kett, *Autobiography*, p. 98; *EMJ, 116* (August 4, 1923), 203; *Oakland Tribune*, June 3, 1960.

99. *EMJ, 104* (December 22, 1917), 1099, *107* (January 3, 1919), 3–4.

100. *EMJ, 109* (January 10, 1920), 102, *112* (November 5, 1921), 737; *M&SP, 116* (June 8, 1918), 806; *M&M, 4* (November 1923), 579.

while a few American engineers continued on, most left, waiting hopefully for more political stability and more encouraging times.[101]

Although they covered the globe, it was probably in Africa —especially South Africa in the latter 1890s—that American mining engineers achieved the greatest recognition for themselves and their work. A few had traveled professionally in the Dark Continent earlier, and during the 1880s a slow influx commenced. Joshua Clayton was apparently offered an African post in 1884, but decided to stay in the West "and fight it out on the line we have chalked out." [102] This offer may have been from the Lisbon-Berlyn Gold Fields, Ltd., a British firm with promoter Baron Albert Grant as the moving spirit, which subsequently hired J. L. Gould of California as its manager, along with half a dozen experienced hydraulic and deep level miners from the same state. A San Francisco agent purchased and shipped to the concern a complete sixty-stamp mill, modeled after that of the Pacific Company in Amador County, and Judd Coppach, an engineer, was sent out to help install it. When Coppach failed to receive his fee, he called on Baron Grant in London. Twice Grant was out; the third time the American put his Colt on the desk, asserting he would wait while the clerk took in a message:

> To Baron Grant, your Lord Almightyness. It is beneath the dignity of an American citizen to call three times for his money. I have done so, and unless I get a cheque at once it would have been better if your mother had been "barren." [103]

In the same year, Gardner F. Williams, one of the earliest University of California graduates, took charge of property in

101. Charles Janin to H. Foster Bain, San Francisco, April 15, 1919, copy, Charles Janin MSS, Box 8; John Hays Hammond to Janin, Washington, D.C., June 12, 1922, ibid., Box 9.

102. Joshua Clayton to unidentified, February 11, 1884, copy, Clayton MSS (Calif.), 2.

103. *EMJ, 38* (December 13, 1884), 396; Alpheus F. Williams, *Some Dreams Come True,* pp. 528–30, 532.

South Africa for the Transvaal Gold Exploration & Land Company, Ltd. The enterprise quickly failed, and Williams was back in California in time for Christmas, 1885. However, Hamilton Smith and Edmund DeCrano of the Exploration Company, Ltd., were convinced that the Transvaal had a great future and pursuaded Williams to return in 1886 as their consulting engineer. Soon Williams met Cecil Rhodes and took over management of the famous De Beers diamond mines, a post he retained until his return to America in 1906.[104]

Gardner Williams was the first of a long line of distinguished American engineers who would form the technical backbone of the South African mineral industry during its crucial formative years. During the 1890s, a period of great consolidation on the Rand, British, French, and German capital predominated and was organized in blocks large enough to be able to afford not only expensive engineering talent, but also costly modern processing plants and the heavy outlay required in developing deep level mines before any returns could be expected. In 1896, it was estimated that at least fifty American engineers were at work in South Africa, among them "some of the brightest oraments of American mining and metallurgy." [105] At the same time, it was asserted that half the mines on the Witwatersrand were being managed by California engineers.[106] Even in 1901, at a Johannesburg dinner honoring W. L. Honnold, of the twenty-six engineers and managers present, fourteen were Americans.[107]

The bulk of these engineers were young men. Ernest Wiltsee, who was not yet thirty when went to Africa, later recounted how he grew a beard in order to look older. George Starr, when not over thirty-five, had "dozens of grey-headed

104. Ibid., pp. 221–27; *DAB, 20,* 261.

105. Rossiter W. Raymond to editor, New York, January 22, 1896, *EMJ, 61* (January 25, 1896), 84.

106. *M&SP, 72* (January 18, 1896), 42; *EMJ, 64* (July 10, 1897), 36.

107. *EMJ, 73* (June 28, 1901), 902. Alpheus F. Williams, ed., *American Mining Engineers in South Africa* (Kimberly, 1902), cited in *EMJ, 74* (July 5, 1902), 17.

mining supts under him" at the Barnato mines, according to one who knew him. Louis Seymour was twenty-eight when he first went to Kimberley; Eugene Hoefer was thirty-six, John Hays Hammond was thirty-eight. Henry C. Perkins at forty-seven and Thomas Mein at fifty-four were the senior engineers in age.[108]

But young though they were, most came with extensive experience in deep level mining. Many, such as Henry Perkins, Hennen Jennings, and Charles Butters—all employed by the Corner House Group—had "cut their wisdom teeth comparatively early in life" at the New Almaden Quicksilver Mining Company; others, including Hammond, Victor Clement, and George Starr—all initially associated with Barnato Brothers in Africa—went with deep level experience at Grass Valley. Many, Hammond and Clement among them, had worked in Mexico, and a number—among them Perkins, Jennings, Louis Seymour and George Webber—had served in the El Callao in Venezuela, a property known as "the nest whence the American engineers flew when they were called upon to institute development of the gold mines of the Witwatersrand." [109]

Englishmen might denounce the "Americanization of British mines" in Africa and show understandable concern that the most lucrative posts were going to American engineers, but they found it difficult to argue with the American's experience in large-scale, deep level production. Cecil Rhodes found them better trained than the British, with a finer balance between theory and practice. An English engineering professor admitted in 1896 that South African gold mining practice was so advanced because it benefitted from American tradition, "which had hitherto been the best." Practice on the Rand, as he described it, was "based upon and modeled after California cus-

108. Wiltsee, "Reminiscences," p. 23; Mary H. Foote to Helena Gilder, Grass Valley, April 19, 1898, M. H. Foote MSS; Rickard, *Interviews*, pp. 413, 420; *DAB, 10,* 55, 56; *M&M, 16* (August & November 1935), 355, 484; *NCAB, 26,* 45.

109. Rickard, ed., *Interviews*, pp. 225–26, 230, 414, 418, 427; Hammond, *Autobiography, 1,* 116, 178; *M&SP, 70* (February 23, 1895), 115; *EMJ, 69* (May 12 & June 23, 1900), 554, 746; *M&M, 1* (May 1920), 19–20; *NCAB, 30,* 567.

tom and experience, modified somewhat by local requirements." [110] Others saw the Americans as being more imaginative, more versatile, and more industrious. When asked to compare American and European engineers on the Rand, Hamilton Smith replied:

> Well, in South Africa the American never lets up. He works from daylight until dark and is thinking about his job in the evening. Our European engineers want to stop at four; the Englishman to play tennis, the Germans for their beer.[111]

High salaries and unparalleled opportunity for promotional or speculative profits attracted men with reputations. John Hays Hammond originally went to work for Barney Barnato at $50,000 a year, but only, he said, until he could convince Barnato that his services were worth more. When Barnato repeatedly ignored his professional advice, Hammond resigned at the end of six months to take charge of Cecil Rhodes's vast holdings at $75,000 a year and a share of the profits—a salary regarded in 1896 as the largest of any engineer in the world.[112] Hammond brought over a full complement of engineers at lesser figures: George Starr received $7,500 the first year and $8,000 the second; but two years later, in Barnato's employ, he was making $25,000 a year.[113] Richard A. Parker came at £2,000 a year, and Pope Yeatman was offered £1,200, but both were assured that these would increase rapidly when the proper openings developed.[114] At the height of the South African fever, those who remained at home were

110. Williams, *Some Dreams Come True,* pp. 551–52; *MI&R, 19* (March 25, 1897), 135; *M&SP, 70* (November 2, 1895), 287, *72* (February 29, 1896), 171.

111. Quoted in A. R. Ledoux, "The American Mining Engineer," *EMJ, 77* (February 25, 1904), 310.

112. Hammond, *Autobiography, 1,* 201, 213–14; *EMJ, 61* (January 18, 1896), 60.

113. John Hays Hammond to George Starr, November 14, 1893, copy, Letterbook 1, Hammond MSS; W. B. Bourn to James D. Hague, San Francisco, June 28, 1893, Hague MSS, Box 11; *M&SP, 71* (October 19, 1895), 255.

114. Hammond to Richard A. Parker, May 29, 1895, & Hammond to Pope Yeatman, May 27, 1895, copies, Letterbook 2, Hammond MSS.

left wishing that "our American friends would get in the way of paying as our English cousins do." [115]

Beyond the basic salaries, possibilities were unlimited. In 1889, Eugene Hoefer, a California Mines graduate, wrote that many Americans were leaving South Africa, some for holidays, some for good. "Most of the older hands have made so much money that work is too troublesome," he said.[116] Hoefer had earlier written that Henry C. Perkins had left Johannesburg "with something over £250,000"; Thomas Mein had gone "with a fair nest egg of £90,000, more or less"; whereas Hennen Jennings "has not got enough money yet for he is still hanging on. Certainly £100,000 ought to satisfy most men." [117] Another engineer reported that Ernest Wiltsee had just cleared $50,000 cash on a deal, and Wiltsee himself admitted a profit on investment in boom property that netted him $187,500 on an input of $250.[118] But John Hays Hammond was the most successful. In an expansive mood in 1895, he wrote a friend concerning his work with Rhodes, whom he called "the greatest Englishman of the age and he has not reached his zenith yet. . . . He suits me to a T." His own work he described as "more commercial and financial than technical." "I am making money fast," he said, "and expect to have about $750,000 well in hand by end of this year—then shall take a rest in London but go on adding to my pile until I reach the million mark—which should not take many months." [119]

American engineers were active in other parts of Africa, but focused particularly in the Transvaal. One occasionally wandered as far afield as French Sudan, and a number pushed into Buluwayo and Matabeleland, an area more fraught with danger from disease and savages than farther south. Charles Jef-

115. Eben Olcott to Thomas Leggett, New York, May 23, 1895, copy, Letterbook 23, Olcott MSS (N.Y.).

116. Eugene Hoefer to Samuel Christy, Pilgrims Rest, February 3, 1898, Christy MSS.

117. Hoefer to Christy, Pilgrim's Rest, September 13, 1896, ibid.

118. *M&SP, 71* (October 19, 1895), 255; Wiltsee, "Reminiscences," p. 68.

119. John Hays Hammond to N. H. Harris, Cape Town, October 8, 1895, Hammond MSS, Box 2.

ferson Clark, who had a reputation as "that American mining man that shoots shillings in the air," amassed a fortune in mining and a variety of other enterprises in Rhodesia, but there contracted a fever that cost him his life.[120] In the uprisings of the Matabele, the last in 1896, several engineers, including Mark Elliot of California, were killed.[121]

It was on the Rand—"the unapproachable Rand," Curle's "Shakespear"—that the American engineer made his great impression. A thousand miles north of Capetown, the center was Johannesburg—"a healthy place to live in," with "all the conveniences of Denver," according to Hammond.[122] Often with their families settled here, the Americans ranged out into the mining field itself.

The American impact was easily discernible. As elsewhere, in the wake of the engineer came mining machinery labeled "Made in U.S.A." Charles Rolker ordered American equipment for South Africa in 1894, and Charles Jefferson Clark purchased $30,000 worth from California in the following year. One of Hammond's subordinates toured the United States, buying electric hoists from General Electric, visiting drum and shaft manufacturers in Akron and the E. P. Allis Company of Milwaukee for electrical transmission equipment, and inspecting the Anaconda works in Montana before his return.[123] British experts reacted negatively and often refused to admit any superiority of American machinery, though Edgar Rathbone chastized his fellow Englishmen in 1896, contending that English manufacturers were only then awakening

120. *M&SP, 65* (August 13, 1892), 197, *71* (October 19, 1895), 255, *72* (February 22, 1896), 152; *EMJ, 68* (July 1, 1899), 14; Lyons, *Herbert Hoover,* p. 64; Charles Jefferson Clark to John Hays Hammond, Ft. Salisbury, Mashonaland, March 19, 1894, Hammond MSS, Box 1. The role of the American engineer in Africa is touched on briefly, but inadequately, in Clarence Clendenen, Robert Collins, & Peter Duignan, *Americans in Africa, 1865–1900,* pp. 91–96.

121. *M&SP, 72* (May 23 & June 6, 1896), 426, 463, *73* (October 31, 1896), 361.

122. John Hays Hammond to Richard A. Parker, May 29, 1895, copy, Letterbook 2, Hammond MSS.

123. Eben Olcott to Enoch Kenyon, New York, October 18, 1894, copy, Letterbook 22, Olcott MSS (N.Y.); *M&SP, 71* (October 19, 1895), 255.

to the potential of the South African market.[124] A good deal of British equipment was used, of course, and some was patterned after the American. One engineer wrote back from Johannesburg in 1894 that American compressors and power drills—especially the Ingersoll-Sergeants—were superior, but were being closely duplicated by the English and the Australians. In the mines, such equipment was tested in races for gross production: "It is a contest in which the managers and staffs down to the miner, are all interested and trying to outdo each other." [125]

To Hennen Jennings, the Rand provided "a splendid field for experience in the handling of men and materiel on a large scale, and also in administration." [126] Hammond, when handling Rhodes's mining interests, had twenty engineers under him—all westerners—and indirectly more than 18,000 men.[127] If it was in some ways a limited, even standardized field, it was one that placed a high premium on efficiency, for the cost of extracting gold was very close to the level of average working costs, especially in depth. One historian believes that the structure of monopolies and concessions fostered by the Transvaal government under Paul Kruger actually compelled the companies "to mine more efficiently and to eliminate waste more resolutely," thus leading to the astounding feats of engineering and administration on the Rand.[128] But such a conclusion ignores the trends of the "new" engineering transplanted by the Americans in the 1890s.

The first companies on the Rand were interested in outcrop mines. As work progressed, it was evident that the veins (or "reefs") dipped at an angle; but before 1890 owners did not suspect the continuance of gold-bearing ore at great depth.

124. *EMJ, 56* (July 15, 1893), 50; *The British and South African Export Gazette* (London), June 5, 1896.

125. Enoch Kenyon to Eben Olcott, Johannesburg, September 10, 1894, Olcott MSS (N.Y.).

126. Quoted in Rickard, ed., *Interviews,* p. 246.

127. *M&SP, 72* (January 18, 1896), 43.

128. C. W. De Kiewiett, *A History of South Africa, Social and Economic,* p. 135.

American engineers were first to establish this point as drill holes sunk by Joseph S. Curtis at the Village Main Reef and by Hennen Jennings at the Rand-Victoria and the Turf Club proved the value at unheard-of depths and set the stage for a row of deep level mines to the south of the outcrops and ultimately for a second row of "deep deeps" even farther south.[129] It was John Hays Hammond who urged Cecil Rhodes to acquire "deep deep" ground, insisting that ore could be mined from at least 5,000 feet down, as was the case in California. By the 1950s, the Robinson Deep, one of the first developed under Hammond's supervision, had reached the 10,000 foot level.[130]

American engineers made great advances in the speed of shaft-sinking, as they began to treat deep level problems on a large unit basis. Shafts were larger, sometimes concreted, and were sunk with growing speed. Under Leslie Simpson, a recent California graduate, crews at the Robinson Deep established world records in shaft-sinking. The Americans also took the lead in electrifying the mines, with Maurice Robeson, of the Corner House group, setting the pace in the introduction of advanced electric hoisting equipment to replace steam.[131]

Another important fruit of the Rand experience was the adoption of a uniform mapping scale—a scale that could be used either for feet or meters and that was more realistic in its proportions (i.e., 1 to 500 or 1 to 1,000) than the awkward one inch to forty feet that was so often used.[132]

Substantial improvements were made in ore treatment. Hennen Jennings soon saw that free gold would not last and that, for continuing profit, sulphide ores must be worked. The popular Jennings thereon called on Charles Butters to erect a chlo-

129. Rickard, ed., *Interviews,* pp. 239–40, 420; Hammond, *Autobiography, I,* 291–94, 296–99, 301; Owen Letcher, *The Gold Mines of South Africa,* pp. 109–10; Frederick H. Hatch & J. A. Chalmers, *The Gold Mines of the Rand,* p. 274; J. Bernard Mannix, *Mines and Their Story,* p. 68.

130. Hammond, *Autobiography, I,* 291–92, 299; *M&SP, 72* (June 6, 1896) 459; June Metcalfe, *Mining Round the World,* p. 103.

131. Rickard, ed., *Interviews,* pp. 239, 242; Hammond, *Autobiography, I,* 300; *M&SP, 72* (June 6, 1896), 459; Letcher, *Gold Mines,* pp. 383–84.

132. Eaton, in Parsons, ed. *Seventy-Five Years of Progress,* pp. 44–45.

rination plant in the Robinson. Butters first came to South Africa in 1891, and his earlier work had not been crowned by success, for his plant at Kennett, California, had been located too far from the ore supply, and he had "played hide and seek with the sheriff" trying to make the enterprise pay. Once the Robinson chlorination mill was in operation, both Jennings and Butters became convinced of the superiority of the new MacArthur-Forest cyanide process for treating the large accumulations of tailings, and Butters proceeded to implement the change at the Robinson, adding numerous improvements and refinements of his own. In 1894, he left the Robinson to organize the Rand Central Ore Reduction Company, a custom chlorination and cyanidation business, "equal in output to any ore working institution in the world."[133] With a large engineering and metallurgical staff at the Rand Central, Butters did much to elaborate and perfect the cyanide process, and when he returned to the United States after eight years, he was one of the foremost experts in the field. One of the founders of the Chemical and Metallurgical Society of South Africa in 1894, he became the organization's president in 1897. At his farewell dinner a year later, the speaker believed that his name "would be remembered and honoured amongst chemists and metallurgists as long as the names of President Kruger and Cecil Rhodes will be remembered in politics."[134]

The mention of politics was no mere accident, for a number of American engineers had been involved in political intrigue in the Transvaal, with near tragic consequences to themselves. The Americans were part of the community of Uitlanders—foreigners—dominated by the Boer government headed by Paul Kruger. The two groups were sharply split over such questions as the franchise, military duty, taxation, and the use of public funds, but it was a more specific series of grievances

133. Samuel Christy to Charles Butters, Berkeley, December 24, 1890, Christy MSS; Williams, *Some Dreams Come True,* p. 551; Rickard, ed., *Interviews,* pp. 120–22, 237; *EMJ, 51* (February 7, 1891), 173; *M&SP, 71* (November 2, 1895), 283; biographical sketch of Charles Butters, Butters Memorial Collection.

134. Rickard, ed., *Interviews,* 242; *Proceedings of the Chemical and Metallurgical Society of South Africa, 2* (February 19, 1898), 253.

that prompted mine owners to seek the overthrow of the Kruger government in the Reform Movement of 1895.

Of particular concern was the government dynamite monopoly, which kept prices high on a commodity vital to deep level mining. Of similar concern was the railway monopoly, which raised transportation costs between the gold mines and the coal fields. Mine owners also complained about the tariff level and the short supply of native labor, and they protested the liquor monopoly granted by Kruger, which permitted the sale of alcohol to native miners against the owners' wishes. A serious grievance was the taxation of mining property even before profits were forthcoming. As a result of such features, mine owners believed "Oom Paul's" policies threatened effective mining operations and were willing to make themselves part of a rebellion. Undoubtedly, they overstated their case. After the movement collapsed, John Hays Hammond, then under arrest, characterized the government in extreme words: "The Transvaal is a small unenlightened retrogressive community under the government of a narrow oligarchy, giving a bad, inefficient administration; monstrous monopolies; corruption rampant,"[135] Hammond was one of a number of American engineers implicated in the Reform Movement—a movement in which Cecil Rhodes and Alfred Beit, men with large control over the two major deep level companies, also played a part. In fall 1895, reform leaders began to smuggle guns into the Johannesburg region. These came from England to Kimberley, where Gardner Williams supervised their loading into Standard Oil drums with false bottoms and consigned them to Hammond to be stored in different mining properties. But the arms accumulated too slowly, and Dr. Leander Starr Jameson, of the British South Africa Company at Bulawayo, was brought into the plot to provide military assistance and a diversion secondary to an uprising within the Transvaal. Jameson moved prematurely and was forced to surrender. In turn, the reformers were forced to relinquish their arms and were arrested on

135. John Hays Hammond to John P. Jones, received January 30, 1895, cable, copy, Hammond MSS, Box 2.

charges of sedition and high treason. Among the sixty-four arrested were Hammond, Victor Clement, Charles Butters, Joseph Story Curtis, and Thomas Mein.[136]

Several months later, Hammond was sentenced to death. Curtis was never brought to trial because of illness, and the other American engineers were sentenced to two years in prison and fines of $10,000 each. Friends of these men sought to enlist the intervention of Secretary of State Olney in Washington, without much success.[137] All readers of the *Engineering and Mining Journal* were urged to write their congressmen, and petitions signed by 78 senators and 172 representatives did go to President Kruger, asking pardon for Hammond.[138] That public sentiment in the United States was at least mildly inflamed was indicated by a bit of doggerel picked up from the American press and republished abroad:

And shall Hays Hammond die?
And shall Hays Hammond die?
There's twenty thousand Yankee boys
Will know the reason why![139]

But soon the Americans' sentences were commuted, and within a few months all were released, Hammond paying a fine of $125,000 and the others $10,000.[140] Gardner Williams was fined about $150 for violating the Cape Colony's arms act, but ironically continued to serve as the American consular agent at Kimberley until 1906. Richard Parker, who

136. Hammond, *Autobiography, 1,* 320–28; Hatch & Chalmers, *Gold Mines of the Rand,* p. 274; Geoffrey Blainey, "Lost Causes of the Jameson Raid," *The Economic History Review, 18* (August 1965), 354–63.

137. Charles M. Rolker to Abram S. Hewitt, Cape Town, July 8, 1896, & Peter B. Olney to Richard Olney (New York, January 17, 1896). These and other photostatic copies of letters from the Olney MSS urging government action are found in Hammond MSS, Box 2.

138. *EMJ, 61* (January 25, 1896), 84; petitions to President Kruger, Washington, D.C., April 28, 1896, copies, Hammond MSS, Box 3; *The Standard and Diggers' News* (London), May 21, 1896.

139. Quoted in *South Africa* (London), March 28, 1896.

140. Hammond, *Autobiography, 2,* 402; Charles Butters to his mother, Pretoria Jail, May 8, 1896, Butters Memorial Collection; *Papers Relating to the Foreign Relations of the United States, 1896,* p. 580.

had also helped smuggle guns, fled the country and escaped arrest.[141]

But a number of American engineers managed to keep out of "the tin pot reform game," as Eugene Hoefer called it.[142] Hennen Jennings, though probably sympathetic, kept aloof, and several, Thomas Leggett and R. E. Brown among them, actually opposed the movement. Brown, known in the profession as "Barbarian" Brown because of his earlier association with an Idaho newspaper, *The Barbarian,* was labeled by Hammond as a paid spy for Kruger and was charged with attempting to break up the American's organizational meeting where the decision was made to support the revolt.[143] Brown subsequently figured in a tangled litigation with the Transvaal government over mining rights in the Wilfontein district and won a judgment of more that $5 million, only to have it overruled by the government as unconstitutional.[144]

Hammond and most of the other guilty engineers withdrew from South Africa after this episode, but conflict between the Boers and the Uitlanders continued. In 1899, when relations between the British government and the South African Republic were tense, a committee of American engineers in Johannesburg, led by Louis Seymour, launched a campaign to "educate" the American public as to conditions in the Transvaal. Friends in the United States badgered President McKinley and Secretary of State John Hay, insisting that, if the Washington administration would support England's demands, President Kruger would "climb down" and war would be averted.[145]

In autumn 1899, cold war gave way to a shooting war in South Africa, and American engineers became involved. The

141. *South Africa,* March 7, 1896; *M&SP, 72* (March 14 & May 30, 1896, 212, 445; *M&M, 16* (August 1935), 355; Williams, *Some Dreams Come True,* pp. 250–51.

142. Eugene Hoefer to Samuel Christy, Pilgrims Rest, September 13, 1896, Christy MSS.

143. Hammond, *Autiography, 1,* 332, 342–43; *M&SP, 144* (March 17, 1917), 360; Rickard, ed., *Interviews,* 347, 420.

144. *EMJ, 67* (January 30, 1897), 112; San Francisco *Chronicle,* March 2, 1899; L. S. Amery, ed., *The Times History of the War in South Africa, 1899–1900, 1,* 146, 228.

145. John H. Ferguson, *American Diplomacy and the Boer War,* pp. 47–48.

Boers immediately laid siege to the British at Kimberley, where Gardner Williams is said to have been "a tower of strength" during the attack.[146] George Labram, the American chief engineer under Williams at De Beers, was also a central figure in the defense of the city. Labram took the lead in making percussion shells and fuses, and was largely responsible for the construction of the famous gun "Long Cecil," built in Kimberley by engineers from adaptation of a design found in a copy of an engineering journal. "Long Cecil" fired 255 shells of twenty-nine pounds each, and today stands in the city as a tribute to De Beers and its American engineers.[147] It was ironic that, when the Boers shelled Kimberley with a long gun of their own brought from Pretoria, Labram was one of nine persons killed. He was buried with full military honors and his widow eventually received the plaudits of the British government and £1,000 in recognition of his "invaluable assistance." [148]

Other engineers also figured in the war: F. R. Burnham was prominent as a scout; Harry Hay, a Berkeley graduate, was an officer in the Rand Rifles Mines Division, organized to protect the deep mines from attack; Louis Seymour, chief organizer and second in command of the Railway Pioneer Regiment, was killed in action at the Zand River, and his name is perpetuated by the Seymour library of technical literature at Johannesburg.[149]

But probably most Americans steered clear of involvement in the conflict; many left South Africa to avoid it. Many were already leaving anyway, and the day when the American engi-

146. Lewis Michell, *The Life and Times of the Right Honourable Cecil John Rhodes, 1853–1902,* p. 279.

147. Gardner F. Williams, *The Diamond Mines of South Africa, 2,* 291, 292–301, 546–47; Amery *Times History, 4,* 560–61. "Long Cecil" is now part of the Honored Dead Memorial honoring those who fell in defense of Kimberley. A. Gordon-Brown, ed., *Guide to Southern Africa,* p. 293.

148. Williams, *Diamond Mines, 2,* 296, 298, 302; Alpheus Williams to Samuel Christy, Kimberley, February 6, 1900, Christy MSS; *Papers Relating to the Foreign Relations of the United States, 1900,* pp. 624–25.

149. *EMJ, 69* (June 23, 1900), 746, *81* (March 10, 1906), 482; *M&M, 19* (November 1938), 508; Amery, *Times History, 1,* 283, *3,* 95, *6,* 311–12; Williams, *Some Dreams Come True,* p. 262; Hennen Jennings to Samuel Christy, H. M. S. *Saxonia,* September 11, 1900, Christy MSS; Gordon-Brown, ed., *Guide to Southern Africa,* p. 253.

neer dominated the Rand, while not yet over, was on the wane. Indeed, the Americans themselves helped to speed the process. Not only did they serve on Transvaal committees to improve mine health and safety and to report on the future of South African mining,[150] some of them took the lead in organizing or improving technical institutions that, in effect, would train their replacements. Hennen Jennings was one of the founders and the first president of the South African Association of Engineers and Architects, a group that became the South African Association of Engineers in 1898. Jennings also helped establish the South African School of Mines and Technology, and many young engineers on the Rand after about 1904 would be trained here or at the Royal School of Mines in London, which Jennings also helped reorganize.[151] With able managers developing on the home scene, American engineers returning to the states were less likely to be replaced by their own countrymen.

And most returned to American affluent and successful and with enhanced reputations that kindled the public's imagination. Even before the reform movement, publicity had projected John Hays Hammond into the spotlight, Eben Olcott believed that Hammond had "made more reputation in So Africa than he would have in the U.S. in twenty years," [152] a comment that contained, perhaps, a touch of professional envy, for the colorful Hammond was not always popular among mining people. In 1898, Mary Hallock Foote could indicate how impressed she was with George Starr, "of South-African mining fame (strictly professional, not newspaper fame, like Jack Hammond)." [153] Hennen Jennings, Henry C. Perkins, and Gardner Williams all returned with reputations as great engineers, but these were modest men and did "not

150. Letcher, *Gold Mines,* pp. 105, 149; Rickard, ed., *Interviews,* p. 246; *DAB, 10,* 56.

151. Letcher, *Gold Mines,* p. 467; Rickard, ed., *Interviews,* pp. 242, 245; *DAB, 10,* 56.

152. Eben Olcott to Robert G. Brown, New York, April 15, 1895, copy, Letterbook 22, Olcott MSS (N.Y.).

153. Mary H. Foote to Helena Gilder, Grass Valley, April 19, 1898, M. H. Foote MSS.

provide much publicity," according to one who knew them well.[154]

But South African experience was a badge of respectability and mark of competence, so much so that, in 1909, the *Pacific Miner* saw fit to caution the industry against accepting "the unsupported statement of every Tom, Dick and Harry, who comes over here and tries to pass himself off for an engineer from the Rand."[155]

Wherever they went, whether Johannesburg or Coolgardie, Cananea or Mount Lyell, beyond the Urals or north of the Great Wall, American mining engineers transplanted western processes and equipment and brought back variations that were often both improved and important for the mineral industry of the West. Moreover, the great variety of experiences gained by a single engineer on four or five continents focused in the West what were probably the most versatile and capable technical men in the mineral world.

154. William Wallace Mein to Howell Wright, San Francisco, March 16, 1937, Howell Wright Collection.

155. *Pacific Miner, 15* (November 1909), 158.

CHAPTER 10

"They Are Very Clever Cosmopolitan Sort of Men"

Now here we have the mining man, in either hand a gun.
He is not afraid of anything, and he's never known a run,
He dearly loves his whiskey, and he dearly loves his beer.
He's a shooting, fighting, dynamiting, mining engineer.
—Colorado School of Mines Song[1]

The mining engineer has always been a romantic figure, and novelists and short-story writers have given him more than his share of glamour. He emerges as a stereotype—a two-fisted, hard-driving outdoor man, able to cope with any situation. High minded, but shrewd, he deals with magnates as easily as with the prospector whose trousers are held in place with a rusty nail. The very personification of manliness, he is at the same time a man of culture and refinement who is as much at home in the drawing room as in an adobe hut in Sonora. Breathtakingly beautiful heroines swoon at his feet, and, after he rescues them, he marries them—or his best friend does.

Invariably, as he swashbuckles his way across the pages of the novel the engineer is "a handsome brute," as Peter B. Kyne describes one of his heroes,[2] and he is likely to have about him "an air of alertness . . . a suggested tireless energy . . . an air of inflexibility, softened with mercy; a rugged honesty that made no compromise with evil-doers."[3] John Stuart Webster, six feet one inch, 190 pounds of bone and muscle, leaves a trail of shattered adversaries in his wake, yet

1. Colorado Mines people insist that the song from which this is taken, "The Mining Engineer," was sung at least ten years before Georgia Tech came into existence. Morgan, *A World School,* pp. 128–129.

2. Peter B. Kyne, *The Understanding Heart,* p. 113.

3. Frank Lewis Nason, *The Vision of Elijah Berl,* p. 4.

frequently cites the Bible and quotes *Paradise Lost* and Cyrano de Bergerac, as he wins a revolution, rich mines, and a fair damsel—in that order.[4] One of Mary Hallock Foote's heroes, a long-striding, cigar-smoking engineer, quotes Bryant's "Hunter of the Prairies" to his lady love and refers to the local spring as "our fountain of Trevi."[5]

Like Gertrude Atherton's Gregory Compton, the engineer of fiction fights for what is rightfully his—and wins—even against the detested Amalgamated.[6] Petty tyrants he takes in stride, and often he has the sanction of the U.S. government behind him. Richard Harding Davis' Robert Clay, patterned after John Hays Hammond, comes straight to the point with his opponents in the South American country of Olancho:

> Try to break that concession; try it. It was made by one Government to a body of honest, decent business men, with a Government of their own back of them, and if you interfere with our conceded rights to work those mines, I'll have a man-of-war down here with white paint on her hull, and she'll blow you and your little republic back up there into the mountains.[7]

A few novelists portray the engineer as somewhat philosophical and introspective,[8] but most are outgoing men of action, troubleshooters, who brook no interference from the scoundrels who seek to do them or their friends in.[9] In one case of blackmail by a rascally lawyer, the engineer concludes "there is only one way to deal with men of his type," then goes into action, shooting the villain "once, twice, thrice" and dead—and is acquitted by a jury on grounds of justifiable

4. Peter B. Kyne, *Webster—Man's Man.*

5. Mary H. Foote, *In Exile, and Other Stories,* pp. 6–10, 14, 20, 26–27.

6. Gertrude Atherton, *Perch of the Devil.*

7. Richard Harding Davis, *Soldiers of Fortune,* p. 63; Fairfax Downey, *Richard Harding Davis & His Day,* p. 128.

8. See, for example, Edwin Lefevre, *The Golden Flood;* C. M. Sublette, *The Golden Chimney;* Frank Waters, *The Dust within the Rock.*

9. See Theodore A. Harper, *Siberian Gold;* Todhunter Ballard, *Westward the Monitors Roar.*

homicide in self-defense, because the deceased had a revolver in his desk.[10]

In most instances, the engineer is a hero figure, but in a few instances, he appears as the knave, generally in a lesser role. In one of Kyne's books he is a philanderer who is "tunneled"—shot dead—by the irate husband.[11] In a novel by Caroline Lockhart, he is portrayed both as a sissy and a crook, characterized by an old-timer thus: "He'd salt his own mother, he'd sell out his grandmother, but in the profession there's none better if he'd stay straight. I knowed him down in Southern Oregon—he was run out."[12] Clyde Murphy's delightful *The Glittering Hill* is built around the exploits of Nick Stryker, a Columbia Mines graduate, undoubtedly suggested from real life by F. Augustus Heinze. The flashy Stryker dresses in a "dark broadcloth suit, his cream-colored vest bearing wide blue checks and, across it, an immense watch chain of solid gold"; involved in million dollar apex lawsuits, and not overly endowed with scruples, he is described by an adversary as "a Simon-pure, eighteen carat 100 per cent son of a bitch, net," but in the end, though obviously a heel, Stryker shows some redeeming features.[13]

Engineers recognized the image of themselves created by the writers of fiction, and indeed by some of the writers of biography designed for the youth of America,[14] and contrasted it with reality. Wrote one in 1916:

> Our type, then, as we find it in the headings and drifts is an individual of the male sex clad in overalls of an incredible degree of dirtiness, with a complexion to match. Unless the rules of the mine forbid it (and even this reservation cannot be taken as of universal application) he smokes a pipe whose external experience is entirely in keeping with its immediate surroundings. He delights in

10. Horace Wilson Bennett, *Silver Crown of Glory,* pp. 159–61.
11. Kyne, *The Understanding Heart.*
12. Caroline Lockhart, *The Man from the Bitter Roots,* 110–11.
13. Clyde F. Murphy, *The Glittering Hill, pp.* 372, 404.
14. See Ingersoll, *In and Under Mexico,* pp. 183–84; Wildman, *Famous Leaders of Industry,* p. 123.

risking his neck by utilizing the hoisting equipment in every conceivable way except sitting in it in the manner intended by the designer, who has labored to render it as safe as human contrivances can be made. His talk is of feet drilled and of rock broken and he is always convinced that the particular part of the workings confided to his care is the "meanest rock in the mine, sir, hard as h--l to drill and 'll come down on your head every five minutes if you don't watch it all the time." [15]

Undoubtedly, there was much material for the novelist in the lives of many engineers—for sometimes there was intrigue: witness Hammond in Africa; and certainly there was danger and adventure in the early West, revolutionary Mexico, or on the steppes of Siberia. But the average engineer was a far cry from the stereotype, and his life, though it had its moments of excitement, was not generally one of overturning governments, rescuing fair damsels, or upholding right and virtue. Rather it tended to be an active, but often prosaic existence, nomadic and restless, at least in the early years, and frequently filled with variety and contrast. For the mining engineer was merely a human being, with all the strengths, weaknesses, differences, and outlooks that characterize the species as a whole.

Perhaps some did resemble the Apollo of tanned complexion and swaggering walk of whom the novelists wrote. Alliene Case was described in 1894 as "a big, soldierly-looking man, with a high-bridge nose, projecting brows, a very marked modeling of the lips & chin; he would make a splendid bronze." Daniel Jackling might qualify, the dashing Bulkeley Wells, who was more than once mistaken for Douglas Fairbanks, and Victor Clement, who, on his white horse, "clad in mouse-colored corduroy, cap to match," was a veritable Galahad to the opposite sex.[16] Handsome genial F. Augustus Heinze was re-

15. "Philosopher" to editor, New York, February 5, 1916, *EMJ, 101* (February 26, 1916), 403.

16. Mary H. Foote to Helena Gilder, Boise, February 19, 1894, M. H. Foote MSS; Cloman, *I'd Live It Over,* p. 64.

membered as a model of fashion, the only man on the editorial staff of the *Engineering and Mining Journal* during his year with that publication "who habitually came to the office wearing a top hat and frock coat."[17]

But Robert Peele described a fellow engineer in 1895 as "round and fat, with a little, high-pitched soprano voice," and Arthur D. Foote was supposed to be "quite stout and looks very knowing in his gold skeleton glasses."[18] Rossiter Raymond was "a short, brisk man . . . with white side whiskers and a sparkling blue eye," who always wore a close-fitting black skullcap which covered "a marvelous bald dome, equal to Shakespeare's in my boyhood imagination," according to one contemporary.[19] Thomas A. Rickard and John Hays Hammond were both little men, Philip Argall resembled Grover Cleveland somewhat, and Louis Ricketts, when decked out in broadcloth and silk hat for the Panama-Pacific Exposition, "bore no small resemblance to the pictures of Lincoln."[20]

In personality, as in physical attributes, they ran the gamut from one extreme to the other. William Ashburner was considered "a jolly good fellow, good at a joke, good-natured, philosophical, and with a great fund of humor and anecdote." Clarence King thought that the witty, urbane Henry Janin was "as champagny and daring a talker" as he had ever heard. Louis Janin was portrayed as "enormously stout but always picturesque and delightful."[21] On the other hand, Watson Goodyear was "a regular old maid"—"he will stand & look at a thing for the longest time before he can make up his mind to

17. *EMJ, 98* (November 14, 1914), 881, *109* (May 15, 1920), 119.

18. Robert Peele to Eben Olcott, New York, November 15, 1895, Olcott MSS (N.Y.); Mary H. Foote to Helena Gilder, Grass Valley, May 8, 1896, M. H. Foote MSS.

19. Edmund Wilson, "Landscapes, Characters, and Conversations from the Earlier Years of My Life," *The New Yorker, 43* (April 29, 1967), 114.

20. Bimson, p. *Ricketts,* 21.

21. Entry for March 9, 1862, Brewer, *Up and Down California,* p. 248; Clarence King to James D. Hague, San Francisco, July 15, 1884, Hague MSS, Box 12; Mary H. Foote to Helena Gilder, Grass Valley, August 2, 1908, M. H Foote MSS.

anything," said Charles Hoffmann.[22] Colonel Gillette, of the Savage mine on the Comstock, stuttered when excited; others had reputations for a lack of humility or an unwillingness to cooperate. After Ernest Wiltsee returned from Africa in 1895, one of his acquaintances remarked: *"Sub rosa,* I am afraid that Wiltsee will soon have to wear a larger sized hat." [23] J. Park Channing had the reputation of being "difficult to get on with" and of chaffing people "because of being *too* energetic, & *too* officious." [24] Mark Requa was described as "quite airy in his manner, and seemed to think he had the world corralled" in 1905, whereas Thomas A. Rickard was known throughout the profession as a man with a high opinion of himself. Charles Janin wrote Edgar Rickard in 1916: "If T.A. wasn't a relative of yours I am afraid I would have to say that he is a conceited ass." [25]

Engineers squabbled among themselves, voiced their displeasure openly, and sometimes aired their dirty linen in public. Robert B. Stanton disliked one of his colleagues from the beginning: "He gave me a pain the first sight I had of him," he said.[26] Another was referred to as "as much of a worm as ever, and probably the most unpopular engineer that I know of." [27] Ernest Wiltsee and Victor Clement shared no love for each other, and Wiltsee sought to undermine the latter's influence with John Hays Hammond.[28] Hammond was referred

22. Charles Hoffmann to Josiah Whitney, May 26, 1870, Hoffmann MSS.

23. Drury, *An Editor on the Comstock Lode,* p. 34; N. H. Harris to John Hays Hammond, San Francisco, July 25, 1896, Hammond MSS, Box 3.

24. Robert Peele to Eben Olcott, New York, August 5, 1895, Olcott MSS (N.Y.).

25. Arthur W. Jenks to Eben Olcott, Salt Lake City, July 22, 1905, ibid.; Charles Janin to Edgar Rickard, San Francisco, January 18, 1916, copy, Charles Janin MSS, Box 6.

26. Entry for December 4, 1906, field notes 22, Stanton MSS.

27. C. T. Nicholson to Charles Janin, January 16, 1917, Charles Janin MSS, Box 6.

28. Ernest Wiltsee to Natalie Hammond, San Francisco, July 30, 1896, Hammond MSS, Box 3. See also entry for August 6, 1905, field notes 23, Stanton MSS; John Hays Hammond to Henry A. Butters, June 5, 1895, copy, Letterbook 2, Hammond MSS; Charles Janin to Chester W. Purington, San Francisco, April 21, 1916, copy, Charles Janin MSS, Box 6.

to as "vain, loud mouthed and a blowheart." "There must be a beastly vulgar vein in the man besides considerable wind." [29]

Like other humans, the mining engineer sometimes gambled, drank too much, or even visited the "light ladies who followed the heavy money." F. Sommer Schmidt marveled at the twenty-four-hour poker games at Ely. "We did our drinking once a month and kept none of it in our shacks," he recalled. Sewell Thomas mentioned a fellow engineer in Nevada who "could drink hard liquor for hours without batting an eye" and who "loved the girls and a great many of them seemed to love him." A few became alcoholics, and some, including Ferdinand Van Zandt, J. H. Ernest Waters, and Bulkeley Wells, committed suicide.[30]

Whatever his personal traits and foibles, the engineer faced a rugged outdoor life, which made a heavy physical demand on him. Until such time as he was well enough established to retire to his office, and leave the active work to his assistants, he had to be able to take to the field in a vigorous way. There were a few exceptions, but in general mining engineering was a career only for the physically fit. Robert B. Stanton practiced the profession despite a deformed arm; both Charles Hoffmann and William Ashburner suffered from bad legs, which sometimes curtailed their fieldwork; and M. H. Burnham attained eminence despite an artificial leg and the "further handicap of an incurable form of tuberculosis," which took his life early.[31] Hale in his younger days, Charles Butters suffered se-

29. Charles M. Rolker to Eben Olcott, New York, August 23, 1893 & August 24, 1894, copies, Letterbook 23, Olcott MSS (N.Y.).

30. Schmidt, "Early Days," p. 20; Thomas, *Silhouettes,* p. 94; Mary H. Foote to Helena Gilder, Boise, November 19, 1891, M. H. Foote MSS; A. H. Wethey to Eben Olcott, Butte, March 15, 1892, & Eben Olcott to R. P. Rothwell, May 10, 1893, copy, Letterbook 20, Olcott MSS (N.Y.); Olcott to Robert G. Brown, New York, September 13, 1894, copy, Letterbook 22, ibid.; Charles Janin to Edgar Rickard, January 2, 1917, copy, and C. T. Nicholson to Janin, London, January 16, 1917, Charles Janin MSS, Box 6; Crampton, *Deep Enough,* pp. 60, 64–65; Gertrude B. Sayre, "Old Smuggler Narratives," typescript, 1940, copy.

31. Smith & Crampton, eds., *Hoskaninni Papers,* p. x; Charles Hoffmann to William Brewer, Carmel, February 1, 1907, Hoffmann MSS; John Hays Hammond, "Character, with Ambition, Industry and Pluck, Gets There," typescript (talk given to wounded veterans at Walter Reed Hospital, May 23, 1919), Hammond MSS, Box 20.

verely from arthritis at the end of his career, but he continued his metallurgical experiments from a folding cot in his laboratory. "Attending business meetings—he would have to go to his car on his hands and knees and—conduct them from an office table where he lay."[32]

But there is no question that the practicing engineer needed to be a tough specimen and in good physical condition to take the punishment demanded by the constant travel and difficult conditions of work he so often faced. Many had been college athletes and continued an interest in sports. F. Sommer Schmidt commented on his acceptance by miners in Nevada and acknowledged that "Being the best billiard player in the County as well as the best second baseman also helped."[33] Edwin Garthwaite was an avid golfer, and Louis Ricketts often slowed down long enough to play around the links. Christopher Corning and Stanley Easton were ardent fly fishermen, whereas Walter Devereux played polo, was a member of the Boone and Crockett Club, and was an expert big game hunter and photographer until a stroke paralyzed his left arm and caused him to retire from active life twenty years prematurely.[34] Robert Livermore was also a polo fan and fished and hunted regularly, often with Bulkeley Wells as a companion. Reports of gunshot accidents would indicate that hunting attracted many. Like a number of other engineers, Livermore sometimes combined business and pleasure. Once he made a trip to examine the Bullard mine in Arizona, and later waxed ecstatic as he recalled the occasion:

> The mine was no good, but the desert at that time of the year was a garden of flowers and teeming with bird life. Every cactus was in bloom, mockingbirds sang, rabbits scurried, and plumed quail piped from every arroyo. I

32. Biographical sketch of Charles Butters, Butters Memorial Collection.

33. Schmidt, "Early Days," p. 6.

34. *EMJ, 118* (September 27, 1924), 512; Len Shoemaker, "Roaring Fork Pioneers," in John J. Lipsey, ed., *The 1962 Brand Book of the Denver Posse of the Westerners,* pp. 63–65; Bimson, *Ricketts,* p. 25; Winfield S. Downes, ed., *Who's Who in Engineering, 1941* (New York, 1941), p. 506.

couldn't resist a day or two with a shotgun and, I believe, I took a few quail home to Gwen [his fiancée].[35]

In their later years, successful engineers might work a typical businessman's cycle. Dozens commuted by ferry daily between Oakland or Berkeley and San Francisco, and many commuted to and from New York. Pope Yeatman, for example, lived in Philadelphia in 1914, and traveled daily to and from his Manhattan office by train, working en route, often with a stenographer.[36] And whether in his office or in the field, the engineer often worked hard, long hours. Fred Bradley always spent several hours on Sunday mornings in his office.[37] Mary Hallock Foote believed her husband worked too hard. "I would like to send a bill in to the Company for his services on Sundays," she said. "He belongs to me on Sundays!"[38] Friends warned Eben Olcott not to "wear yourself out for other people by working from 5 AM to 12 PM. It won't pay."[39] A San Francisco engineer working at the Oathill mine in California acknowledged a friend's letter, but said he had "been too busy to answer it in the day time and at night I am too tired to do anything but sleep."[40] But most felt like Charles Janin, who, in 1912, remarked on how busy he had been: "But I would rather die of work than to starve to death."[41]

If the life of the engineer was busy and vigorous, it was also nomadic. "The mining engineer is like a soldier or sailor," wrote Edward McCarthy. "He has to pack his traps and say

35. Gressley ed., *Bostonians and Bullion,* p. 140. For accidents, see *San Francisco Call,* November 24, 1904; *M&SP, 95* (December 7, 1907), 697. For other combined mining and fishing trips, see Fred L. Morris to Philip Wiseman, July 10, 1916, copy, Morris MSS, Box 1.

36. *McClure's, 42* (April 1914), 111.

37. Interview with James P. Bradley, San Francisco, May 18, 1964.

38. Mary H. Foote to Helena Gilder, New Almaden, July 4, 1877, M. H. Foote MSS.

39. Robert Peele to Eben Olcott, New York, May 10, 1894, Olcott MSS (N.Y.)

40. Ralph B. Newcomb to Walter W. Bradley, Middletown, California, February 5, 1914, Walter W. Bradley MSS.

41. Charles Janin to F. Ayer, Seattle, August 20, 1912, Charles Janin MSS, Box 5.

good-bye to loved ones at a moment's notice, and when requested to do so." [42] Thomas A. Rickard estimated that from 1889 to 1902 he averaged 35,000 miles a year, including two voyages round the world. During the first year after his marriage, he was "at home only once for a whole week." [43] The secretary of the AIME pointed out in 1903 that, of a membership of 3,300, he had made 2,250 changes of address in the previous twelve months, with most of these confined to about 20 percent, mainly younger men, of the total subscription. Thus, he said, "there are about 600 men who are as nomadic as the Kirghiz of the steppes." [44]

But even this does not tell the whole story, for an engineer was not likely to change his mailing address for a relatively short trip. 'Letters to the above address will reach me, though I am on the go most of the time and only at home for short intervals" Robert Booraem wrote in 1889.[45] John Hays Hammond maintained a permanent address, but according to a friend, he was on the go so much that he was "almost as bad as the Irishman's flea." [46]

In good times, consulting might keep an engineer almost constantly on the move. Eben Olcott wrote in 1891 from Gilman, Colorado, that he would be back in Denver later in the week, "then hasten on to El Paso" to inspect a mine, after which he would look at another in New Mexico before heading for California to examine a third. After several stops in the Sierra foothills, he would be in Salt Lake City and elsewhere in Utah on mining business, before returning to Colorado.[47]

42. McCarthy, *Further Incidents,* p. 58.

43. Rickard, *Retrospect,* p. 53; Rickard, ed., *Interviews,* pp. 535–36. Philip N. Moore sometimes traveled as many as 50,000 miles in twelve months. *DAB, 13* 135.

44. *EMJ, 75* (January 24, 1903), 143.

45. Robert E. Booraem to Eben Olcott, Jersey City, November 7, 1889, Olcott MSS (N.Y.).

46. Daniel M. Barringer to Charles Denby, July 12, 1912, copy, Letterbook R, Barringer MSS.

47. Eben Olcott to Euphemia Olcott (Gilman, November 29, 1891), Olcott MSS (Wyo.), Box 2.

Even in times when the profession was relatively depressed, the engineer displayed a restlessness as he took whatever jobs came along, always looking for something better:

> Underground with rod and transit—
> Shaky roof? You blithely chance it!
> Squinting, scowling through a level,
> Where it's blacker than the devil;
> Wallowin' in mud and slickens;
> Just a-rustlin' like the dickens
> To end the job
> And join the mob
> In search of other pickin's!
>
> Scalin' porphyritic ledges—
> Seekin' hand-holes on the edges—
> Tracin' fault and vein and croppin'—
> Poundin' samples—never stoppin';
> On some lonesome expedition
> Workin' with the sole ambition
> To end the job
> And join the mob
> To find a new position!
>
> Polin' sluggish tropic rivers—
> Eatin' quinine for the shivers—
> Hackin' with a dull machete
> Till your very boots are sweaty—
> Diggin' out a wily chigger—
> Shoutin' curses at a nigger—
> To end the job
> And join the mob
> In search of "something bigger"! [48]

For many, there was something attractive about the constant travel. To a mining engineer, it was worth something to be introduced, as Mrs. Foote did one of her "dear old Lead-

48. Walter H. Gardner, "The Mining Engineer," *EMJ, 113* (February 11, 1922), 250.

ville boys," as "Will Fisher of Boston—and all over the West."[49] Not only did the itinerant life bring the engineer into "close contact with nature," it brought him in touch with "all sorts of people, educated, uneducated, refined, coarse, poor and rich, of both sexes, of different nationalities and temperaments."[50] For the young man, "who is by temperament vigorous, adventurous, and aggressive, who loves a roving life in quest of new things, whose instincts are curious, inventive and creative," John Hays Hammond could think of "*no* profession more inspiring."[51]

But, frequently, after years of such an existence, when families arrived, this aspect of the life became less inspiring, and engineers sought a more sedentary existence. Reuben Rickard of Berkeley complained about the migratory nature of his work in 1885 and found "This being away from home so much is anything but pleasant." From 1890 to 1894, he gave up travel, in favor of a quiet home life, but when his wife died, he again took to the field.[52] As early as 1881, Eben Olcott said he was "trying to swear off entirely" from long trips out of the country, but would still consider short ones.[53] Probably none felt the impact of the roving nature of the engineer's work more than his wife. While preparing to move to Grass Valley in 1895, Mary Hallock Foote expressed a sentiment that must have been common. "How much simpler it is to die than move," she said. "To the Other Country we can go without a trunk. I suppose civilization means that: a trunk!"[54]

Engineers did not take their families with them on consult-

49. Mary H. Foote to Helena Gilder, Grass Valley, April 15, 1896, M. H. Foote MSS.

50. C. O'Brien to editor, San Francisco, March, 1910, *Pacific Miner, 16* (March 1910), 94.

51. John Hays Hammond, "The Mining Engineer," typescript, January 2, 1933, Hammond MSS, Box 16.

52. Reuben Rickard to Samuel Christy, Chicago, September 18, 1885, Christy MSS; *EMJ, 61* (April 4, 1896), 327.

53. Eben Olcott to John Hays Hammond, April 22, 1891, copy Letterbook 17, Olcott MSS (N.Y.).

54. Mary H. Foote to Helena Gilder, Boise, August 30, 1895, M. H. Foote MSS. See also Augusta Garthwaite, "Addenda to Reminiscences of a Mining Engineer," typescript, preface.

ing trips, although sometimes their wives went along.[55] On managerial posts, families often went, even abroad, unless circumstances were simply too oppressive—and even danger did not deter some. When Guido Küstel came to the Santa Rita mines in Arizona in the late 1850s, he was not alone.

> His sister and niece, the fraulein Kline,
> Were company for those at the mine.[56]

But Charles Hoffmann was separated nearly two years from his family while managing a property in Mexico in the early 1880s, having decided that Mexico was no place to raise his four boys. When Eben Olcott went to Peru in 1890, he left his wife and child in Butte, but admitted that "Butte is a pretty bad place for both of them." [57] More pleasant was the winter in Phoenix arranged by Daniel Barringer for his family in 1901, while he did professional work south of the border.[58] Many of those in South Africa in the 1890s brought their families to Johannesburg, but sent their children to schools in Europe. Chester Purington was in Yokohama in 1923, waiting for an opportunity to return to Siberia, when he died in the great earthquake, trying to save his two children.[59]

Western mining towns were not necessarily pleasant places for wives and families. Virginia City, Nevada, with its reputation as a "city of stovepipes and single men's wives," was considered "a miserable place to live in and very expensive" by Charles Hoffmann, whose wife was with him there in 1877.[60] Another engineer complained bitterly about New Virginia in Arizona.

55. *EMJ, 109* (May 15, 1920), 1111; *M&M, 4* (November 1923), 578.

56. From Charles D. Poston, *Apache-Land,* p. 53.

57. Charles Hoffmann to Josiah Whitney, Oakland, August 26, 1883, Hoffmann MSS; Eben Olcott to Euphemia Olcott, Phara, Peru, June 29, 1890, Olcott MSS (Wyo.), Box 2.

58. Daniel M. Barringer to W. M. Morgan, November 11, 1901, Letterbook F, Barringer MSS.

59. *M&M 4* (November 1923), 578, 579.

60. Charles Hoffmann to Josiah Whitney, Virginia City, February 5, 1877, Hoffmann MSS; *Territorial Enterprise,* October 24, 1873.

> One might as well be in Hell, I would never think of bringing my family *and* that is all I live for. . . . There is not a stick of wood within 20 miles of us to the north, *and* within 50 to 100 in any other direction. Dry, Desert, barren, Hot.[61]

One from New York was equally unimpressed with Ely, Nevada at the turn of the century. "I must say that the town of Ely contains some of the meanest types of humanity that it has been my misfortune to meet," he wrote.[62] Another could describe the copper camp of Bingham, Utah, as "a sewer five miles long." [63]

After he was married, Robert Livermore joined a prospecting crew high on the Taylor Fork of the Gunnison, installing his bride in a one-room log cabin, formerly a cowboy line camp, where she cooked for four hungry men until the job played out and the Livermores moved back to "civilization" at Telluride. There they became part of the "gay crowd of Eastern emigres," who adoringly followed the social lead of Bulkeley Wells—that amazing bon vivant who drove his coach and four through the streets, played poker with sheepherders while hunting, and kept a string of polo ponies, a valet, and a cook in a private railroad car.[64]

Many wives were able to take the environment in stride, accepting philosophically Arthur Foote's dictum that "an engineer's climate is where his work is" and that we "must prepare ourselves to encounter what we cannot prevent." [65] In isolated mining regions, social life might be limited to the few families living at the mine,[66] while in more established camps,

61. Charles B. Dahlgren to Adolph Sutro, New Virginia, February 24, 1878, Sutro MSS, Box 5.

62. Frederick D. Smith to James B. Orr, Elmira, New York, April 8, 1902, copy, Read MSS.

63. *EMJ, 91* (January 28, 1911), 204.

64. Gressley, ed., *Bostonians and Bullion,* pp. 140–42; Sayre, "Old Smuggler Narratives."

65. Mary H. Foote to Helena Gilder, Grass Valley, December 22, 1897, M. H Foote MSS; Cloman, *I'd Live It Over,* p. 216.

66. E. H. Clausen to George Louderback, Camp Seco, California, April 19 1913), Louderback MSS.

although such cultured men as Alexis Janin complained of "the lack of intellectual recreation," [67] engineers and their wives developed surprisingly refined social circles.

The fact is that many mining engineers were remarkably sophisticated men. Probably no other group in the West was as well traveled and as well educated. That they were concerned with more than technical matters is readily apparent. Observers of the social milieu in early Central City, Colorado, were impressed, especially with "our young Baron, only a year or two out of Freiberg, in his native Saxony—accepted critic of music and the arts." [68] The house of Arthur and Mary Foote, whether in California, Colorado, or Idaho, was always a center for engineers who could talk on many subjects. When a German engineer visited their New Almaden home in 1876, dinner was followed by a lively and serious discussion of Wagner.[69] The Footes' log cabin above Leadville a few years later was a frequent focus for the "most attractive and intelligent set of young technical men" of the camp, who came to discuss literary, philosophical, and artistic as well as scientific, questions.[70]

Mrs. Foote often described their engineer visitors in significant terms. She considered Hamilton Smith "very cultivated," and she characterized the triumverate of James Hague, Louis Janin, and William Ashburner as "very clever cosmopolitan sort of men." [71] William S. Ward, whom she met at the Leadville Assembly Ball that she immortalized in fiction, was "very agreeable and cultivated." The son of a missionary in

67. Alexis Janin to Juliet C. Janin, Pioche, March 19, 1872, Janin family MSS, Box 5.

68. Frank C. Young, *Echoes from Arcadia,* pp. 111–12; see also Grace Greenwood, *New Life in New Lands: Notes of Travel,* pp. 92–93.

69. Mary H. Foote to Helena Gilder, New Almaden, November 15, 1876, M. H. Foote MSS.

70. Mary H. Foote to unidentified, Leadville, July 8, 1879, ibid.; Thomas Donaldson, "Idaho of Yesterday," typescript, 1941, pp. 389–91; Philip N. Moore, "My Reminiscences of Leadville," *M&M, 5* (January 1924), 10.

71. Mary H. Foote to Helena Gilder, New Almaden, December 7, 1876 & December 31, 1877, M. H. Foote MSS. A man of eclectic interests, Hague corresponded with Darwin over a period of years on the life habits and methods of communication of ants. *NCAB, 23,* 164.

India, educated at Princeton and Columbia, and in 1879 manager of the Evening Star mine, Ward erected in Leadville "an elegant residence with spacious grounds and adorned with the various virtu which a refined taste can suggest and wealth supply."[72] Charles J. Moore, another of the technical group at Leadville, also maintained a fine home:

> he did not leave his aesthetic tastes behind him, but took with him his very large and well-selected library of classical music, for which he is passionately fond; his residence also contains many excellent publications and works of art of high character, together with Chinese and Japanese bric-a-brac, such as one would not expect to find in the heart of the Rocky Mountains, nor elsewhere, except in the homes of the cultivated in eastern cities.[73]

Edwin Garthwaite was an excellent violinist and cellist; his wife was an accomplished pianist, and when they went to Mexico in 1891, they took a piano with them.[74] John C. F. Randolph, who had traveled professionally for more than thirty-five years, had a wide acquaintance with many lands and peoples "and knew their great songs and music and pictures and scenes and customs as well as their mines." Richard A. Parker had a passionate interest in art and in his later years was one of the directors of the Denver Art Association.[75]

Many mining engineers read widely. In Japan in 1873, Alexis Janin instructed his brother to send him some books:

> I want something "meaty" in the Herbert Spencer and Buckle style. I have read a good deal of Herbert latterly and like him immensely. A volume or two of John Stuart Mill on Political Economy or such like subjects, a little

72. Mary H. Foote to Helena Gilder, Leadville, May 28, 1879, M. H. Foote MSS; Griswold, *Carbonate Camp,* p. 238.

73. R. G. Dill, "History of Lake County," in *History of the Arkansas Valley, Colorado* (Chicago, 1881), p. 354.

74. Augusta L. Garthwaite, "Addenda," p. 2; Augusta L. Garthwaite, "Our Life Together," pp. 4–5, 14–16, 71, typescript, n.d.

75. *EMJ, 91* (February 18, 1911), 363; *M&M, 16* (August 1935), 344.

> Husby, some Tyndall, and a volume or two showing that whatever is, is wrong, will be about the ticket.[76]

Janin achieved an excellent knowledge of Japanese and spoke "the German, French, Italian and Spanish languages with great fluency."[77]

William Hague was an avid reader of history, and a number of engineers in the Southwest regularly perused the Sunday edition of *The New York Times*.[78] His diaries indicate that Frank Sizer methodically read three or four books a month during his career. Among those he completed in 1900 were J. D. More's *John Quincy Adams,* Ignatius Donnelly's *Atlantis,* Thackeray's *Vanity Fair,* and writings of Bret Harte, Wilkie Collins, and Nathaniel Hawthorne. An undated listing in the papers of Arthur D. Foote, probably written late in his career, shows him to have been a member of twenty-six clubs and societies, including not only the standard professional organizations, but also the Franklin Institute, the American Association for the Advancement of Science, the National Economic League, the National Conservation Association, the American Asiatic Association, and the American Historical Association. In addition, Foote subscribed to twenty newspapers and periodicals, among them *The New York Times, Atlantic Monthly,* and *Foreign Affairs*.[79] His eyesight was perennially poor, but his wife read to him aloud down through the years.

Rossiter W. Raymond was referred to in 1881 as "a very clever fellow, who is engineer, poet, novelist, editor, man of business, musician, composer and Sunday-School teacher, all at the same time."[80] For years, the columns of the *Engineering and Mining Journal* were graced with Raymond's pungent edi-

76. Alexis Janin to Henry Janin, Tokyo, November 5, 1873, Janin family MSS, Box 5.

77. San Francisco *Bulletin,* January 14, 1897.

78. Arthur D. Foote, "William Hague," n.d., copy, A.D. Foote MSS; *M&SP, 112* (June 17, 1916), 891.

79. Undated list, A. D. Foote MSS.

80. Sidney Gilchrest Thomas, quoted in the address of John Fritz, *Rossiter Worthington Raymond* (1910), p. 56.

torials, his fascinating travel descriptions of the West, and his witty poetry, both serious and comic.[81] He taught several generations of children in the Sunday school of Henry Ward Beecher's Plymouth Church in Brooklyn, and, on Beecher's death, was offered the pastorate of the church. Annually, he wrote a children's Christmas story, and he turned out an occasional novel (*Brave Hearts,* 1873), a book of travel, and a life of Peter Cooper. His wit was proverbial: when he visited a mine in Canada where the miners were permitted to carry out mica for their own use, he quipped "Mica is one of the minor profits, I see." [82] He was equally at home "on a pile of skins and blankets," chatting science and politics with Clarence King in the Utah wilderness, and in the study of his country home, "Hilltop," in Connecticut, discussing classical literature.[83] He was adept at whist, and once played the great Steinitz to a draw in chess. His professional trips he spiced with lectures and sermons: at Treasure City, Nevada, in 1869, he was billed as "one of the most interesting and instructive speakers living," and when he took to the pulpit, his sermons were regarded as "able." [84]

Nor was Raymond the only engineer with a literary flair. Lewis Parson's short story, "Dick Tresco and the Yellow Streak," appeared in the *American Magazine* in 1920 and was well received. On occasion, Courtenay DeKalb wrote both fiction and poetry; Mark L. Requa was the author of *Grubstake* (1933), a novel; and Frank L. Nason published at least three novels, two of them in mining settings.[85]

81. An example is one of Raymond's "Metallurgical Mother Goose" series, which ran in 1879:

"Lucy Locket struck a 'pocket,
Took the stuff, and ground it;
Not a penny was there in it,
Save the 'gouge' around it."

EMJ, 27 (July 28, 1879), 461.

82. *EMJ, 109* (June 26, 1920), 1415; *DAB, 15,* 414–15.

83. *EMJ, 8* (July 27, 1869), 51; Wilson, "Landscapes," *The New Yorker* XLIII (April 29, 1967), p. 114.

84. *White Pine News,* June 28, 1869, quoted in Jackson, *Treasure Hill,* p. 88; *M&SP, 77* (September 24, 1898), 307, *119* (October 11, 1919), 504, 505.

85. For Parsons, see "Dick Tresco and the Yellow Streak," *The American Magazine, 89* (March 1920), 62–65, 270–78; *EMJ, 109* (March 6, 1920), 590.

Mining engineers displayed a surprisingly strong sense of history. A number were members of historical societies; [86] others were active contributors to the historical literature of the period. James Douglas wrote at least three books dealing with early Canadian history.[87] Ernest A. Wiltsee, head of the Emperor Norton Memorial Society in California, was the author of a volume on Frémont, another on early pack animal transportation, and a third on gold rush steamers.[88] A. B. Parsons' *The Porphyry Coppers* (1933) and Ira Joralemon's *Romantic Copper* (1934) are both heavily historical, and Thomas A. Rickard, the author of at least twenty-five books, gave us one of our few real histories of American mining.[89] Rickard, Hammond, Hoover, Henry C. Morris, and a few others published memoirs that are more interesting than modest; [90] Wiltsee, Garthwaite, and Livermore left behind unpublished reminiscences; and a few, such as Joshua E. Clayton and George W. Maynard, had strong, but unfulfilled, desires to record what they had seen.[91]

Few other mining engineers emulated Raymond in taking to

For DeKalb, see his poem "Redemption," and the humorous story, "Mrs. Bumper's Investment," in *Out West, 23* (January–June 1905), 378, *23* (July–December 1905), 160–68. Nason's *The Vision of Elija Berl* deals with a mining engineer who turns to irrigation in California; *The Blue Goose* (New York, 1903), is set in the Colorado mines; he also wrote *To the End of the Trail* (Boston, 1902). For Nason's career, see *Who's Who in America, 1928–1929* p. 1549.

86. See *M&M, 10* (December 1929), 580, *14* (September 1933), 392; *DAB, 6,* 87; *Who's Who in the West,* p. 205; *Quarterly of the California Historical Society, 12* (December 1933), 371.

87. *Canadian Independence, Annexation and British Imperial Federation* (1894); *Old France in the New World* (1905); *New England and New France* (1913). See *DAB, 5,* 397.

88. *A Partial List of Organizations in California Interested in California History,* p. 85. Wiltsee wrote. *The Pioneer Miner & the Pack Mule Express* (1931); *The Truth about Fremont; an Inquiry* (1936); *Gold Rush Steamers of the Pacific* (1938).

89. *A History of American Mining* (1932). His *Man and Metals* (1932) and *The Romance of Mining* (1945) are also historical. Herbert and Lou Hoover are well known for their translation of Agricola's *De Re Metallica.*

90. One critic insisted that Hammond's *Autobiography* was "full of lies"—that Hammond incorporated into it as his own experiences what had actually happened to other engineers. William Wallace Mein to Howell Wright, San Francisco, October 27, 1937, Howell Wright collection.

91. *EMJ, 55* (June 3, 1893), 505, *95* (February 22, 1913), 433.

the pulpit, although occasionally one left the profession for the ministry,[92] but probably most had an average interest in religion. Some, such as Eben Olcott, were stalwart churchgoers, but likely the nomadic nature of their work reduced their regular attendance. A few, including Lionel Nettre of Montana and Ross E. Browne, were free thinkers. Browne not only published an attack on the Einstein theory of relativity and wrote an essay, "How to Grow Old," he produced a book *Views of an Agnostic* (1915), in which he showed a rationalistic sympathy with the scientific school of Spencer, Darwin, and Huxley.[93]

Because he was a man of some culture and education, and because he was often torn between his work in a remote, undeveloped area and the pull of urban civilization with its amenities, the mining engineer lived a life of contrasts. One moment he might be in raw wilderness "forty miles from no place," sleeping in an insect-infested hut or in a haystack in a thunderstorm, or drinking water full of "wiggle-tails";[94] the next might find him moving in sophisticated circles in San Francisco, New York, or London. To those fresh from three or four years of school in Europe, adaptation to the crude mining camps of Nevada or Montana must have come as a shock. Not long out of Freiberg, Louis Janin described his life on the Comstock in 1864:

> I wish you could take a peep at me and see how a fellow "accustomed to the refined society of Europe" can exist. I and two companions sleep in a small room 10 x 12 ft, which serves at the same time for sitting room and office. The room is one-half of a little shanty made of rough boards with "intervals of 15 minutes" between them, so we have plenty of air. We have one chair, one little table,

92. *EMJ, 94* (September 21, 1912), 562.

93. Ross E. Browne, *Views of an Agnostic*, pp. 5–9, 28; *M&M, 17* (July 1936) 368; Granville Stuart to Lionel R. Nettre, Fort Maginnis, Montana, January 23, 1882, copy, Letterbook 1, Granville Stuart MSS.

94. Frank E. Johnesse to Mayme Patten, Lewiston, Idaho, November 15, 1899, Johnesse family MSS; Kett, *Autobiography*, pp. 44–45; Gressley, ed., *Bostonians and Bullion*, p. 124.

> a tin wash basin and one tumbler between us. . . . One of my companions is a carpenter and the other is the clerk of the mine. This latter one & myself dine together at the workmen's lodging house, but we are still so imbued with aristocratic notions that we dine after the men are finished, and have a table cloth (to be sure, a very dirty one) set for us. We eat our hash and watery potatoes with the greatest contentment, although hash in fly time is not the most desirable food.[95]

In 1935, Robert Livermore commented on the contrasts that had helped make his life so interesting.

> Our greatest delight was after months of hard work at high altitudes, to start horseback, clad in digging clothes, descend to town,—bath and a shave,—then hie us Eastward, achieving more luxury with every step. At Chicago, we doffed Stetsons and purchased, "2 hard hats." In New York, nothing but a "topper" and all that goes with it would go.

Livermore recalled another contrast:

> On a visit to New York, as usual seeking solace from underground grime, clad in Broadway togs, we received word to examine a new strike 100 miles south of Hudson Bay;—three days later, in furs, in a dog team, under Northern lights wondering whether we would get somewhere before we froze to death.[96]

Louis Ricketts, one of the most successful engineers, exemplified the ever-present contrast. Frequently, he appeared in public in dirty slouch hat, clay-stained shirt, and beltless pants held up by two nails. In this garb—*mal cinto,* "the badly belted," the Mexican miners called him—he supposedly once

95. Louis Janin to Juliet C. Janin, Ormsby County, Nevada, August 2, 1864, Janin family MSS, Box 10. A few years later, Henry Janin reported that brother Alexis had "introduced into his partner's house the unheard of luxury of sheets." "How effeminate," said Henry. Henry Janin to Juliet C. Janin, Hamilton, Nevada, January 30, 1870, ibid., Box 6.

96. Gressley, ed., *Bostonians and Bullion,* pp. 192–93.

appeared in one of the leading banks of San Francisco asking if a telegram had arrived transmitting funds to close out a major oil deal. After he had been studiously ignored for an hour or so, the wire arrived—for $12 million—and bank officials, three vice presidents among them, scurried to complete the transaction.[97]

Thomas A. Rickard agreed that the contrasts of the life provided not only interest, but also helped to shape the engineer's basic outlook:

> If the profession of mining-engineering has reached no higher philosophy than that of Wall Street and Throgmorton Avenue, it is obtuse indeed; if varied experiences, their own and others, the hardships of the trail and the luxuries of the city; the great silence of the mountains and the unresting noisiness of the streets; the poverty of the *peon* and the wealth of high finance; if all of these, in constant contrast, do not make a mining man something of a philosopher on his own account, then he is indeed as unimpressionable as the wooden Indian of the tobacconist.[98]

The profession was a fairly close-knit fraternity and many engineers came to know each other occupationally and socially. At times, they displayed considerable esprit de corps, whether under congenial working conditions such as those at Grass Valley where James Hague, John Hays Hammond, and others "spent our time with quartz by day and pints by night," as Hague put it, or whether on a more solemn occasion, such as the funeral of Francis Vinton at Leadville in 1879, where the "civil and mining engineers and assayers on horseback" made up part of an impressive cortége.[99] Because they moved in the same social circles, it is not surprising that a number of engineering families intermarried—the Brownes and the Hoffmanns, the Bradleys and the Parkses, or the Meins and the

97. Bimson, *Ricketts,* pp. 19–20.
98. "To the Mining Student," *EMJ, 78* (October 6, 1904), 539.
99. Hammond, *Autobiography, 1,* 179; *Leadville Reveille,* quoted in *EMJ, 28* (October 25, 1879), 298.

Williamses, for example.[100] They formed their own groups —the Engineers' Club in San Francisco and the Rocky Mountain Club in New York City—and it was such men as Rossiter Raymond, William P. Blake, George Maynard, and Raphael Pumpelly—engineers of western experience—who would be the prime movers in creating the American Institute of Mining Engineers in 1871.[101] Moreover, they supported the AIME well over the years, and the West played an increasingly larger role. Its annual meetings were held periodically in such cities as Denver, Helena, or San Francisco, and the papers read and discussed tended to emphasize problems of mining and milling gold, silver, copper, and lead in the region beyond the Mississippi. In 1908, the Mining and Metallurgical Society of America came into being, but the AIME remained the more important group.[102]

Many engineers felt some special obligation. They regarded themselves as a kind of elite, somehow ordained to share their knowledge within the profession, but also to uplift society in general. Mining engineers were expected to cultivate "a certain facility of expression and of good manners," for as one said, "The engineering professions should be followed by gentlemen." [103] The president of the Colorado School of Mines spoke of the successful "artist-engineer," who in speech, conduct and dress must be able to move in influential circles: "He cannot afford to be a curiosity blown down from the mountains." [104] Engineers were activists and very conscious

100. San Francisco *Bulletin,* October 13, 1903; Wolfe, *Men of California,* p. 55; Brewer, *Up and Down California,* p. 132.

101. *EMJ, 13* (January 23, 1872), 51.

102. *EMJ, 34* (September 23 & October 7, 1882), 158, 184, *85* (April 25 1908), 871; *M&SP, 58* (May 18, 1889), 360; *Helena Herald* July 14 & 15, 1887. Western-oriented engineers who were at one time president of the AIME included Rossiter W. Raymond, Richard Pearce, James Douglas, Eben Olcott, John Hays Hammond, David W. Brunton, Benjamin B. Thayer, Louis Ricketts, Philip N. Moore, Horace V. Winchell, Herbert Hoover, Edwin Ludlow, Authur S. Dwight, Frederick Bradley, Robert E. Tally, Daniel Jackling, Donald Gillies, Harvey S. Mudd, and Louis Cates. Parsons, ed., *Seventy-Five Years of Progress,* pp. 495–511.

103. *SMQ, 31* (April 1910), 207.

104. Victor C. Alderson, "Artist Engineer," *Mining World, 31* (September 4, 1909), 516.

of the high status of their profession. Perhaps their general attitude was epitomized by one who learned in 1924 that his alma mater was graduating forty lawyers and only one mining engineer that year: "Forty parasites and future blockheaded Congressmen to one doer and producer." [105]

Successful mining engineers lived well. Social prestige was high, and some maintained a patrician outlook. Of Henry Janin, it was said by a contemporary that he "lives well, on the best of the land; drinks only the best of wines, and smokes only the best of cigars." Janin was "aristocratic in his tastes, gentlemanly in demeanor, and careless of the opinions of those he does not esteem." [106] Even those who were only moderately successful often kept servants. When Mary Hallock Foote was in New Almaden, she "lived like 'a lady' as they say—never lifting a finger in the way of work." [107] At Telluride, Robert Livermore provided his wife with "a maid of all work"—a "lady with a past" who had successfully graduated from "the Row." [108] Clarence King usually had a Negro body servant with him, who in time became a knowledgeable assistant in underground work. According to one story, another engineer, possibly Louis Janin, also employed a valet, and on one occasion when the two experts were preparing for an apex case, King observed his servant doubling up with laughter.

> "What's the joke?" asked King. "Say, Mr. King," was the reply, "you know Mistah Janin's nigger? Well, dat

105. N. E. Guyot to editor, San Francisco, n.d., *EMJ, 117* (May 31, 1924), 892. A recent historian of mechanical engineers points out that the AIME "exhibited almost no interest in the questions of professionalism or professional status at all", and attributes this to low membership standards and the fact that the mining engineer was more entrepreneurial than other engineers. Monte A. Calvert, *The Mechanical Engineer in America, 1830–1910,* p. 213. Perhaps the AIME as an organization showed little concern, but the pages of the technical journals clearly indicate that the mining engineers as individuals were indeed aware of professional questions and status and worried a good deal about such issues, as their struggle against the unqualified would attest. See pages 76–77, this book.

106. *Pacific Coast Annual Mining Review* (1878), p. 51.

107. Mary H. Foote to Helena Gilder, January 17, 1878, M. H. Foote MSS.

108. Gressley, ed., *Bostonians and Bullion,* p. 142.

> nigger is the most ignorantist nigger I ever see: yessah, he don't know nuffin. Wy dat nigger—ho,ho,ho,ho—dat nigger, Mistah King, he so ignorant he don't even know what hawnblende andesite am!" [109]

If success can be measured in terms of wealth, then a number of mining engineers indeed must be considered successful. A survey of deans of mining and technical schools in the United States indicated that in 1921 the most highly esteemed mining engineers in the country were, in descending order, Herbert Hoover, John Hays Hammond, Josiah Spurr, J. Parke Channing, Daniel C. Jackling, and Pope Yeatman.[110] Most of these men, with perhaps the exception of Spurr, who was named for his editorial reputation, were extremely wealthy.

Hoover, Hammond, Jackling, Hennen Jennings, H. H. Webb, James Douglas, and a number of others were full-fledged millionaires who often displayed their affluence accordingly. In 1897, Ernest Wiltsee, who later became the owner of "the richest oil well in the world," purchased the seventy-eight-foot *Aggie,* then with one exception considered "the largest and finest yacht on San Francisco Bay." [111] But this was a mere dinghy alongside Dan Jackling's steam-driven 300-foot, 2,500 ton *Cyprus,* which had a crew of fifty, accommodations for thirty guests, and a host of appointments including a movie theater, miniature golf course, and two brass guns to ward off pirates.[112] Jackling also kept a private railroad car, as did Douglas, Rossiter Raymond, and Bulkeley Wells. Hammond owned a Newport mansion and one of the finest residences in Washington, D.C., which he sold in 1913 to the Russian Embassy.[113]

But these were exceptional men, atypical of the profession

109. Josiah Spurr, "Prospecting Deluxe," *EMJ, 115* (May 26, 1923), 919
110. *EMJ, 111* (April 30, 1921), 757.
111. *M&SP, 75* (December 4, 1897), 530; *San Francisco Call,* July 8, 1900; *Sacramento Union,* November 9, 1947.
112. *EMJ, 101* (June 10, 1916), 1040.
113. *EMJ, 96* (November 15, 1913), 945.

as a whole. The average mining engineer probably maintained a respectable standard of living. Like Ross E. Browne in Oakland, he might have a house "with a big billiard room and tennis court," and he was able to give his children a college education.[114] And yet the profession had its risks and its ups and downs. There were lean years and fat years, and some never prospered, and some who were once well off slipped into financial adversity later in their careers. Philip Deidesheimer filed a petition of bankruptcy in 1877, listing liabilities of $534,600 and no assets.[115] Joshua Clayton constantly complained about his money troubles and the "doubtful living" he made in mining engineering.[116] At the end of a long career in 1895, Melville Attwood was "sadly in need of a little money," and borrowed funds while he sought to sell his mineral collection.[117] Even George Maynard and Louis Janin, two of the foremost West Coast engineers, felt financial pinches in their later years, and Robert Brewster Stanton in 1910 recorded his dire situation. "Can not figure a possible dollar for the first of month to pay board, etc.," he wrote, "I never came to such an absolute dead lock & stone wall before."[118]

But these, too, were exceptions. By and large, the profession paid reasonably well, although the laws of prudent investment and supply and demand undoubtedly caused frequent fluctuations in the income level of many. In 1865, for example, Louis Janin complained that business was dull and that "Mining Companies are getting mean" and reducing salaries. But if Janin were complaining, commented Josiah Whitney, "the others of the profession do not find anything to do at all, I

114. Charles Hoffmann to William Brewer, Oakland, April 4, 1908, Hoffmann MSS.

115. Herbert L. Smith, "The Bodie Era," p. 3, Herbert L. Smith Scrapbooks, 5.

116. Joshua Clayton to Robert A. Kirker, Salt Lake City, January 24, 1883, copy, Clayton MSS (Calif.), 2; Clayton to Edward J. Curtis, June 26, 1884, copy, ibid., 4; Goodwin, *As I Remember Them,* p. 239.

117. Melville Attwood to Adolph Sutro, Berkeley, January 14, 1895, Sutro MSS, Box 2.

118. Eben Olcott to Enoch Kenyon, New York, October 18, 1894, copy, Letterbook 22, Olcott MSS (N.Y.); W. B. Bourn to James D. Hague, San Francisco, March 7, 1905, Hague MSS, Box 11; entry for July 25, 1910, field notes 24, Stanton MSS.

imagine."[119] But the industry prospered and in 1872 a San Francisco promoter noted that Janin and other good engineers were "all *monopolized,"* even at fees that were considered expensive.[120]

In the aftermath of the depression of 1873, William Ashburner and James Hague both lamented the slowness of business: "No reports, no fees," said Hague,[121] but neither suffered very long. Again in 1888, Carl Stetefeldt wrote that mining matters were "fearfully dull" and that mining engineers "would do better to open candy stores, or blacken boots."[122] On his return from Africa in 1893—another depression year—Charles Rolker complained that New York engineers charging "cut rates" had "absolutely demoralized" income.[123] Yet Eben Olcott noted that he had cleared $6,000 in 1894, "an unusually dull" year compared with some in which he had netted more than $9,000.[124]

But the profession boomed in the middle 1890s, giving reputable engineers "about all we could attend to." Demand was greater than supply in 1906, and about half the graduating class at Berkeley had found positions as soon as they completed their examinations.[125] Then came an era of dislocation as the result of the Mexican Revolution and World War I, with a substantial number of engineers drifting back to the western part of the United States from areas in turmoil. "Mining is rotten dull out here," wrote Charles Janin from

119. Louis Janin to Louis Janin, Sr., Virginia City, September 10, 1865, Janin family MSS, Box 10; Josiah Whitney to William Brewer, San Francisco, December 10, 1866, Brewer-Whitney MSS, Box 2.

120. George S. Roberts to Asbury Harpending, San Francisco, March 18 1872, Harpending MSS, Box 1.

121. William Ashburner to George Brush, San Francisco, April 3, 1875 Brush family MSS; James D. Hague to S. F. Emmons, San Francisco, April 9, 1875, copy, Letterbook 6, Hague MSS.

122. Carl A. Stetefeldt to Samuel Christy, New York, May 11, 1888, Christy MSS.

123. Charles M. Rolker to Eben Olcott New York, August 23, 1893 & August 24, 1894, Olcott MSS (N.Y.).

124. Eben Olcott to W. H. Radford, New York, July 29, 1895, copy, Letterbook 23, ibid.

125. Louis Janin to John Hays Hammond, San Francisco, July 30, 1896, Hammond MSS, Box 3; *EMJ, 82* (August 18, 1906), 319.

California in 1913.[126] About the same time, Charles Munro wrote an engineer friend, W. H. Lanagan, soon after Woodrow Wilson had been inaugurated: "Say Harry you ought to be glad to have a job—they say things are awful tight in the states without much improvement for the future. They do not know what the new President is going to do." [127]

Other engineers condemned the low wages, insisting that ordinary miners were being paid more and that bona fide engineers were being driven from the field by inadequate salaries.[128] Some predicted that employers were reducing the engineer to "professional peonage" and that, in self-defense, the engineer might be driven to organize.[129] The early 1920s saw the same situation, and New York engineers with "empty pockets and bellies" were accused of "cutting fees from their previous high prices." [130] A Utah engineer in 1922 grumbled about the glut of technical men on the market and cited an instance in which an engineer of more than twenty years' experience was asked to manage two mines for a fee of $200 a month, less office expenses. "Are mining engineers a commodity?" he asked, "If so, why not have us listed on the Stock Exchange, along with wheat and corn?" [131] Clearly, at the end of the era, it was more difficult for a young engineer to get a start, and salaries, in comparison with others, were not so good as in the more golden years. But, in the long run, most mining engineers maintained a respectable upper middle class existence or better.[132]

A skilled professional man of better-than-average education, if not income, the mining engineer was a highly respected citi-

126. Charles Janin to Chester W. Purington, San Francisco, February 3, 1913, copy, Charles Janin MSS, Box 5.

127. Charles Munro to W. H. Lanagan, aboard the S.S. *Guatemala,* March 28, 1913, Morris MSS, Box 2.

128. *M&SP, 110* (February 6 & March 13, 1915), 228, 414, *113* (October 28 1916), 618.

129. *M&SP, 116* (May 11, 1918), 641–42.

130. Charles Janin to H. Foster Bain, San Francisco, April 27, 1921, copy Charles Janin MSS, Box 9.

131. "The Guilty One" to editor, Salt Lake City, n.d., *EMJ, 113* (April 8, 1922), 566.

132. *Mining Science, 69* (April 1914), 17–18.

zen whose opinion even on nonmining matters carried considerable weight. Yet, at any level of political life—local, state, and national—despite the fact that some engineers were always active, it would be difficult to characterize the engineer as a political animal. The roving nature of his work and his frequent inability to sink roots deeply into a community often prevented his building either a strong interest or a significant political following. In 1887, Hennen Jennings, then manager of the New Almaden mine, testified in a case involving charges of vote intimidation by his company, that, though "perhaps a little tainted" in the direction of Mugwumpism, he was too busy to take much notice of politics.[133] In response to a questionnaire asking about his politics and political life in 1897, Daniel M. Barringer wrote: "No experience in either. Have been absent from Philadelphia so much on professional work that have had no opportunities to take an active interest in political affairs." [134]

Journal editors were fond of deploring the tendency of mining engineers to retreat into the background when the "political buzzsaw" "commenced to whirl," leaving farmers and lawyers to dominate legislative bodies.[135] On the other hand, some editors believed it an error for the professional engineer to get caught up in "the crazy whirlpool of misappreciation that excites the American public, yet such men follow one another into the fatal flame like so many fickle moths after sundown," only to be swamped by political specialists.[136]

Being human beings, mining engineers did express themselves on political questions, and they represented almost every shade of opinion. Some were Southern sympathizers during the Civil War, and G. C. Swallow had been imprisoned as such in St. Louis in 1862, although his friends insisted that "skullduggery" was at the bottom of it. The Janin brothers had

133. *San Jose Daily Mercury,* April 23, 1887, clipping, Christy MSS.

134. Information for "Universities and Their Sons," undated copy, Letterbook B. Barringer MSS.

135. See *EMJ, 55* (May 6, 1893), 420, *57* (June 30, 1894), 601, *109* (April 17, 1920), 913; *M&SP, 101* (July 30, 1910), 136.

136. *Northwest Mining Journal, 6* (October, 1908), 41.

grown up in New Orleans, but found it best to remain quiet in the West. "When Papa refused to let Henry and myself go South, I determined to think and to say as little as possible about the war, and have found the plan to work admirably," confided Louis in 1865.[137]

In 1858, Joshua Clayton was a popular sovereignty Democrat, but in 1877 he viewed the election of Rutherford Hayes as "good news." [138] James Hague supported the election of McKinley in 1896, expressing the hope that California would "join the expected majority and throw Bryan overboard." [139] After the turn of the century, his interest in national politics now kindled, Daniel Barringer thought McKinley "will go down in history as one of our ablest Presidents." [140] He also thought well of Theodore Roosevelt, whom he had known for many years, and wrote Roosevelt after the 1904 election that "you are the first Republican nominee that I ever voted for." He also urged Roosevelt to do something for the Negro and suggested the burning of sulphur to rid Panama of mosquitoes and curb malaria while the canal was being built.[141] Barringer also badgered President Taft and his secretary of state, Philander Knox, urging the appointment of his friend Charles Denby as minister to China and suggesting that former President Porfirio Díaz be invited to Washington to advise on Mexican policy! [142]

Barringer was also a friend and classmate of Woodrow Wilson, and had once advised Wilson, then at Princeton, against any additional investment in Georgia gold property. In

137. Merrill, *The First Hundred Years,* p. 426; Louis Janin to Juliet C. Janin, Virginia City, March 9, 1865, Janin family MSS, Box 10.

138. Joshua Clayton, "Kansas Affairs," copy, written for the Mariposa *Democrat,* March 22, 1858, in Clayton MSS (Yale); entry for March 2, 1877, Clayton diary, 1877, Clayton MSS (Calif.), 5.

139. James D. Hague to C. W. Howard, New York, October 6, 1896, copy Letterbook 21, Hague MSS.

140. Daniel M. Barringer to John Brockman, September 16, 1901, copy Letterbook E, Barringer MSS.

141. Barringer to Theodore Roosevelt, December 15, 1904, copy Letterbook J, & July 14, 1905, copy, Letterbook K, ibid.

142. Barringer to William Howard Taft, April 16, 1909, copy, Letterbook O, & Barringer to Philander C. Knox, April 27, 1912, copy, Letterbook P, ibid.

the campaign of 1912, he worked actively on behalf of Wilson, contributed funds, and became disenchanted with Theodore Roosevelt, whom he came to regard as "a good man who has allowed his egotism to run away with him." [143] Ironically, while he was furthering "Tommy" Wilson's cause, he was cooperating in some Austrian mining business with John Hays Hammond, who was actively campaigning for Taft. Hammond, by his own admission, knew every president from Grant to Hoover, with the exception of Arthur. He was a pallbearer of Cleveland, a college friend of Taft, with whom he "enjoyed a flutter in the direction of the vice-presidency," and was later appointed by Taft as special ambassador at the coronation of George V in England. He turned down an appointment as minister to China, but was enough of an unofficial adviser to Taft that Colonel House could subsequently be referred to as "the John Hays Hammond of the Wilson Administration." [144]

When World War I was over, Charles Janin had returned from his wartime work in Washington and become increasingly critical of problems on the domestic scene, particularly organized labor, high taxes, and the specter of radicalism:

> Bolshevists in this country must be handled without gloves and promptly or there will be lots of trouble. The scum of Russia and Germany that didn't leave the country or were not interned are trying to stir up some trouble especially in Chicago where Berger who should be shot if given just deserts is raising cain.

Janin knew who was to blame—"and it ain't Colonel House," he said.[145]

143. Barringer to S. McClure Lindsay, May 8, 1900, copy, Letterbook D, Barringer to Woodrow Wilson, October 13, 1903, copy, Letterbook I, Barringer to Wilson, September 9, 1912, copy, Letterbook R, & Barringer to John King, July 22, 1912, copy, Letterbook R, ibid.

144. Barringer to Charles Denby, July 5, 1912, copy, Letterbook I, ibid.; Hammond, *Autobiography, 2,* 527–31, 534, 633.

145. Charles Janin to Ira Joralemon, November 27, 1918, copy, Charles Janin MSS, Box 7; Janin to H. Foster Bain, San Francisco, October 24, 1919, copy, ibid., Box 8.

On the other hand, Fred Bradley supported Wilson to the end. "While history is going to record the President's great accomplishments, it should in some way scourge all those who have deserted him and pounced on him in his time of sickness and political weakness," insisted Bradley.[146] Some engineers were already by 1920 advocating Herbert Hoover for president.

> Attention all who hear her call!
> No trifle so can move her.
> There's one best mate for the Ship of State,
> "His name is Herbert Hoover." [147]

The call was premature. Hoover turned down a full partnership with Daniel Guggenheim in 1921 to become secretary of commerce, a post he was to retain until 1928, the year of his presidential election—an event that prompted much rejoicing on the part of the engineering fraternity, both Republican and Democrat.[148]

Many of his fellow engineers shared Hoover's conservative political philosophy. Rossiter W. Raymond never had much sympathy for government remedies, whether they be suggestions for a federally built silver reduction plant in the 1860s or the Interstate Commerce Law of 1887.[149] The profession split on the question of free silver, but probably the majority favored bimetallism.[150] Like Robert Brereton, who emphasized that on "the cricket-field of life," "what scores I have made have been won off my own bat," most engineers were advocates of "rugged individualism," and saw state action as smacking of socialism.[151]

146. Fred Bradley to Bernard Baruch, December 28, 1920, copy, Bradley MSS.

147. From "1920," *M&M, 1* (February 1920), 6; *EMJ, 109* (April 17, 1920) 913.

148. Hoover, *Memoirs, 2,* 186; Fred Bradley to Bernard Baruch, San Francisco, November 5, 1928, copy, Bradley MSS.

149. *AJM, 4* (November 23, 1867), 328; *EMJ, 45* (February 4, 1888), 86.

150. *EMJ, 56* (July 8, 1893), 25, *58* (November 24, 1894), 483.

151. Brereton, *Reminiscences* 5; Ernest A. Wiltsee to Eleanor Bancroft, Calistoga, California, April 14, 1940, Wiltsee MSS, Part 2; *EMJ, 109* (January 24, 1920), 278. See also Mark L. Requa, *The Relation of Government to Industry,* pp. 227–36.

Throughout the entire period, mining engineers did hold political offices at various levels, but by no means in keeping with the full measure of influence they might have wielded in the West had they been more politically inclined. At the muncipal level, Charles Goodale took considerable pride in having been alderman of the second ward in Butte, 1888–90; Thomas Rickard was mayor of Berkeley; and William M. Ferry, a Colorado Mines graduate, was elected mayor of Salt Lake City in 1915.[152]

Mining engineers served as members of the legislatures in practically all western states and territories.[153] William Ralston became speaker pro tem of the California assembly, subsequently sat in the state senate, and was for seven years assistant treasurer of the United States at San Francisco, giving up this post in 1913 "to make room for a Democrat," as he put it, and to run unsuccessfully for the governorship.[154] Engineers sat as members of constitutional conventions,[155] and a few became governors of their states.

James B. Grant, one of the West's outstanding metallurgical engineers, was elected the first Democratic governor of Colorado in 1882.[156] Another Colorado engineer, Jesse F. McDonald, served as mayor of Leadville and as a member of the state senate before being elected lieutenant governor in 1904. But the governor resigned a few hours after being

152. Harper, ed., *Who's Who on the Pacific Coast,* pp. 222–23, 300; *EMJ, 91* (April 15, 1911), 778; *M&M, 19* (March 1938), 169. For other examples of engineers as mayors of small towns, see *Who's Who in Engineering, 1922–23,* pp. 1000, 1412; *M&SP, 106* (April 12, 1913), 563.

153. These include Sherman Day and Thomas Farish in California; A. J. Seligman and Alexander Burrell in Montana; P. H. Hunt and George W. McCaskell in Utah; Thomas Davis, John Christy, and David Morgan in Arizona; Charles Christien in Colorado; Cony Brown in New Mexico; and Frank E. Johnesse in Idaho. *M&SP, 16* (January 4, 1868), 8, *120* (May 15, 1920), 728; *EMJ, 103* (March 10, 1917), 429; *M&M, 11* (March 1930), 189, *12* (March 1931), 171, *16* (June 1935), 280; Harper, ed., *Who's Who on the Pacific Coast,* p. 191, 300; Wolfe, ed., *Men of California,* p. 21; Bunje, et al., *Careers,* 29; Charles G. Christien dictation, n.d.

154. *Who's Who in Engineering, 1922–23,* p. 1026; *W. C. Ralston,* pp. 3–7; *EMJ, 98* (October 31, 1914), 804.

155. *M&M, 11* (December 1930), 601.

156. *EMJ, 92* (November 18, 1911), 1004; Alice Polk Hill, *Colorado Pioneers in Picture and Story,* pp. 489–91.

sworn in, and McDonald served out the term, though he was defeated in his bid for re-election in 1908.[157] Emmet D. Boyle, a graduate of the Nevada School of Mines and consulting engineer to many western enterprises in the early twentieth century, was successively Nevada state engineer, tax commissioner, and—from 1915 to 1923—governor.[158] George H. Dern served two terms as governor of Utah and became secretary of war under President Franklin D. Roosevelt.[159]

A number of mining engineers, including William Ashburner and Daniel Jackling,[160] aspired to sit in the national halls of Congress, but only a few ever achieved that goal. Possibly the first, if judged by the work he did, was Nathaniel P. Hill, who was one of Colorado's senators from 1879 to 1886. A chemistry professor at Brown University, Hill had organized the Boston & Colorado Smelting Company in 1867 and built a pioneering mill in the Rockies.[161] Perhaps William S. Rosecrans would also qualify, for he did both civil and mining engineering work in the post-Appomattox era, and sat in Congress as representative from California, 1881–85.[162] It seems likely, however, that there was no technically trained mining engineer in Congress until 1907, when William F. Englebright, of Nevada City, California, took his seat in the House. He was not re-elected in 1910 and returned to engineering, but his son, Harry L. Englebright, a California Mines graduate, served in the house from 1927 to 1943.[163] Samuel S. Arentz, South Dakota School of Mines, 1904, was elected representative from Nevada in 1920. Defeated in the 1922 pri-

157. *M&M, 23* (June 1942), 364; *EMJ, 79* (March 23, 1905), 596; *Who's Who in Engineering, 1922–23,* p. 811.

158. *M&M, 7* (February 1926), 89; *Who's Who in Engineering, 1922–23,* p. 177.

159. *M&M, 6* (January 1925), 40, *17* (October 1936), 506.

160. Josiah Whitney to William Brewer, San Francisco, August 27, 1867, Brewer-Whitney MSS, Box 3; *EMJ, 88* (August 28, 1909), 418, *114* (September 2, 1922), 424; *M&SP, 116* (May 11, 1918), 667.

161. *DAB, 9,* 43; *EMJ, 27* (January 25, 1879), 55.

162. James L. Harrison, ed., *Biographical Directory of the American Congress, 1774–1949,* p. 1759.

163. Ibid., p. 1133; "Mining Engineers and Politics," *M&SP, 101* (July 30, 1910), 136.

mary for a Senate appointment, he returned to private practice in Reno momentarily, but was in the same year re-elected to the House, where he continued to serve until 1933.[164] Lewis W. Douglas, trained at MIT in metallurgy and geology, was representative from Arizona, 1927–33, after which he became Franklin Roosevelt's director of the budget, deputy administrator of the War Shipping Administration, and, in 1947, ambassador to Great Britain.[165]

In addition to the handful who sat in Congress, a few engineers also figured in national party deliberations. Stanley A. Easton, of the Bunker Hill & Sullivan, was one of the Idaho delegation to the Republican National Convention in Chicago in 1920. Earlier that same year, Will Hays, chairman of the Republican National Committee, had appointed Daniel Jackling, Mark L. Requa, and John Hays Hammond among those serving on an advisory subcommittee on policies and platform.[166] Hammond's "flutter for the vice presidency," as a possible runningmate of Taft in 1908 has been mentioned, but it is doubtful if he was a serious contender because of his close association with "The Trusts"—the Guggenheims.[167] However, in 1924, John Campbell Greenway, an Arizona engineer with a distinguished record in the Spanish American War and World War I and the husband of a Democratic National Committeewoman, received the ballots of eight states in the Democratic Convention, but was passed over in favor of Charles W. Bryan, brother of the illustrious William Jennings.[168]

If the mining engineer made only a minimal impact in national politics, he did make an impressive record in public af-

164. Harrison, ed., *Biographical Directory,* p. 788; *Black Hills Engineer, 16* (1928), 188–90.

165. Harrison, ed., *Biographical Directory,* pp. 1098–99; *M&M, 13* (October 1932), 465, *14* (September 1933), 387.

166. *EMJ, 109* (May 22, 1920), 1180; *M&M, 1* (March 1920), 48. Requa would serve as campaign manager for Coolidge and later Hoover in California and for a time was Republican national committeeman from that state. L. K. Requa to author, Salt Lake City, May 11, 1964.

167. *EMJ, 85* (June 12, 1908), 1215; *Northwest Mining Journal, 5* (June 1908), 96.

168. *EMJ, 118* (July 19, 1924), 105; *M&M, 7* (February 1926), 88.

fairs, especially during World War I. A number had fought in the Civil War,[169] and several, including Greenway, Chester A. Thomas, Arthur L. Williams, and R. H. Channing, had been with Roosevelt's Rough Riders in Cuba in 1898.[170] During 1914–17, the period of American neutrality in the great war in Europe, a German engineer with experience in the West was implicated in a plot to dynamite ships leaving New York with munitions for the Allies, and at least one American mining engineer urged the support of the German cause for economic reasons,[171] but the bulk of engineering sentiment in the United States was overwhelmingly on the side of the Allied powers.

No doubt this was strengthened by the fact that American mining engineers were aboard ships torpedoed by German submarines on the high seas. One went down on the *Falaba* off the Scully Islands in 1915; another was lost when the *Persia* was sunk in the Mediterranean a year later. At least three went down with the ill-fated *Lusitania,* and Scott Turner was fished out of the water after six hours, suffering from a broken nose, dislocated shoulder, and bruises from head to toe.[172]

The work of Herbert Hoover and fellow engineers in providing aid to stranded Americans and to starving Belgians also reinforced the pro-Allies position of the profession. In the summer of 1914, Hoover, Edgar Rickard, John L. Hoffmann, and John B. White—all mining engineers—organized a refugee committee to succor American travelers in Europe. A few

169. James D. Hague had served a year with the U.S. navy during the war; Rossiter Raymond had served in the federal army from 1861 to 1864, part of the time as aide de camp on the staff of General Frémont. Almon D. Hodges was with the Massachusetts regiment, and several others connected with mine management, including Fitz John Porter and Napoleon Buford, had been Union officers. James B. Grant had served in the Confederate army. *DAB, 3,* 125, *6,* 87, *15,* 91; *EMJ, 52* (July 18, 1891), 71, *92* (November 18, 1911), 1004, *107* (January 18, 1919), 135–36.

170. *EMJ, 66* (July 9 & September 17, 1898), 44, 344, *104* (December 15 1917), 1060, *110* (November 27, 1920), 1053; *Who's Who in Arizona,* pp. 465–66; Theodore Roosevelt, *The Rough Riders,* p. 40.

171. *EMJ, 48* (November 21, 1914), 254, *100* (November 13, 1915), 815.

172. *EMJ, 99* (May 15, 1915), 877; *M&SP, 110* (April 4 & May 8, 15, & 29, 1915), 505, 711, 749, 777, 853, *112* (February 12, 1916), 254; Edgar Rickard to Charles Janin, London, May 20, 1915, Charles Janin MSS, Box 5.

years later, one of the profession noted that American engineers were uniquely qualified for such a task:

> Many of them had themselves been stranded, lost in darkest Africa, near death's door in half a dozen places, and had tackled and handled successfully much worse situations than the feeding and shipping of a lot of stray sheep in a place where there was plenty of food and lots of shipping.[173]

Next, Millard Shaler, an American engineer, took the lead in organizing and Hoover agreed to direct, the Commission for Relief of Belgium. White, Millard Hunsiker, William Honnold, and Edgar Rickard all worked with this endeavor for no pay, and Rickard, who made at least one speaking tour to raise funds in the United States, found the experience exhilarating.[174]

American mining engineers were quite aware of the plight of their British counterparts in the service, and as a New Year's gift in 1916, sent boxes of Havana cigars to more than a hundred members of the Institution of Mining and Metallurgy then in the trenches in France.[175] American engineers also joined the British and Canadian forces. Harold Rickard went from Colorado, Lionel Lindsay from California, Fred Reece from Arizona. Morton Webber of New York, who had ranged professionally through the West, was twice wounded as an officer in the Royal Field Artillery, and Peter Nisson, "well known in Canada and the West," became a major in the Royal Engineers, where he devised a sectional steel military hut that would be widely used in two world wars.[176]

173. F. F. Sharpless, "Mining Engineers in the War," *EMJ, 107* (January 3, 1919), 1.

174. Thomas A. Rickard, "Mining Engineers and Belgian Relief," *M&SP, 114* (April 28, 1917), 574–75; *EMJ, 118* (October 25, 1924), 644; Edgar Rickard to Charles Janin, London, March 23, 1915, Charles Janin MSS, Box 5; Rickard to Janin, Chicago, 1916, ibid, Box 6.

175. *EMJ, 101* (March 25, 1916), 575.

176. *M&SP, 111* (August 14, 1915), 258, *112* (April 29, 1916), 620, *113* (August 5, 1916), 220, *114* (May 12, 1917), 649–53, *115* (August 25, 1917), 293, *120* (March 20, 1920), 401; *EMJ, 103* (March 10, 1917), 436.

When the United States entered the conflict, western mining engineers flocked to the colors. Some were disappointed. While Congress was still debating, Robert Brewster Stanton wrote in his field book: "Offered my services (through Wm. Barclay Parsons) for the War, in Engineering work. But who wants a man 71 years old—for anything." [177] But so many mining men enlisted that a shortage of skilled engineers and miners was evident in the West before the end of 1917.[178]

These men were scattered throughout the American expeditionary forces, but many were with the Eleventh Engineers, one of the first American units into action, and especially the Twenty-seventh Engineers, which was regarded as "the only strictly mining engineer regiment organized during the war." Under the command of O. B. Perry, one of the Guggenheims' dredge experts, the Twenty-seventh was trained for front-line work—tunneling, preparation of underground shelters, demolition, and saw its share of action in France.[179]

> Here's to our gallant regiment, whose spirit knows no fears—
> The glorious Twenty-seventh of mining engineers.
> They set the guns that kill the Huns. Then here's three rousing cheers!
> They'll surely win, they'll reach Berlin, our mining engineers.[180]

A number of young engineers, including Forbes Rickard, Jr., of Denver and William Hague of California, died in France; the *Transactions* of the AIME for 1919 ran obituaries on twenty-three mining engineers who had perished in service.[181]

177. Entry for April 3, 1917, field notes 29, Stanton MSS.

178. *EMJ, 103* (June 2, 1917), 999; *Mining American, 75* (October 13, 1917), 17.

179. *M&M, 3* (March 1922), 17, *5* (December 1924), 572; *EMJ, 104* (November 3, 1917), 807, *109* (January 17, 1920), 139.

180. By Mrs. George M. Taylor of Colorado Springs, quoted in *EMJ, 106* (September 28, 1918), 590.

181. *M&SP, 116* (February 2, 1918), 145–46, *117* (August 10, 1918), 201; AIME *Trans., 61* (1919), 724–50.

It was on the home front that western engineers played their most significant part in the war effort. The AIME did its bit by expelling twenty-one Germans and one Austrian from membership, and Rossiter Raymond's poetic wit was enlisted in the struggle:

> "Wilhelm, what blacked your eagle eye?"
> "O curious journalist!
> It was this eye, now swollen high,
> That smote old Foch's fist." [182]

The Rocky Mountain Club, a group of New York-based engineers who had already donated $500,000 saved for a new building to Hoover's Belgian relief program, entertained soldiers from western states as they passed through Manhattan to or from Europe.[183]

But more importantly, dozens of mining engineers became a part of the national machinery to mobilize the economy for the war. Even before American entry, William Truran and Charles Butters of California had been called to the British Ministry of Munitions to advise on the production of explosives,[184] but after spring 1917, countless others moved into Washington as "dollar a year men," or at greatly reduced salaries to work with the Wilson administration. According to J. Parke Channing, Washington quickly discovered "that a mining engineer has had such varied experience, and has driven so little in ruts, that at a minute's notice he can jump from 6-in. projectiles to bailed hay." [185]

Philip N. Moore, a western engineer and president of the AIME, did important work in coordinating production of strategic metals through the War Minerals Committee and also helped organize the American Engineering Council, which cooperated with the Council of National Defense in stepping up the output of vital minerals. In 1919, Moore became chairman of the War Minerals Relief Commission, established to

182. "The German Explanations," *EMJ, 106* (September 14, 1918), 507.
183. *EMJ, 103* (June 30, 1917), 1169, *107* (March 15, 1919), 504.
184. *M&SP, 113* (September 16, 1916), 439; *M&M, 15* (January 1934), 75.
185. J. Parke Channing, "Man-Power," *M&SP, 116* (June 1, 1918), 762.

indemnify those who had undertaken mineral production at government solicitation, but who had not recouped their expenditures by the time of the armistice.[186] Ernest Wiltsee, for example, one of the leading West Coast producers of chrome, had invested more than $100,000 in new concentrators in 1918, only to see prices fall when the war ended and government price supports were relaxed.[187] The War Minerals Committee itself was headed by William Y. Westervelt, an iron and copper expert.[188]

Pope Yeatman quelled an urge to serve in the army and brought his quarter of a century's engineering experience to bear as head of the nonferrous metals section of Bernard Baruch's War Industries Board. Hennen Jennings served on the platinum committee of the WIB, then was appointed chairman of a special committee to study the gold situation. Mark L. Requa, a Nevada copper engineer, spent some time with Hoover's Food Administration, but in 1918 became head of the petroleum division of the Fuel Administration.[189]

Fred Bradley served on a committee to aid in procurement of lead for the army and the navy, and at least five mining engineers, including David Brunton and B. B. Thayer, were on the Naval Consulting Board. Two others, F. A. Eustis and F. Huntington Clark, of the Emergency Fleet Corporation, became embroiled with General Goethals in a controversy over the practicality of constructing wooden ships and were ousted.[190]

William Wallace Mein, an old African hand and President of the International Nickel Company of Canada, was in 1918 put in charge of the fertilizer branch of the U.S. Department of Agriculture. Daniel Jackling would win a Distinguished Service Medal for his wartime services in supervising the con-

186. *DAB, 13,* 135.

187. Wiltsee, "Reminiscences," pp. 255–58.

188. *Mining American,* 75 (August 18, 1917), 3.

189. *EMJ, 109* (March 27, 1920), 761, *110* (October 16, 1920), 769; *M&SP, 116* (January 19, & May 18, 1918), 75, 675; *Who's Who in Engineering, 1922–23,* p. 1426; *M&M, 1* (May 1920), 22; *DAB, 10,* 56.

190. Bernard Baruch to Fred Bradley, April 7, 1917, Bradley MSS; *EMJ 103* (June 16, 1917), 1073, *107* (January 3, 1919), 3.

struction of federal explosives plants, and David W. Brunton served as chairman of a war department committee on devices and inventions useful to the war effort.[191]

Herbert Hoover headed the Food Administration and to it attracted West Coast engineers in particular. Among the "California Contingent," or the "Frisco Fusileers," as they were called, was Robert E. Cranston, one of the leading dredge experts in the county, and Edgar Rickard, who would later take charge of the Food Administration after Hoover was assigned to direct global relief efforts. Another was Charles W. Merrill, one of the foremost metallurgists in the West, who was asked by Hoover to head the Division of Collateral Commodities—a request received with mixed emotions, according to Merrill:

> I was annoyed because I didn't want the interference with my professional and business career, which was then extremely active. And I was indignant because I knew my respect and admiration for Hoover was so great I couldn't possibly refuse his request. . . . And now, as I look back on it, it is one of the two things I have done in a rather long life that brought me the greatest real satisfaction.

Hoover believed that Merrill did "a magnificent job," and tried to induce him into the government again in 1928, without success.[192]

Once the war had ended, most mining engineers were mustered out of government service, both military and civilian. Hoover, with Edgar Rickard as his right-hand man, was engaged in direct relief operations abroad; Philip Moore continued the War Minerals Relief Commission, with a number of western engineers as consultants;[193] but in general, there was

191. *M&SP, 116* (April 13 & June 8, 1918), 531, 806; *EMJ, 104* (December 22, 1917), 1102, *109* (February 21, 1920), 537.

192. Ryder, *The Merrill Story,* xi (foreword by Herbert Hoover), 64; *EMJ, 107* (May 17, 1919), 891; Charles Janin to Edgar Rickard, San Francisco December 23, 1918, copy, Charles Janin MSS, Box 7.

193. *DAB, 13,* 135; *EMJ, 111* (January 1, 1921), 27; *M&SP, 124* (January 13, 1922), 49.

an exodus from Washington. This release of engineering talent, just as wartime mining operations were being cut back, in a period of re-adjustment, created a glut of technical men on the market. On the machinery of the American engineering service, which had provided technical specialists to the war department during the emergency, was erected a new Engineering Societies Employment Bureau to help relocate engineers.[194]

It is certain, however, that the mining engineer made a substantial contribution to the mobilization of resources on the home front, far out of keeping with his numerical importance. It is equally as certain that the mining engineer in wartime service in the nation's capital would hardly fit the stereotype set down by romantic novelists and popularizers.

That he was a public spirited individual in peacetime is also clear. He might be found in the midst of various reform or civic causes, ranging from the distribution of relief funds in the aftermath of the San Francisco earthquake and fire to the simplified spelling movement.[195] His papers indicate that John Hays Hammond, a conservative in many ways, gave public addresses in favor of women suffrage, improved accident compensation laws, and the economic effects of foreign policy. Stanley Easton was a district worker for the Boy Scouts for many years, and in 1913 was a member of a commission appointed to draft a workingmen's compensation act for the State of Idaho.[196] Another engineer, a Michigan Mines graduate, gave up his professional career to found and operate a school for poor orphan boys in Honduras.[197]

Many—Hamilton Smith, Fred Bradley, and John Hays Hammond among them—established university fellowships, and donated funds for buildings or libraries.[198] Numerous others served on the boards of trustees of their own alma ma-

194. *EMJ, 109* (January 17, 1920), 139.

195. *EMJ, 76* (August 8, 1908), 295, *110* (August 14, 1920), 313.

196. Downs, ed., *Who's Who in Engineering, 1941,* p. 506; *EMJ, 96* (November 8, 1913), 898.

197. Kett, *Autobiography,* pp. 170–71.

198. *EMP, 70* (July 14, 1900), 34, *79* (April 13, 1905), 714, *112* (December 17, 1921), 890; *M&SP, 90* (April 8, 1905), 219.

ters or other institutions of higher learning.[199] In general, they were active contributors to philanthropic and social causes, and the generalizations made by Raymond Merritt in a recent study of engineering and culture from 1850 to 1875 would have to apply. Like members of the profession at large, they were leaders of the technological revolution, instruments of managerial change, and men of cosmopolitan perspective who were committed to the educational process. Their work influenced not only the advance of their own professionalism, but the progress of American civilization in a broader sense.[200]

199. Among those so serving were William Ashburner (University of California and Stanford University); Charles W. Merrill (University of California); Arthur Dwight (Columbia University); Robert E. Talley and Louis D. Ricketts (University of Arizona); A. A. Blow (Colorado School of Mines); and Stanley Easton (Whitman College and University of Idaho). *M&SP, 40* (March 6, 1880), 145, *54* (April 23, 1887), 268; *EMJ, 105* (January 19, 1918) 153, *112* (September 25, 1921), 503; *M&M, 18* (February 1937), 133; Bimson, *Ricketts,* p. 20; *Who's Who in America, 1962–63, 32,* 884; Frank Probert, "Mining Alumni," *California Monthly, 18* (April 1925), 448; Columbia University, *Honorary Degrees Awarded in the Years 1902–1945,* p. 121.

200. Raymond Harland Merritt, "Engineering and American Culture, 1850–1875," p. iv, Ph.D. dissertation, University of Minnesota, 1968.

CHAPTER 11

"A Most Important Missionary of Civilization"

When I die (said the mining engineer) do not bury me at all;
Cache me on the bottom level, with a pick beside my pall;
Leave a candlestick and matches, then cave the stopes and drifts,
And I'll be a tommy-knocker for a hundred thousand shifts.

—Samuel B. Ellis [1]

Although not a political being especially, the mining engineer did have some impact on government and public affairs, and he was, in the long run, an important figure to the West. Despite his early fight to establish himself and despite his nomadic way of life, he was invariably a man of stature at the local level, generally a member of the local elite and in a position to exert influence in the community. As an educated professional man, he helped to shape the western cultural, legal, and economic base, and he served as a significant link, tying the West to the rest of the world.

In his own way, the mining engineer did much to popularize and advertise the West and its resources. He sometimes took charge of western mineral displays at national or international expositions,[2] and he often carried the glories of the West to

1. From "The Tommy-Knockers," *EMJ, 110* (October 2, 1920), 675. A "tommy-knocker" is the mining equivalent of the gremlin.

2. Guido Küstel, for example, set up such a display at the Vienna Exposition of 1873. James Hague, Henry and Louis Janin, William Ashburner, John Boalt, and Winfield Keyes were all involved in bringing together the AIME display for the Philadelphia centennial in 1876. Hague was offered a similar post at the Paris exposition two years later, whereas Joshua Clayton represented Utah and John Church, Arizona in collecting minerals for the Denver mining exposition of 1882. *Report of Guido Küstel, Commissioner to the Vienna Exposition of 1873, appointed by the Mechanics' Institute of San Francisco, Cal., M&SP, 27* (August 23, 1873), 120, *32* (March 18, 1876), 177, *36* (February 9, 1878), 81, *44* (June 24, 1882), 416; Clayton to A. Hanauer, Salt Lake City, December 27, 1882, copy, Letterbook 1882, Clayton MSS (Calif.), 2.

eastern cities by private conversation or public addresses. Rossiter W. Raymond announced that, as of January 1, 1869, he would be available to lecture on an assortment of topics including "Wonders of the West," "The Art of Mining," "The History of Mining," "Artesian Wells," "The Steam Engine," and "The Catacombs of Egypt, Sicily, Naples, Rome and Paris."[3] Raymond, Joshua Clayton, John A. Church, and other engineers discussed western mining before the Bullion Club, a New York group formed in 1878 to advance legitimate precious mining interests,[4] and John Hays Hammond, John B. Farish, and Thomas Leggett were among those organizing the Rocky Mountain Club thirty years later, in part to further the interest of western communities among eastern capitalists.[5] Others carried the gospel of the mineral Southwest to Philadelphia or Harrisburg, speaking eloquently from hired halls.[6] Especially after 1880, they pushed the West and its resources to the forefront at professional meetings of the AIME.

Because of the nature of their work and interests, mining engineers were among the early organizers and supporters of technical and scientific societies in the West. They were charter members of and frequent contributors to the California Academy of Sciences, where meetings "were sometimes more grotesque and amusing than scientific, and were not always harmonious," according to one constituent.[7] Augustus Bowie and Melville Attwood were two of the founders of the California State Geological Society in 1876, and Attwood read several papers on mining and milling topics before meetings of the association.[8] Bowie, Ross Browne, Charles Hoffmann, and

3. *AJM, 6* (December 19, 1868), 392.

4. *United States Annual Mining Review and Stock Ledger* (New York, 1879) pp. 27–28; *M&SP, 38* (April 26, 1879), 265, *39* (August 9, 1879), 86

5. *EMJ, 85* (January 25, 1908), 201.

6. *Mining World, 2* (March 20 & April 20, 1882), 182, 196.

7. Edward Bosqui, *Memoirs,* p. 87. See also Bancroft Scrapbooks, *17,* 2–3; *51,* no. 3, 1046; *M&SP, 3* (May 18, 1861), 5, *12* (May 26, 1866); 321, *45* (December 23, 1882), 408, *72* (January 25, 1896), 76.

8. *Articles of Incorporation and By-Laws of the California State Geological Society,* p. 1; *Alta California,* September 15, 1878.

A. D. Hodges were active in helping to establish the Technical Society of the Pacific Coast in 1884, bringing together civil, military, mechanical, and mining engineers, along with architects, geologists, and other professional men.[9]

A number of mining engineers helped form the Salt Lake Mining Institute in 1883, with the object of setting up a museum, library, and agency for the discussion and publication of information about the mineral resources of Utah.[10] Already, the technical men at Leadville had sought to organize themselves, and out of their efforts, in 1882, came the Colorado Scientific Society, which in 1895 was said "to embrace in its membership almost every mining expert in Colorado who can lay claim to being a scientific expert." At its inception, the society had 12 members; by 1896, 160; and in 1923, it opened a new technical library in Denver, the funds provided by fees collected under the unpopular licensing act.[11]

Mining engineers wrote skillfully on both technical and nontechnical subjects and in the process transmitted their ideas far and wide. It was often contended that an excellent way for a young engineer to become known was to write about his work or his profession "in some high-class periodical,"[12] and countless numbers took this advice to heart. Among other writings, Joshua Clayton gave advice to prospectors through the columns of the *Mariposa Gazette* in 1861 and contributed an article on fissure veins for a Salt Lake newspaper in 1879.[13] A veteran Leadville editor, Carlyle Davis, recalled that many mining engineers, including A. A. Blow, Max Boehmer, and Louis S. Noble, had written articles for his papers to

9. *M&SP, 28* (May 9, 1874), 289, *44* (March 18, 1882), 176, *48* (March 8, 1884), 171, *65* (July 16, 1892), 45.

10. *EMJ, 36* (October 27, 1883), 257; Robert W. Sloan, ed., *Utah Gazeteer and Directory of Logan, Ogden, Provo and Salt Lake Cities, for 1884,* p. 604.

11. *EMJ, 33* (March 4, 1882), 117, *57* (February 24, 1894), 180, *115* (May 12, 1923), 835; *MI&R, 16* (November 14, 1895), 204; *17* (January 2, 1896), 295. For a brief but excellent discussion of the Colorado Scientific Society, see Rodman W. Paul, "Colorado as a Pioneer of Science in the Mining West," *The Mississippi Valley Historical Review, 47* (June 1960), 47–50.

12. *EMJ, 51* (April 18, 1891), 464.

13. *M&SP, 3* (June 1, 1861), 3; entry for January 15, 1879, Clayton diary, 1879, Clayton MSS (Calif.), 6.

"prove up" the Leadville mineral belt.[14] It was John Hays Hammond who contributed the article on "Gold" for the 1900 edition of the *Encyclopedia Britannica.*[15]

Hundreds of engineers wrote detailed accounts of their experiences in the West and all over the world for the columns of the technical journals. These ranged all the way from Joshua Clayton holding forth on the "pocket veins" of Esmeralda District, Nevada, or Winfield S. Keyes on "How to Make Mining Investments Pay," to P. M. Randall writing about vein formations on the Island of Aruba, T. A. Rickard on the Broken Hill Mines of New South Wales, and Charles Butters describing the cyanide process in South Africa.[16] As an up-to-date compendium of what was being done in the profession, the technical press came to be regarded as "the greatest of the mining schools." [17] But it was the willingness of the individual engineer to contribute that made it so.

Nor was the mining engineer any less hesitant to publish his professional knowledge in book form. Guido Küstel's *Nevada and California Processes of Silver and Gold Extraction* (1863)—a volume "written for the people in good plain English"—became a standard reference in his day, and mining men hurried to get copies of his new work on ore concentration when it was published a few years later.[18] *The Explorers', Miners' and Metallurgists' Companion,* written by J. S. Phillips, became the veritable bible of many self-educated mining men in the quarter of a century after its initial publication in 1871.[19] Subsequently, it would receive stiff competition from Richard H. Stretch's *Prospecting, Locating and Valuing*

14. Davis, *Olden Times,* p. 333.
15. MS article, Hammond MSS, Box 10.
16. See *M&SP, 8* (March 5, 1864), 145, *28* (May 23, 1874), 328; *AJM, 2* (December 22, 1866), 201–02; *EMJ, 52* (November 7, 1891), 530–32, *54* (October 51, 1892), 365–66.
17. "The Mining Schools of the United States," Rothwell, ed., *Mineral Industry, 2* p. 811.
18. *M&SP, 7* (December 7, 1863), 8; Henry Guyer to J. Guyer, Argenta, Montana, September 16, 1868, Henry Guyer MSS.
19. *EMJ, 13* (April 30, 1872), 281; *M&SP, 45* (July 8, 1882), 27.

Mines (1899), a volume recommended by Daniel Barringer to a young friend interested in becoming a mining engineer.[20] And after 1918, no engineer would be without a copy of Robert Peele's compendium, the *Mining Engineers' Handbook*. Augustus J. Bowie's *Hydraulic Mining in California* (1878) had gone into eleven editions by 1910, and was especially important because it preserved on paper the details of standard hydraulic techniques after hydraulicking had been curtailed in California as a result of the bitter debris controversy. When Eben Olcott went to Canada on a professional trip in 1889, he sent back for his copy.[21] In the same vein, Robert B. Stanton took pains to copy out an outline of T. A. Rickard's *Sampling and Estimation of Ore in a Mine* (1904) in one of his field books for use "on the go." [22]

To detail the technical literature produced by mining engineers would serve no useful purpose, but a few additional examples might be noted. James D. Hague is best remembered for his contribution to the mining volumes published in connection with Clarence King's Survey of the Fortieth Parallel and in connection with the Census of 1880. Watson Goodyear translated the German classics on assaying and wrote on western coal mining. Hamilton Smith wrote on hydraulics, David Brunton on tunneling, William H. Storms on timbering, Daniel Barringer on mining law, and Charles Janin on dredging.[23]

20. Daniel M. Barringer to Guy P. Bennett, July 29, 1899, copy, Letterbook C, Barringer MSS.

21. *M&SP, 64* (February 27, 1897), 170; Eben Olcott to J. P. Carmichael, Ottawa, July 1889, Olcott MSS (N.Y.).

22. Field notes 20, pp. 61–63, Stanton MSS.

23. James D. Hague, "Mining Industry," in Clarence King, *United States Geological Exploration of the Fortieth Parallel, 3* (Washington, D.C., 1870); Watson A. Goodyear, *A Treatise On the Assaying of Lead, Silver, Copper, Gold, and Mercury,* trans. T. Bodemann (New York, 1865); Goodyear, *The Coal Mines of the Western Coast of the United States* (San Francisco, 1877); Hamilton Smith, *Hydraulics* (New York, 1886); David W. Brunton & John A. Davis, *Modern Tunneling* (New York, 1914); William H. Storms, *Timbering and Mining* (New York, 1909); Daniel M. Barringer & John S. Adams, *Law of Mines and Mining in the United States* (2 vols. Boston, 1897–1911); Charles Janin, "Gold Dredging in the United States," U.S. Bureau of Mines, *Bulletin,* no. 127, Washington, D. C., 1918.

But the most important contribution of the mining engineer in the West was not merely that he passed on his technical knowledge and experience in printed form, but that he put them into practice himself, and in so doing helped advance the mineral industry through superior management and innumerable improvements in mining and milling practice. From the engineer came better planning and greater preliminary development; the increased use of power, particularly electrical; advanced engineering and equipment for tunneling, boring, hoisting, draining, and ventilating, not to mention countless innovations in the working of all types of ore. With the application of technology, mining became a modern industry, with increased production per unit of capital and manpower. And in this transition, the engineer moved in the vanguard.

In fact, said the editor of *Mining and Scientific Press* in 1905, speaking of the previous thirty years, "Nearly all the advance in mining science is directly attributable to mining engineers." [24] That is not to say that the engineer was always the inventor of techniques or equipment, though often he was, but he was quick to see what could be applied in a given situation and what could not. Deidesheimer's square-set timbering, for example, was one of the most valuable contributions in the history of mining, but engineers learned that it was not the answer to every problem. It was too costly in some mines or could not be employed in the face of shifting ground; hence methods of stoping and filling using only one-fifth as much timber were introduced or block-caving or open-cut techniques were employed that revolutionized the industry.[25] Yet practical men, particularly the Cornish miners, who sang "The Wreck of the Arethusa" and "Trafalgar's Bay" on paydays and who long remained a mainstay in western underground mining,[26] often refused to give up the square-set even when not needed. "Have you ever had to deal with any of the old

24. "Reforms Needed in Mining," *M&SP, 90* (April 1, 1905), 198.

25. Storms, *Timbering and Mining,* pp. 140–42, 242; Walter R. Crane, *Gold and Silver,* p. 354.

26. For a detailed, but sympathetic treatment of the "Cousin Jacks" in the United States, see Arthur Cecil Todd, *The Cornish Miner in America* (1967).

'Cousin Jack' miners from Grass Valley?" asked one of the modern engineers, with the square-set in mind.[27]

One common characteristic of the American mining engineer was his ability to initiate and to adapt. "If he thinks he has a scheme which will be an improvement over old-time methods, either in mining practice or in the reduction of ores, he is not afraid to try it." [28] Thus, Fred Bradley could institute a finely honed system of operation at the Spanish mine in California to achieve what mining men called "the best record ever made on gold quartz" prior to the late 1880s.[29] Such engineers as Dan Jackling, Louis Ricketts, Felix MacDonald, and Louis Cates defied tradition with their mass approach to low-grade copper ores; [30] but by 1916, in the words of one informed observer, "it makes a mining engineer feel old to recall the days when a 2 or 3% copper orebody made capitalists sneeze." [31] Three vital factors in the enduring success of the great Homestake mine in the Black Hills were an ability through a combined open-cut–deep-level approach to work gold ore on a large scale, Charles Merrill's advances in cyanide technology, and Donald McLaughlin's geological success in locating the vein in folds.[32]

Progressive engineers took the lead in introducing the latest equipment. Fred Bulkley and Walter B. Devereux pioneered in 1888 with an electric hoist installed in the workings of the Aspen Mining and Smelting Company property in Colorado, an innovation that caught on rapidly. By 1901, at Cripple Creek,

27. John Tyssowski to editor, New York, February 20, 1911, *EMJ, 91* (February 25, 1911), 403.

28. *M&SP, 89* (July 30, 1904), 67.

29. Samuel Christy to the president & board of regents of the University of California, Berkeley, June 28, 1892, copy, Christy MSS.

30. McDonald, for example, was responsible for applying a system of under cut block caving borrowed from the Michigan iron mines to Bingham copper in Utah. This revolutionary new approach spread rapidly and cut costs tre mendously. Eaton, "Seventy-Five Years of Progress in Metal Mining," Parsons, ed., *Seventy-Five Years of Progress*, p. 63.

31. *M&SP, 113* (September 30, 1916), 483.

32. Interview with Donald McLaughlin, San Francisco, March 28, 1964.

> the miner may go up to his work from the town on an electric car, go down in the mine by an electric hoist, operated by electric signals, the shaft being kept dry by an electric pump, do his work by an electric light, talk to the town and thence to the world by an electric telephone, run a drill electrically, and fire his shots by an electric blaster.[33]

James Douglas believed in 1893 that the West had been "the most fertile field" for technological innovation, especially in the modification and adaptation of machines and methods.[34] American engineers could take the gold dredge, a New Zealand contribution, and within a few years build up the most advanced dredge technology in the world; they could take the cyanide process with its special development on the Rand or the flotation process from Australia and advance these to their full potential. From South Africa, Yankee engineers borrowed the idea of the Kimberley skip for vertical shafts and for a uniform, realistic mapping scale, and in a thousand different ways brought to the West the fruits of their experience in other mining environments.

But they were more than borrowers and imitators. Not only were American mining engineers responsible for conceiving that drink known as the Daiquiri—a direct result of the Cuban experience—[35] they contributed numerous inventions bearing even more directly on the mineral industry. Josiah S. Phillip's "Wee Pet Assayer" and his patented "testing machine" for gold, silver, and lead were designed for the general prospector of the 1870s, but Stetefeldt's furnace, Arents' "siphontap," Butters' leaf filter, or Blatchly's rock drill had more professional application. Henry C. Perkins perfected a hydraulic nozzle deflector; George C. Rice, a novel mine rescue cage; Peter Nisson, an improved circular mortar and gravity

33. *M&SP, 82* (January 19, 1901), 47; *EMJ, 53* (January 23, 1892), 134–35; M. S. Holt, "Electricity in Mining as Applied to the Aspen Mining and Smelting Company, Aspen, Colo.," AIME *Trans., 20* (1891), 316, 318–21; *NCAB 25,* 256.

34. James Douglas, "Summary of American Improvements in Ore-Crushing and Concentration," AIME *Trans., 22* (1893), 321, 344.

35. Downey, *Richard Harding Davis & His Day,* p. 128.

stamp, which was widely used.[36] Arthur R. Wilfley, who learned his engineering at the knee of Victor Hills in Colorado, in 1897 invented the Wilfley concentrating table, a "side-jerk" ore-dressing device, which, despite its tendency to "vibrate and shimmy like an old negro mammy at a southern hoedown," soon became standard throughout the mining world; by 1927, more than 23,000 had been manufactured.[37]

But perhaps the most active and versatile engineer in this respect was David W. Brunton, whose fertile imagination brought forth one idea after another. Among Burton's many inventions were a quartering shovel used for sampling, a mining pump, an electric hoist, a "Mine and Tunnel Velocipede," a pocket alidade, a car coupling, a fire control, a system of round timber framing, and a mechanical sampler that automatically selected 1/625 of the ore passing through it.[38] Undoubtedly his most important contribution, however, was Brunton pocket transit, which he patented in the 1890s and which came to be used "all over the civilized world." This was a simple hand surveying instrument, easily carried and sighted by one man rather than two. Equipped with a mirror reflector, it became basic in warfare as well as engineering, and by 1968 the original manufacturers, Wm. Ainsworth & Sons of Denver, had produced nearly 150,000 of them.[39] The Brunton transit, remarked one mining engineer, "was the symbol of office" in the profession. "It was like being given your sword by King Arthur of Round Table fame." Columbia-trained Gelasio Caetani, hero of the Italian armies in World War I, commented in 1923: "Fortunately I had with me the Brunton compass which had followed me in my wanderings from the Taku Inlet of Alaska to the deserts of Arizona." Not only was the device vital in driving a tunnel under Austrian lines,

36. *M&SP, 21* (December 24, 1870), 433, *38* (April 19, 1879), 259; *SMQ, 35* (November 1913), 100; *EMJ, 80* (October 7, 1905), 648; Augustus J. Bowie, Jr., *A Practical Treatise on Hydraulic Mining in California,* pp. 50, 183–84.

37. *EMJ, 112* (October 8, 1921), 577; *NCAB, 21,* 360; Otis A. King, *Gray Gold,* p. 66.

38. *EMJ, 51* (June 20, 1891), 718, *56* (July 15, 1893), 57; *100* (December 11, 1915), 967; Rickard, ed., *Interviews,* p. 78; *NCAB, 23,* 99.

39. Robert F. Bauer, Production Manager, Wm. Ainsworth & Sons, Inc., to author, Denver, August 12, 1968.

without it Caetani wondered "how I could have shaved in the dugouts." [40]

As American engineers carried their technology throughout the West and to the far corners of the world, they brought a better balance to the mineral industry; through more expert evaluation and more judicious management, they provided additional protection for capital and the investor. They influenced mining law, although sometimes they exported the bad along with the good. They were instrumental, for example, in writing extralateral rights—with all their complications—into the Rhodesian mining code.[41] But the good—deep level techniques in the Transvaal, chlorination in Mexico, or dredging in Siberia—frequently came back improved and more advanced.

For all the world, the American West served as a kind of gigantic post graduate school of mines; it was a major crossroads for mining engineers of all nations, where ideas were exchanged, tested against the rugged western environment, sometimes discarded, sometimes adapted or even re-exported in modified form. By the 1920s, the American engineer was being regarded as "a most important missionary of civilization in opening up new territory." [42] Through him, the West indirectly made its indelible imprint on the mineral industry around the globe.

In a larger sense, too, the engineer took with him the basic American philosophy that practical problems, both technical and human, could be solved by rational processes. Both knowledge and application were interchangeable: an ability to handle effectively the physical side of mining implied the capability to deal likewise with social, political, or economic questions of broader moment. Understandably, the American public took the mining engineer to its bosom as one of the heroes of the industrial era.

40. *EMJ, 59* (May 18, 1895), 465; *Northwest Mining News, 1* (July 1907), 8; Ingersoll, *In and Under Mexico,* p. 54; *M&M, 4* (March 1923), 134; "The Evolution of Mine-Surveying Instruments," AIME *Trans., 29* (1899), 952–55.

41. *M&SP, 112* (June 3, 1916), 809.

42. John Hays Hammond, "The Engineer in Public Life," *M&M, 10* (March 1929), 115.

BIBLIOGRAPHY

MANUSCRIPT COLLECTIONS

Adelberg and Raymond MSS, New York Public Library.

Ashburner, William, MSS, Bancroft Library, University of California, Berkeley.

Barlow, Samuel L. M., MSS, Huntington Library, San Marino.

Barringer, Daniel Moreau, MSS, Western History Research Center, University of Wyoming, Laramie. Eventually, these manuscripts will be deposited at Princeton University. For purposes of convenience, the following designations have been given to various letterbooks:

DATE OF LETTERBOOK	DESIGNATION
December 17, 1894–March 2, 1896	Letterbook A
September 5, 1898–December 30, 1898	Letterbook B
January 3, 1893–November 16, 1899	Letterbook C
November 25, 1899–March 5, 1901	Letterbook D
March 5, 1901–November 9, 1901	Letterbook E
November 9, 1901–October 23, 1902	Letterbook F
February 1, 1902–May 15, 1902	Letterbook G
December 30, 1902–October 2, 1903	Letterbook H
October 12, 1903–August 13, 1904	Letterbook I
August 15, 1904–July 8, 1905	Letterbook J
July 10, 1905–March 1, 1906	Letterbook K
March 1, 1906–April 27, 1907	Letterbook L
April 29, 1907–April 29, 1908	Letterbook M
April 27, 1908–November 24, 1908	Letterbook N
November 25, 1908–June 4, 1909	Letterbook O
September 29, 1911–May 3, 1912	Letterbook P
May 3, 1912–January 7, 1913	Letterbook R
Commonwealth Exploration Company, February 19, 1909–August 10, 1910	Commonwealth Exploration Company Letterbook I.

Bayley, William S., MSS, University of Illinois Archives, Urbana.

Bradley, Frederick Worthen, MSS, in the possession of Phillip R. Bradley, Jr., 2801 Oak Knoll Terrace, Berkeley, California.

Bradley, Walter W., MSS, Bancroft Library, University of California, Berkeley.

Brewer, William, and Josiah D. Whitney, MSS, Bancroft Library, University of California, Berkeley.
Brown, D. R. C., MSS, Western History Collection, University of Colorado Libraries, Boulder.
Browne, Spencer C., MSS, in possession of Mrs. Spencer C. Browne, 1 Eagle Hill Road, Kensington, California.
Brush family MSS, Sterling Library, Yale University.
Butters, Charles and Jessie, Memorial Collection, Bancroft Library, University of California, Berkeley.
Christy, Samuel Benedict, MSS, University of California Archives, Berkeley.
Clayton, Joshua E., MSS, Bancroft Library, University of California, Berkeley.
———, MSS, Western Americana Collection, Yale University.
———, address to the Quartz Miners of Mariposa County, single MS letter, Western Americana Collection, Yale University.
Coffey, George Thomas, MSS, Bancroft Library, University of California, Berkeley.
Elkins, Stephen B., MSS, West Virginia University Library.
Ellwood, Isaac L., MSS, Western History Research Center, University of Wyoming, Laramie.
Farish, Frederick G., photograph collection, Western History Research Center, University of Wyoming, Laramie.
Foote, Arthur DeWint, MSS, Bancroft Library, University of California, Berkeley.
Foote, Mary Hallock, MSS, Bancroft Library, University of California, Berkeley. Five reels of microfilm on deposit.
Goldfield (Nevada) Mining Companies, MSS, Bancroft Library, University of California, Berkeley.
Guyer, Henry, MSS, Western Americana Collection, Yale University.
Hague, James Duncan, MSS, Huntington Library, San Marino.
Hammond, John Hays, MSS, Sterling Library, Yale University.
Harpending, Asbury, MSS, California Historical Society, San Francisco.
Hawley, James H., MSS, Idaho State Historical Society, Boise.
Hoffmann, Charles, MSS, Bancroft Library, University of California, Berkeley.
Ingersoll, Robert G., MSS, manuscripts division, Library of Congress.
Janin, Charles, MSS, Huntington Library, San Marino.
Janin family MSS, Huntington Library, San Marino.
Johnesse family MSS, Idaho State Historical Society, Boise.
John Taylor and Sons MSS, index book, 2 White Lion Court, London.
King, Clarence, MSS, Huntington Library, San Marino.
Lamoureaux, T. J., MSS, Huntington Library, San Marino.
Louderback, George D., MSS, University of California Archives, Berkeley.

Manhattan Silver Mining Company MSS, Bancroft Library, University of California, Berkeley.

Morris, Fred Ludwig, MSS, Bancroft Library, University of California, Berkeley.

Noel family MSS, Bancroft Library, University of California, Berkeley.

Olcott, Ebenezer E., MSS, New York Historical Society, New York City.

———, MSS, Western History Research Center, University of Wyoming, Laramie.

Prince, L. Bradford, MSS, State Museum of New Mexico, Santa Fe.

Read, William M., MSS, Bancroft Library, University of California, Berkeley.

Reed, Simeon G., MSS, originals in Reed College Library. Microfilm courtesy of the Idaho State Historical Society, Boise.

Renehan, A. B., and Carl H. Gilbert, MSS, State Museum of New Mexico, Santa Fe.

Roberts, George, MSS, Bancroft Library, University of California, Berkeley.

Roberts, Stephen, MSS, Bancroft Library, University of California, Berkeley.

Rogers, William King, MSS, Bancroft Library, University of California, Berkeley.

Sanders, Wilbur Edgerton, MSS, Montana Historical Society, Helena.

Sizer, Frank Leonard, MSS. In possession of F. M. Sizer, 2960 Magnolia Avenue, Berkeley, California.

Stanton, Robert Brewster, MSS, New York Public Library, New York City.

Stuart, Granville, MSS, Western Americana Collection, Yale University.

Sturges, George S., MSS, Bancroft Library, University of California, Berkeley.

Sutro, Adolph, MSS, Bancroft Library, University of California, Berkeley.

Titcomb, Harold A., MSS, Western History Research Center, University of Wyoming, Laramie.

Weld, Christopher Minot, MSS, Photoduplication. New York Public Library.

Wiltsee, Ernest Abram, MSS, Bancroft Library, University of California, Berkeley.

Wood, Henry E., MSS, Huntington Library, San Marino.

Wright, Howell, collection, Sterling Library, Yale University.

Yerington, Henry M., MSS, Bancroft Library, University of California, Berkeley.

INTERVIEWS AND AUTHOR'S CORRESPONDENCE

Bauer, Robert F., to author, Denver, August 12, 1968.

Bradley, James Parks, interview, 620 Market Street, San Francisco, May 18, 1964.

Browne, Mrs. Spencer C., interviews, 1 Eagle Hill Road, Kensington, California, March 8 & 15 & April 10, 1964.

Carlisle, Henry C., interview, 2511 Broadway, San Francisco, October 28, 1963.

Foote, Francis Seeley, interview, 2601 Shasta Road, Berkeley, California, May 8, 1964.

Hurst, George, interview, Bethlehem Steel Company, 20th & Illinois, San Francisco, June 5, 1961.

Joralemon, Ira B., interview, 315 Montgomery Street, San Francisco, April 6, 1964.

McLaughlin, Donald H., interview, 100 Bush Street, San Francisco, March 28, 1964.

Requa, L. K., to author, Salt Lake City, May 11, 1964.

GOVERNMENT PUBLICATIONS

FEDERAL PUBLICATIONS

Census Reports, twelfth, 1900.

"Coeur D'Alene Mining Troubles," U.S., *Senate Document,* no. 142, 56th Cong., 1st Sess., 1899–1900.

Congressional Record.

"Emma Mine Investigation," U.S., *House Report,* no. 579, 44th Cong., 1st Sess., 1875–76.

"Investigation of Mexican Affairs," U.S., *Senate Document,* no. 285, 66th Cong., 2d Sess., 1919–20.

Jones, William A., "Report upon the Reconnaissance of Northwestern Wyoming, Made in the Summer of 1873," U.S., *Executive Document* 285, 43d Cong., 1st Sess., 1874.

King, Clarence, *United States Geological Exploration of the Fortieth Parallel,* Washington, D.C., Government Printing Office, 1870.

"Meeting of the Mining and Metallurgical Society of America," U.S., *Senate Document,* no. 233, 64th Cong., 1st Sess., 1915–16.

Papers Relating to the Foreign Relations of the United States, 1861–.

"Report of the Commission to Amend the General Mining Laws," U.S., *House Report,* no. 639, 63d Cong., 2d Sess., 1913–14.

"Report of the Public Lands Commission," U.S., *House Executive Document,* no. 46, 46th Cong., 2d Sess., 1879–80.

Statistics of Mines and Mining in the States and Territories West of the Rocky Mountains, 1867–75. Published annually.

U.S. Bureau of Mines, *Bulletins,* 1910– .
——, *Technical Papers,* 1911– .
United States Statutes at Large.

STATE PUBLICATIONS

Aubury, Lewis E., "Gold Dredging in California," California State Mining Bureau, *Bulletin,* no. 57. Sacramento, 1910.

Galloway, John Debo, "Early Engineering Works Contributory to the Comstock," University of Nevada, *Bulletin, 41,* no. 5, Geology and Mining Series, no. 45.

COURT TESTIMONY AND REPORTS OF JUDICIAL DECISIONS

Arizona Reports, 1866– .
Colorado Appeals, 1891– .
Federal Cases, 1789–1880.
The Federal Reporter, series 1. 300 vols. 1880–1924.

Opinion of Referee. August 22, 1864. Gould & Curry Silver Mining Co., Plaintiff, v. North Potosi Gold & Silver Mining Co., Defendant. Virginia City, Territorial Enterprise, 1864.

The Pacific Reporter, 1883– .

The Richmond Mining Company of Nevada et al., Appellants, v. the Eureka Consolidated Mining Company, and the Richmond Mining Company of Nevada, Appellants, v. the Eureka Consolidated Mining Company. Appeals from the Circuit Court of the United States for the District of Nevada, n.d.

Sawyer, *Reports* . . . Circuit and District Courts, 9th Circuit. 14 vols. 1870–91.

Testimony of Clarence King, in the Case of Eureka Consolidated Mining Company v. Richmond Mining Company, of Eureka Nevada, at the March Term of the Sixth Judicial District Court, 1873. Eureka, Eureka Sentinal, 1873.

NEWSPAPERS

Alta California (San Francisco), August 29, 1864, September 6, 1869, November 17, 1874.

The Amador Record (Sutter Creek, California), special mining edition, April 1897.

The Black Hills Daily Pioneer (Deadwood), January 15–31, 1878.

Bulletin (San Francisco), June 5, 1878, August 27, 1879, March 5, 1880, March 15, 1880.

The Butte Inter Mountain, December 30, 1905.

The Chicago Tribune, April 10, 1869.

Chronicle (San Francisco), October 14, 1879, November 4, 1879, December 27, 1907, April 28, 1936.
The Colorado Miner (Georgetown), June 25, 1887.
The Double Jack (Dillon, Wyoming), May 13, 1905.
Financial Times (London), December 8, 1900.
Georgetown Courier (Colorado), January–July 1883.
Helena Herald, 1870–90.
Oakland Tribune, June 3, 1960.
Reese River Reveille, March 27, 1873.
Sacramento Union, November 9, 1947.
San Francisco Call, July 8, 1900, August 30, 1903, November 24, 1904.
Silver Standard (Silver Plume, Colorado), 1887.
South Africa (London), March 7, 1896.
The Standard and Diggers' News (London), May 21, 1896.
Stock Report (San Francisco), December 22, 1879.
Territorial Enterprise (Virginia City, Nevada), February 6, 1868, October 24, 1873.
The Times (London), December 2, 1872, January 7, 1887.
Virginia Evening Chronicle (Virginia City, Nevada), August 3 & 8, 1877, June 28, 1880, July 1, 1880.

CONTEMPORARY PERIODICALS

The American Mining Gazette, and Geological Magazine (New York), April 1, 1964–December 31, 1865.
Anglo-Colorado Mining and Milling Guide (London), October 29, 1898.
The British and South African Export Gazette (London), June 5, 1896.
Colorado School of Mines Magazine (Golden), October 1910–December 1930. Title varies.
The Economist (London), August 27, 1904.
The Engineer (San Francisco), October 1, 1876.
The Engineer of the Pacific (San Francisco), September 1877–March 1880. Series broken.
The Engineering and Mining Journal (New York), 1866–1932. Title is *American Journal of Mining, Milling, Oil Boring, Geology, Mineralogy, Metallurgy, etc.* until July 1869.
The Engineering Magazine (New York), May 1912–March 1916.
Engineering News and American Railway Journal (New York), October 6, 1892, November 3, 1892.
McClure's Magazine (New York), 1913–14.
The Miner: A Monthly Magazine, Devoted especially to the Promotion of the Mining Interests of the Pacific States and Territories (San Francisco), March–July 1866.

Mining and Metallurgy (New York), 1920–40.
Mining and Scientific Press (San Francisco), 1860–March 25, 1922. Title is *The Scientific Press* until 1872.
The Mining Industry and Review (Denver), July 18, 1895–June 30, 1898. Title is *The Mining Industry and Tradesman* until August 15, 1895.
The Mining Journal (London), January 13, 1872.
Mining Magazine (New York), July 1904–June 1906.
The Mining Magazine: Devoted to Mines, Mining Operations, Metallurgy, &c., &c. (New York), July 1853–June 1857.
The Mining Record (New York), October 17, 1878–December 25, 1880.
The Mining Record (San Francisco), May 21, 1881.
Mining Reporter (Denver), January 5, 1905–April 29, 1918. Title becomes *Mining Science* in December 1907, and *Mining American* in October 1915.
Mining World (Las Vegas, New Mexico), September 1880–August 1882, September–October 1885.
Mining World (London), February 8 & August 2, 1873.
The Mining World (Chicago), July 1905–December 1916. Title becomes *Mining and Engineering World* in July 1911.
The Nevada Monthly (Virginia City), February 1880–October 1880.
Northwest Mining Journal (Seattle), January 1906–June 1909.
Northwest Mining News (Spokane), March 1907–March 1913.
The Pacific Miner (Portland), June 1, 1902–November 1906; June 1909–August 1911. Published in San Francisco after June 1909.
The Pahasapa Quarterly (Rapid City), February 1912–December 1937. Title becomes *Black Hills Engineer* in January 1923.
Salt Lake Mining Review (Salt Lake City), April 15, 1912–February 15, 1920.
The School of Mines Quarterly (New York), 1879–1914.
Science, n.s. (New York), November 25, 1904.
The Statist (London), March 25, 1899.

TRANSACTIONS OF SOCIETIES

Journal of the Franklin Institute (Philadelphia), November 1883.
Report of the Proceedings of the American Mining Congress, 1903–24.
Transactions of the American Institute of Mining Engineers, 1871–1932.
Transactions of the American Society of Civil Engineers, 1934.
Transactions of the Institution of Mining & Metallurgy, (London), 1893–94.

COMPANY PUBLICATIONS AND REPORTS

Adelberg & Raymond, *Manhattan Silver Mining Company of Nevada,* New York, William C. Bryant, 1865.

Alturas Gold, Ltd., *Directors' Report,* February 18, 1888.

Arizona Copper Company, Ltd., *Annual Report,* year ending September 30, 1900.

Brunckow, Frederick, *Report of Frederick Brunckow, Geologist, Mineralogist, and Mining Engineer, To a Committee of the Sonora Exploring & Mining Co.,* Cincinnati, Railroad Record Printing, 1859.

Colorado United Mining Company, Ltd., *Annual Report,* year ending July 16, 1879.

Exploration Company, Ltd., *Memorandum and Articles of Association,* November 1, 1886, London, Board of Trade, Companies Registration Office, Bush House.

———, *Report for Year ending December 31, 1901.*

Foote, Arthur D., *The Idaho Mining and Irrigation Company,* New York, Theodore L. DeVinne, 1884.

Montana Company, Ltd., *Final Report of the Committee of Inquiry To Be Presented at the Extraordinary General Meeting of the Shareholders, To Be held On Thursday, the 19th Day of March, 1885,* London, 1885.

———, *Report of Messrs. N. Story-Maskelyne and J. R. Armitage to the Board of Directors,* London, November 12, 1884.

New Colorado Silver Mining Company, Ltd., *Directors' Report,* November 15, 1892–May 31, 1894.

Poorman Gold Mines, Ltd., *Circular to Shareholders,* June 28, 1901.

Rickard, Thomas A., *Report upon the Seven Stars Group of Mines, Yavapai County, Arizona,* August 10, 1892.

Riotte, Eugene N., *Report on the Havilah Mining Co's Property in Kern County, California,* San Francisco, Cuberly, 1875.

Seventh Annual Report of the Gould & Curry Silver Mining Co., San Francisco, Towne & Bacon, 1866.

Silliman, Benjamin, *Report on the Newly Discovered Auriferous Gravels of the Upper Rio Grande del Norte, in the Counties of Taos and Rio Arriba, New Mexico,* Omaha, Herald Publishing House, 1880.

Stratton's Independence, Ltd., *Directors' Report,* May 1, 1899–June 30, 1900.

———, *Annual Report,* year ending June 30, 1901.

———, *Prospectus,* May 11, 1899.

Symington, William N., *Report on the Quantity and Value of the Ore in Sight in the Robinson Mine, Summit County, Colo.,* December 26, 1881.

CONTEMPORARY BOOKS AND PAMPHLETS

Among the Silver Seams, Georgetown, Colorado, Georgetown Courier, 1880.

Annual Mining Review and Stock Ledger, San Francisco, July 1876.

Articles of Incorporation and By-Laws of the California State Geological Society, San Francisco, Alta California, 1877.

Bloomer, John G., *Pacific Cryptograph, for the Use of Operators in Mining Stocks, Mining Superintendents, Bankers and Brokers,* 2d ed. San Francisco, J. G. Bloomer, 1874.

Bowie, Augustus J., Jr., *Hydraulic Mining in California,* New York, privately printed, 1878.

———, *A Practical Treatise on Hydraulic Mining in California,* New York, Van Nostrand, 1885.

Bulletin of the University of California, no. 5, November 1874.

"Burgher" (Edmund Randolph?), *The New Almaden Mine,* San Francisco, Daily National Office, 1859.

Catalogue of the Colorado School of Mines, Golden, Colorado, 1898–99.

Church, John A., *Mining Schools in the United States,* New York, Waldron & Payne, 1871.

Considerations in Reference to the Establishment of a National School of Mines as a Means of Increasing the Product of Gold and Silver Bullion, Washington, D.C., Intelligence Printing House, 1867.

Criminal Record of the Western Federation of Miners from Coeur d'Alene to Cripple Creek 1894–1904, Colorado Springs, Colorado Mine Operators' Association, 1904.

Cushman, Samuel, and J. P. Waterman, *The Gold Mines of Gilpin County, Colorado,* Central City, Register, 1876.

Goodyear, Watson A., *The Coal Mines of the Western Coast of the United States,* San Francisco, Bancroft, 1877.

———, *A Treatise on the Assaying of Lead, Silver, Copper, Gold, and Mercury,* trans. T. Bodemann, New York, Wiley, 1865.

Hinton, Richard J., *The Hand-Book to Arizona: Its Resources, History, Towns, Mines, Ruins and Scenery,* New York, American News, 1878.

Ingersoll, Ernest, *Knocking Round the Rockies,* New York, Harper, 1883.

Ingham, G. Thomas, *Digging Gold among the Rockies,* Philadelphia, Cottage Library, 1881.

Janin, Henry, *A Brief Statement of My Part in the Unfortunate Diamond Affair,* San Francisco, privately printed, 1873.

Kent, Lewis A., *Leadville,* Denver, Daily Times, 1880.

Küstel, Guido, *Nevada and California Processes of Silver and Gold Extraction,* San Francisco, Carlton, 1863.
Langdon, Emma F., *The Cripple Creek Strike, 1903–1904,* Victor, Colorado, Daily Record Press, 1904.
Lists of the Alumni of Columbia College School of Mines, 1892, New York, Columbia College, 1892.
McCoy, Amasa, *Mines and Mining in Colorado: A Conversational Lecture, Delivered in the Lecture Room of Cosby's Opera House, to the International Mining and Exchange Company,* Chicago, International Mining and Exchange, 1871.
MacKnight, James Arthur, *The Mines of Montana,* Helena, Wells, 1892.
National Mining and Industrial Exposition Association, *The Mining Industry,* Denver, News Printing, 1881.
Opinions of the Press and of Eminent Public Men on the Importance of Our Mineral Resources and the Advantages To Be Derived from the Establishment of a National School of Mines, Washington D.C., Moore, 1869.
Pacific Coast Annual Mining Review, San Francisco, Francis & Valentine, 1878.
Poston, Charles D., *Apache-Land,* San Francisco, Bancroft, 1878.
Register of the University of California, 1894–95, Berkeley, 1895.
Report of Guido Küstel, Commissioner to the Vienna Exposition of 1873, Appointed by the Mechanics' Institute of San Francisco, Cal., San Francisco, Sterett, 1874.
Richthofen, Baron Ferdinand, *The Comstock Lode: Its Character, and the Probable Mode of Its Continuance in Depth,* San Francisco, Sutro Tunnell, 1866.
Robertson, J. R., *The Life of Hon. Alex. Del Mar, M.E.,* London, Gooch, 1881.
Rothwell, Richard P., ed., *The Mineral Industry, Its Statistics, Technology and Trade in the United States and Other Countries from the Earliest Times to the End of 1893,* New York, Scientific, 1894.
Rusling, James F., *Across America: or, the Great West and the Pacific Coast,* New York, Sheldon, 1874.
School of Mines, Columbia College, editions for 1865–66, 1866–67, 1872–73, 1876–77, 1883–84.
Shuck, Oscar, ed., *The California Scrap-book: a Repository of Useful Information and Select Reading,* San Francisco, Bancroft, 1869.
Sloan, Robert W., ed., *Utah Gazetteer and Directory of Logan, Ogden, Provo and Salt Lake Cities, for 1884,* Salt Lake City, Herald, 1884.
Smart, Stephen F., *Leadville, Ten Mile, Eagle River, Elk Mountain, Tin Cup and All Other Noted Colorado Mining Camps,* Kansas City, Ramsey, Millett and Hudson, 1879.
Smith, Hamilton, *Hydraulics,* New York, Wiley, 1886.

Taylor, Bayard, *Colorado: A Summer Trip,* New York, Putnam, 1867.
Triplett, Colonel Frank, *Conquering the Wilderness,* New York, Thompson, 1883.
United States Annual Mining Review and Stock Ledger, New York, Mining Review, 1879.
University of California, *Reports to the President of the University, from the Colleges of Agriculture and the Mechanic Arts,* Sacramento, California State Printer, 1877.
W. C. Ralston; A Business Man for Governor, San Francisco, Pacific Printing, 1914.
Whitney, James P., *Colorado, in the United States of America,* London, Cassell, Petter & Galpin, 1867.

MEMOIRS AND REMINISCENT ACCOUNTS

Bennett, Russell H., *Quest for Ore,* Minneapolis, Denison, 1963.
Bosqui, Edward, *Memoirs,* San Francisco, 1904.
Bowles, Samuel, *Across the Continent: A Summer's Journey to the Rocky Mountains, the Mormons, and the Pacific States, with Speaker Colfax,* New York, Hurd & Houghton, 1865.
Brereton, Robert Maitland, *Reminiscences of an Old English Engineer 1858–1908,* Portland, Irwin-Hodson, 1908.
Brewer, William H., *Up and Down California in 1860–1864,* ed. Francis P. Farquhar, New Haven, Yale University Press, 1930.
Browne, J. Ross, *Adventures in the Apache Country: A Tour through Arizona and Sonora, with Notes on the Silver Regions of Nevada,* New York, Harper, 1871.
Christien, Charles, dictation, Bancroft Library, University of California, Berkeley.
Cloman, Flora, *I'd Live It Over,* New York, Farrar & Rinehart, 1941.
Corning, Frederick Gleason, *A Student Reverie,* New York, privately printed, 1920.
———, *Papers from the Notes of an Engineer,* New York, Scientific, 1889.
Crampton, Frank A., *Deep Enough,* Denver, Sage Books, 1956.
Curle, J. H., *The Shadow-Show,* London, Methuen, 1912.
———, *This World of Ours,* New York, Doran, 1921.
Davis, Carlyle Channing, *Olden Times in Colorado,* Los Angeles, Phillips, 1916.
Davis, Richard Harding, *The West from a Car-Window,* New York, Harper, 1892.
Dawson, Thomas Fulton, *Life and Character of Edward Oliver Wilcott,* 2 vols. New York, Knickerbocker Press, 1911.
Deidesheimer, Phillip, dictation, San Francisco, December 8, 1886, Bancroft Library, University of California, Berkeley.

Donaldson, Thomas, "Idaho of Yesterday," typescript, 1941, Bancroft Library, University of California, Berkeley.

Drury, Wells, *An Editor on the Comstock Lode,* New York, Farrar & Rinehart, 1936.

Durbrow, William, interview conducted by Willa K. Baum, Berkeley, 1958, University of California Regional Cultural History Project.

Ellis, Amanda M., *The Strange, Uncertain Years,* Hamden, Connecticut, Shoe String Press, 1959.

Garthwaite, Augusta Lowell, "Our Life Together," typescript, in possession of Mrs. Otis Marston, 233 Vine, Berkeley, California.

———, "Addenda to Reminiscences of a Mining Engineer," typescript, ca. 1937, in possession of Mrs. Otis Marston.

Garthwaite, Edwin Hatfield. "Reminiscences of a Mining Engineer," typescript, ca. 1936, in possession of Mrs. Otis Marston.

Goodwin, Charles C., *As I Remember Them,* Salt Lake City, 1913.

Gorham, Harry M., *My Memories of the Comstock,* Los Angeles, New York, Suttonhouse, 1939.

Greenwood, Grace (Mrs. Sara Jane Lippencott), *New Life in New Lands: Notes of Travel,* New York, Ford, 1873.

Gressley, Gene M., ed., *Bostonians and Bullion: the Journal of Robert Livermore,* Lincoln, University of Nebraska Press, 1968.

Haren, LeRoy R., and Ann W., eds., *The Diaries of William Henry Jackson,* Glendale, Clark, 1959.

Hammond, Isaac B., *Reminiscences of Frontier Life,* Portland, privately printed, 1904.

Hammond, John Hays, *The Autobiography of John Hays Hammond,* 2 vols. New York, Farrar & Rinehart, 1935.

Harland, Hester Ann, *Reminiscences,* mimeographed, n.d.

Harris, Theodore D., ed., *Negro Frontiersman: The Western Memoirs of Henry O. Flipper,* El Paso, Texas Western Press, 1963.

Haskell, Charles W., dictation, Grand Junction, January 18, 1887, Bancroft Library, University of California, of Berkeley.

Hoover, Herbert, *The Memoirs of Herbert Hoover,* 3 vols. New York: Macmillan, 1951.

Horsley, Albert, *The Confessions and Autobiography of Harry Orchard,* New York, McClure, 1907.

Ingersoll, Ralph M., *In and Under Mexico,* New York, Century, 1924.

———, *Point of Departure,* New York, Harcourt, Brace, 1961.

Kett, William Francis, *Autobiography of William Francis Kett,* n.p., 1951.

McCarthy, Edward T., *Further Incidents in the Life of a Mining Engineer,* New York, Dutton, n.d.

———, *Incidents in the Life of a Mining Engineer,* New York, Dutton, 1918.

McLaughlin, Roy P., *The Tenderfoot Comes West,* New York, Exposition Press, 1968.

Morris, Henry Curtis, *Desert Gold and Total Prospecting,* Washington, D.C., privately printed, 1955.

———, *The Mining West at the Turn of the Century,* Washington, D.C., privately printed, 1962.

Nevins, Allan, and Milton Halsey Thomas, eds., *The Diary of George Templeton Strong,* 4 vols. New York, Macmillan, 1952.

Parsons, George W., *The Private Journal of George Whitwell Parsons,* Phoenix, Arizona Statewide Archival and Records Project, 1939.

Pumpelly, Raphael, *My Reminiscences,* 2 vols. New York, Holt, 1918.

Rathbone, Basil, *In and Out of Character,* Garden City, Doubleday, 1962.

Richards, Robert Hallowell, *Robert Hallowell Richards: His Mark,* Boston, Little, Brown, 1936.

Rickard, Thomas A., *Across the San Juan Mountains,* San Francisco, Dewey, 1907.

———, ed., *Interviews with Mining Engineers,* San Francisco, Mining and Scientific Press, 1922.

———, *Retrospect,* New York, McGraw-Hill, 1937.

Sala, George Augustus, *America Revisited: From the Bay of New York to the Gulf of Mexico, and From Lake Michigan to the Pacific,* 6th ed., London, Vizetelly, 1886.

Sales, Reno H., *Underground Warfare at Butte,* Caldwell, Caxton, 1964.

Sayre, Gertrude B., "Old Smuggler Narratives," typescript, 1940, xerox copy at Western History Research Center, University of Wyoming, Laramie.

Schmidt, F. Sommer. "Early Days at the Nevada Consolidated Copper Company, Ely, Nevada," September 23, 1949, courtesy of Prof. Russell R. Elliott, University of Nevada, Reno.

Shaler, Nathaniel Southgate, *The Autobiography of Nathaniel Southgate Shaler,* Boston, Houghton Mifflin, 1909.

Siringo, Charles A., *A Cowboy Detective,* Chicago: Conkey, 1912.

Skinner, Emory Fiske, *Reminiscences,* Chicago: Vestal, 1908.

Stanton, Robert Brewster, *The Hoskaninni Papers: Mining in Glen Canyon, 1897–1902,* ed. C. Gregory Crampton & Dwight L. Smith, University of Utah Anthropological Papers, no. 54, Glen Canyon Series, no. 15, Salt Lake City, 1961.

Thomas, Sewell, *Silhouettes of Charles S. Thomas,* Caldwell, Caxton, 1959.

Vigouroux, George E., ed., *The Diary of a Mining Investor,* New York, Quick News, 1910.

Wagner, Henry R., *Bullion to Books,* Los Angeles, Zamorano Club, 1942.

Williams, Alpheus F., *Some Dreams Come True,* Capetown, Timmons, 1948.

Wilson, Edmund, "Landscapes, Characters, and Conversations from the Earlier Years of My Life," *The New Yorker, 43,* (April 29, 1967), 50–131.

Wyman, Walker, ed., *California Emigrant Letters,* New York, Bookman, 1952.

Young, Frank C., *Echoes from Arcadia,* Denver, privately printed, 1903.

SECONDARY ACCOUNTS

Amery, L. S., *The Times History of the War in South Africa, 1899–1900,* 7 vols. London, Low, Marston, 1900–09.

Ashbaugh, Don, *Nevada's Turbulent Yesterday,* Los Angeles, Westernlore Press, 1963.

Bailey, Robert G., *Hell's Canyon,* Lewiston, Idaho, Bailey, 1943.

Bancroft, Hubert H., *The Works of Hubert Howe Bancroft,* 39 vols. San Francisco, History, 1882–90.

Barringer, Daniel Moreau, and John S. Adams, *The Law of Mines and Mining in the United States,* 2 vols. Boston, Little, Brown, 1897–1911.

Bates, J. C., ed., *History of the Bench and Bar of California,* San Francisco, Bench & Bar, 1912.

Beatty, Bessie, ed., *Who's Who in Nevada,* Los Angeles, Home, 1907.

Bernstein, Marvin D., *The Mexican Mining Industry, 1890–1950,* Albany, State University of New York Press, 1964.

Bimson, Walter R., *Louis D. Ricketts,* New York, Newcomen Society, 1949.

Blainey, Geoffrey, *The Rush That Never Ended,* Melbourne, Melbourne University Press, 1963.

———, "Lost Causes of the Jameson Raid," *The Economic History Review, 18* (August 1965), 354–63.

Boyer, Mary G., ed., *Arizona in Literature,* Glendale: Clark, 1934.

Brown, Wesley A., "Eleven Men of West Point," *The Negro History Bulletin, 19* (April 1956), 147–57.

Browne, Ross E., *Views of an Agnostic,* Oakland, privately printed, 1915.

Brunton, David W., & John A. Davis, *Modern Tunneling,* New York, Wiley, 1914.

Bunje, E. T. H., F. J. Schmitz, and D. W. Wainright, *Careers of University of California Mining Engineers, 1865–1936,* Berkeley, California Cultural Research Project, 1936.

Byers, William N., *Encyclopedia of Biography of Colorado,* Chicago, Century, 1901.

Calvert, Monte A., *The Mechanical Engineer in America, 1830–1910,* Baltimore, Johns Hopkins Press, 1967.

Chalfant, William A., *Outposts of Civilization,* Boston, Christopher, 1928.

Clendenen, Clarence, Robert Collins, and Peter Duignan, *Americans in Africa, 1865–1900,* Stanford, Hoover Institution on War, Revolution & Peace, 1966.

Clifford, Henry B., *Rocks in the Road to Fortune,* New York, Gotham Press, 1908.

Columbia University, *Honorary Degrees Awarded in the Years 1902–1945,* New York, Columbia University Press, 1946.

Crane, Walter R., *Gold and Silver,* New York, Wiley, 1908.

Croft, Helen Downer, *The Downs, the Rockies—and Desert Gold,* Caldwell, Caxton, 1961.

Crook, Thomas, *History of the Theory of Ore Deposits,* New York, Van Nostrand, 1933.

Curle, J. H., *The Gold Mines of the World,* 3d ed., New York, Engineering & Mining Journal, 1905.

Curti, Merle, & Kendell Birr, *Prelude to Point Four,* Madison, University of Wisconsin Press, 1954.

Defenbach, Bryan, *Idaho,* 3 vols. New York, American Historical Society, 1933.

De Kiewiett, C. W., *A History of South Africa, Social and Economic,* Oxford, Oxford University Press, 1957.

Dobson, Charles W., "Mine Salting," *Cosmopolitan, 24* (April 1898), 575–83.

Downey, Fairfax, *Richard Harding Davis & His Day,* New York, Scribner's, 1933.

Elliott, Russell R., *Nevada's Twentieth-Century Mining Boom,* Reno, University of Nevada Press, 1966.

Ferguson, John H., *American Diplomacy and the Boer War,* Philadelphia, University of Pennsylvania Press, 1939.

Festschrift zum hundertjährigen Jubiläum der Königl. Sächs. Bergakademie zu Freiberg, Dresden, Meinhold, 1866.

Finch, James K., *The Story of Engineering,* Garden City, Doubleday, 1960.

Glasscock, C. B., *The War of the Copper Kings,* New York and Indianapolis, 1935.

Goetzmann, William H., *Exploration and Empire,* New York, Knopf, 1966.

Gordon-Brown, A., ed., *Guide to Southern Africa,* London, Robert Hale, 1967.

Grew, Edwin Sharpe, *Field-Marshal Lord Kitchener: His Life and Work for the Empire,* 4 vols. London, Gresham, 1917.

Griswold, Don L., & Jean H., *The Carbonate Camp Called Leadville,* Denver, University of Denver Press, 1951.

Gudde, Erwin, *California Place Names,* 2d ed., Berkeley, University of California Press, 1960.

Hammond, John Hays, & Jeremiah W. Jenks, *Great American Issues,* New York, Scribner's, 1926.

Harper, Franklin, ed., *Who's Who on the Pacific Coast,* Los Angeles, Harper, 1913.

Harrison, James L., ed., *Biographical Directory of the American Congress, 1774–1949,* Washington, D.C., Government Printing Office, 1950.

Hatch, Frederick H., and J. A. Chalmers, *The Gold Mines of the Rand,* New York, Macmillan, 1895.

Hawley, James H., "Steve Adams's Confession and the State's Case against Bill Haywood," *Idaho Yesterdays,* 7 (Winter 1963–64), 16–27.

Hill, Alice Polk, *Colorado Pioneers in Picture and Story,* Denver, privately printed, 1915.

History of the Arkansas Valley, Colorado, Chicago, Baskin, 1881.

Hoover, Herbert, *Principles of Mining,* New York, McGraw-Hill, 1909.

Jackson, W. Turrentine, *Treasure Hill,* Tucson, University of Arizona Press, 1963.

Jensen, Vernon H., *Heritage of Conflict,* Ithaca, Cornell University Press, 1950.

Johnson, Allen, & Dumas Malone, eds., *Dictionary of American Biography,* 22 vols. New York, Scribner's, 1943–58.

Joralemon, Ira B., *Romantic Copper,* New York, Appleton-Century, 1934.

Kelley, Robert L., *Gold vs. Grain,* Glendale, Clark, 1959.

Kessler, Camillus, *At the Bottom of the Ladder,* Philadelphia, Lippincott, 1926.

King Memorial Committee of the Century Association, *Clarence King Memoirs,* New York, Putnam's 1904.

King, Joseph L., *History of the San Francisco Stock and Exchange Board,* San Francisco, privately printed, 1910.

King, Otis Archie, *Gray Gold,* Denver, Big Mountain Press, 1959.

Lavender, David, *The Story of Cyprus Mines Corporation,* San Marino, The Huntington Library, 1962.

Leckenby, Charles H., *The Tread of Pioneers,* Steamboat Springs, Steamboat Pilot Press, 1945.

Leith, Charles K., *Economic Aspects of Geology,* New York, Holt, 1921.

Letcher, Owen, *The Gold Mines of South Africa,* London, Waterlow, 1926.
Lewisohn, Sam A., "Industrial Leadership and the Manager," *Atlantic Monthly, 126* (September 1920), 414–18.
Lyons, Eugene, *Herbert Hoover: A Biography,* Garden City, Doubleday, 1964.
Mangam, William D., *The Clarks: An American Phenomenon,* New York, Silver Bow Press, 1941.
Mannix, J. Bernard, *Mines and Their Story,* Philadelphia, Lippincott, 1913.
Marcosson, Isaac F., *Anaconda,* New York, Dodd, Mead, 1957.
Merrill, George P., *The First One Hundred Years of American Geology,* New York, Hafner, 1924.
Merritt, Raymond Harland, "Engineering and American Culture, 1850–1875," Ph.D. dissertation, University of Minnesota, 1968.
Metcalfe, June, *Mining Round the World,* New York, Oxford University Press, 1956.
Michell, Lewis, *The Life and Times of the Right Honourable Cecil Rhodes, 1853–1902,* New York, M. Kennerley, 1910.
Michelson, Miriam, *The Wonderlode of Silver and Gold,* Boston, Stratford, 1934.
Miller, Joseph, ed., *Arizona Cavalcade,* New York, Hastings House, 1962.
Morgan, Jesse R., *A World School: The Colorado School of Mines,* Denver, Sage Books, 1955.
National Cyclopedia of American Biography, 49 vols., New York, White, 1898–1966.
Newton, Harry J., *Pitfalls of Mining Finance,* Denver, Daily Mining Record, 1904.
———, *Yellow Gold of Cripple Creek,* Denver, Nelson, 1928.
Parsons, Arthur B., *The Porphyry Coppers,* New York, AIME, 1933.
A Partial List of Organizations in California Interested in California History, Los Angeles, California State Historical Association, 1942.
Paul, Rodman W., "Colorado as a Pioneer of Science in the Mining West," *The Mississippi Valley Historical Review, 47* (June 1960), 34–50.
———, *Mining Frontiers of the Far West, 1848–1880,* New York, Holt, Rinehart & Winston, 1963.
Peele, Robert, ed., *Mining Engineers' Handbook,* 2 vols., 2d ed., New York, Wiley, 1927.
Probert, Frank H., "Mining Alumni, An Appraisal and Appreciation," *The California Monthly, 18* (April 1925), 447–51.
Progressive Men of the State of Montana, Chicago, Bowen, n.d.
Read, Thomas T., *The Development of Mineral Industry Education in the United States,* New York, AIME, 1941.

Reeks, Margaret, *Register of the Associates and Old Students of the Royal School of Mines, and History of the Royal School of Mines,* London, Royal School of Mines (Old Students') Association, 1920.

"Report of the Committee of Inquiry on the Colorado School of Mines," *Bulletin of the American Association of University Professors, 6* (May 1920), 19–40.

Requa, Mark L., *The Relation of Government to Industry,* New York, Macmillan, 1925.

Rice, George Graham, *My Adventures with Your Money,* Boston, Gorham Press, 1913.

Rickard, Thomas A., *A History of American Mining,* New York, McGraw-Hill, 1932.

———, ed., *Rossiter Worthington Raymond,* New York, AIME, 1920.

———, *The Sampling and Estimation of Ore in a Mine,* New York, Engineering and Mining Journal, 1904.

Roberts, Clarence N., *History of the University of Missouri School of Mines and Metallurgy, 1871–1946,* Rolla, 1946.

Rolle, Andrew, *California,* New York, Crowell, 1963.

Roosevelt, Theodore, *The Rough Riders,* New York, Scribner's, 1920.

Rossiter Worthington Raymond, n.p., 1910.

Ryder, David W., *The Merrill Story,* privately printed, 1958.

St. Johns, Adela Rogers, *First Step Up toward Heaven,* Englewood Cliffs, Prentice-Hall, 1959.

Sales, Reno H., "Geophysical Mining Claims," *Rocky Mountain Mineral Law Institute,* Albany, Bender, 1957.

Savage, W. Sherman, "The Negro on the Mining Frontier," *The Journal of Negro History, 30* (January 1945), 30–46.

Shamel, Charles H., *Mining, Mineral and Geological Law,* New York, Hill, 1907.

Shoemaker, Len, "Roaring Fork Pioneers," in John J. Lipsey, ed., *The 1962 Brand Book of the Denver Posse of the Westerners,* Denver, Johnson, 1963.

———, *Roaring Fork Valley,* Denver, Sage Books, 1958.

Shuck, Oscar T., ed., *History of the Bench and Bar of California,* Los Angeles, Commercial, 1901.

Sprague, Marshall, *Money Mountain,* Boston, Little, Brown, 1953.

Storms, William H., *Timbering and Mining,* New York, McGraw-Hill, 1909.

Stretch, Richard H., *Prospecting, Locating and Valuing Mines,* New York, McGraw-Hill, 1909.

Sutcliffe, George, ed., *Who's Who in Berkeley 1917,* n.p., n.d.

Todd, Arthur Cecil, *The Cornish Miner in America,* Glendale, Clark, 1967.

Wentworth, Frank L., *Aspen on the Roaring Fork,* Denver, World Press, 1950.

Weatherbe, D'Arcy, *Dredging for Gold in California,* San Francisco, Mining and Scientific Press, 1907.

Who Was Who in America, 1951–1960, Chicago, Marquis, 1960.

Who's Who in America, 1928–29, Chicago, Marquis, 1928.

Who's Who in America, 1962–63, Chicago, Marquis, 1962.

Who's Who in Arizona, n.p., n.d.

Who's Who in Engineering, 1922–23, Brooklyn, Lewis Historical, 1922.

Who's Who in Engineering, 1931, New York, Lewis, 1931.

Who's Who in Engineering, 1941, New York, Lewis, 1941.

Who's Who in Mining and Metallurgy, London, Mining Journal, 1908.

Wildman, Edwin, *Famous Leaders of Industry,* Boston, Page, 1921.

Wilkins, Thurman, *Clarence King,* New York, Macmillan, 1958.

Williams, Gardner F., *The Diamond Mines of South Africa,* 2 vols. New York, Buck, 1906.

Wolfe, Wellington C., ed., *Men of California, 1900–1902,* San Francisco, Pacific Art Company, 1901.

WORKS OF FICTION

Atherton, Gertrude, *Perch of the Devil,* New York, Stokes, 1914.

Ballard, Todhunter, *Westward the Monitors Roar,* Garden City, Doubleday, 1963.

Bennett, Horace Wilson, *Silver Crown of Glory,* Philadelphia, Winston, 1936.

Davis, Richard Harding, *Soldiers of Fortune,* New York, Scribner's, 1897.

Foote, Mary Hallock, *In Exile, and Other Stories,* Boston, Houghton Mifflin, 1894.

———, *The Ground-Swell,* Boston, Houghton Mifflin, 1919.

———, *The Last Assembly Ball and the Fate of a Voice,* Boston, Houghton Mifflin, 1889.

Greenleaf, Lawrence N., *King Sham, and Other Atrocities in Verse; Including a Humorous History of the Pike's Peak Excitement,* New York, Hurd & Houghton, 1868.

Harper, Theodore A., *Siberian Gold,* Garden City, Doubleday, 1927.

Hurt, Walter, *The Scarlet Shadow: A Story of the Great Colorado Conspiracy,* Girard, Kansas, The Appeal to Reason, 1907.

Kyne, Peter B., *The Understanding Heart,* New York, Grosset & Dunlap, 1926.

———, *Webster—Man's Man,* New York, Grosset & Dunlap, 1917.

Lefevre, Edwin, *The Golden Flood,* New York, McClure, Phillips, 1905.

Lockhart, Caroline, *The Man from the Bitter Roots,* New York, Burt, 1915.

Murphy, Clyde F., *The Glittering Hill,* New York, World Publishing Company, 1944.
Nason, Frank Lewis, *The Blue Goose,* New York, McClure, Phillips, 1903.
———, *The Vision of Elijah Berl,* Boston, Little, Brown, 1905.
Raymond, Rossiter W. (Robertson Gray, pseud.), *Brave Hearts,* New York, Ford, 1873.
Requa, Mark L., *Grubstake,* New York, Scribner's 1933.
Sublette, C. M., *The Golden Chimney,* Boston, Little, Brown, 1931.
Vaile, Charlotte M., *The M. M. C.: A Story of the Great Rockies,* Boston, Wilde, 1898.
Waters, Frank, *The Dust within the Rock,* New York, Liveright, 1940.

SCRAPBOOKS

Bancroft, Hubert Howe, Scrapbooks, Bancroft Library, University of California, Berkeley.
Hayes Scrapbooks, Bancroft Library, University of California, Berkeley.
Montana clippings, Scrapbook, Western Americana Collection, Yale University.
Smith, Herbert L., Scrapbook, Bancroft Library, University of California, Berkeley.

INDEX